Ron Burke

INTRODUCTORY GEOMETRY

INTRODUCTORY GEOMETRY

MARY KAY HUDSPETH

THE PENNSYLVANIA STATE UNIVERSITY

ADDISON-WESLEY PUBLISHING COMPANY
Reading, Massachusetts ■ Menlo Park, California
London ■ Amsterdam ■ Don Mills, Ontario ■ Sydney

Reproduced by Addison-Wesley from camera-ready copy supplied by the author.

Copyright © 1983 by Addison-Wesley Publishing Company, Inc.

All rights reserved. No part of this publication may be reproduced, stored in a retrieval system, or transmitted, in any form or by any means, electronic, mechanical, photocopying, recording, or otherwise, without the prior written permission of the publisher. Printed in the United States of America. Published simultaneously in Canada.

ISBN 0-201-10690-6
ABCDEFGHIJ-AL-89876543

In Memory of Robert

PREFACE

An instructor working with developmental students quickly realizes that many of them are talented and ambitious, yet they lack the fundamentals of geometry that are necessary for success in technical math, trigonometry, and/or calculus. This text, based on eight years' experience, is written specifically for such students. It assumes <u>no</u> prior knowledge of geometry; it does not emphasize deductive proof; rather it builds concepts on intuitive understanding using informal, yet precise, language. The specific features of the text are:

1. The text is less formal than most mathematics texts so that the students can <u>read</u> it.

2. To help the student focus on key concepts and relationships, all major procedures, formulas, and relationships have been set off in boxes. Several charts are included in the exercises to help the student gain a better understanding of the interrelationships among basic concepts.

3. The text contains numerous exercises so that the students will have ample opportunity to practice and develop their skills. Each set of exercises begins with problems reinforcing the basic concepts, then moves to problems that are progressively more challenging. In general the problems are paired so that every even-numbered problem is preceded by a similar odd-numbered problem for which the answer is given.

4. Answers to the odd-numbered problems in the Exercises and to all the problems in the Chapter Reviews are in the back of the book. For construction problems, a full-scale drawing is given so that the students may more easily check the accuracy of their constructions. The answer key contains detailed proofs for the sections on proofs.

5. Each geometric term and symbol is discussed in detail in the text. Because so much new terminology is encountered in geometry, a glossary of terms begins on page 496. A list of symbols and abbreviations used in the text is inside the front cover, and a summary of formulas and relationships is inside the back cover.

6. The text may be used with or without a calculator. A calculator is strongly recommended for problems marked by "C", and would be particularly useful when covering Chapters 7 and 8. When a calculator is not recommended, the problem states "Give an exact answer."

7. Radian measure is introduced parallel to degree measure so that the student may become thoroughly familiar with it prior to more technical courses.

8. Because developmental students in general are not skilled in recognizing algebra problems when they appear in a different context, this text is designed to give these students experience with algebra as it can occur in geometry problems. While the text presumes a knowledge of first-year algebra (through simultaneous solution of two linear equations in two variables), problems that require a knowledge of factoring are starred (*), as are other more challenging problems.

9. Pre-calculus problems are included to provide experience for students who are preparing for calculus. These transition problems, which are double starred (**), give the students practice in determining the appropriate equation or expression for word problems presenting geometry as it will be used in calculus.

The text contains more than enough material for a semester course. Because mathematics is based on careful and logical thought and because many developmental students are unaware of the structure of logical thought, formal logic is covered in Chapter 9 and may be studied first. Sections on proofs are included where appropriate. Either the logic and/or the sections on proof may be omitted without loss of continuity. Either Chapter 7, Introduction to Trigonometry, or Chapter 8, Three-Dimensional (Solid) Geometry, or both, may be omitted. Those problems and examples in Chapter 8 that require trigonometry are preceded by a "T" so that they may be omitted if Chapter 7 has not been covered.

The author owes much to the many students and tutors, particularly Gary Svetz, who have so carefully and clearly commented on the text as it developed. The book would not have been possible without the support of the mathematics faculty of the Developmental Year Program who have taught much of the material and who have contributed much to its development, with special credit due the eagle eye of Fay Jester. Especially helpful were the comments of the reviewers: Professor Pauline Jenness of William Rainey Harper College; Mrs. Carletta Elich of Logan High School, Logan, Utah; and Professor Frederick Strauss of the University of Texas at El Paso.

Special thanks go to Pat Mallion, Executive Editor for Mathematics at Addison-Wesley for her encouragement, and to the typist, Bonnie Randolph, and the draftsman, Delwyn Knott, for their diligent and careful work. The support of my husband has been indispensable.

MKH

TABLE OF CONTENTS

Chapter 1	POINTS, LINES, AND ANGLES	
1.1	Points, Curves, and Planes	1
1.2	Measurement of Line Segments	14
1.3	Angles and Their Measurement	27
1.4	Constructing and Bisecting Angles	41
	Chapter 1 Review	56

Chapter 2	TRIANGLES	
2.1	Basic Definitions	61
2.2	Triangles and Their Classification	63
2.3	Right Triangles and the Pythagorean Theorem	70
2.4	Construction of Triangles	91
2.5	Congruence	102
2.6	Proofs of Congruence	115
2.7	More Proofs of Congruence	125
	Chapter 2 Review	132

Chapter 3	PARALLEL LINES AND SIMILAR TRIANGLES	
3.1	Parallel Lines	138
3.2	Proofs	151
3.3	Similar Triangles	155
3.4	Parallel Lines and Transversals	169
3.5	Proofs of Similarity	175
	Chapter 3 Review	178

Chapter 4 POLYGONS

4.1	Definitions of Quadrilaterals	183
4.2	Trapezoids and Parallelograms	190
4.3	Rectangles, Rhombuses, and Squares	199
4.4	Proofs	211
4.5	Other Polygons	214
	Chapter 4 Review	222

Chapter 5 PERIMETER AND AREA OF POLYGONS

5.1	Squares and Rectangles	225
5.2	Parallelograms and Triangles	235
5.3	Rhombuses and Trapezoids	245
5.4	Regular Polygons and Composite Figures	255
	Chapter 5 Review	261

Chapter 6 CIRCLES

6.1	Basic Definitions	265
6.2	Chords and Tangents	283
6.3	Areas, Sectors, Composite Figures	297
6.4	Constructions	308
6.5	Proofs	329
	Chapter 6 Review	332

Chapter 7 INTRODUCTION TO TRIGONOMETRY

7.1	Sine, Cosine, and Tangent for Acute Angles	340
7.2	Applications	352
7.3	Other Trigonometric Functions	360
7.4	Trigonometric Functions for $90° \leq \theta \leq 360°$	363
	Chapter 7 Review	371

Chapter 8   THREE-DIMENSIONAL (SOLID) GEOMETRY

    8.1   Basic Definitions   374

    8.2   Surface Area of Prisms and Pyramids   384

    8.3   Surface Area of Cylinders, Cones, and Composite Figures   398

    8.4   Volume   409

    Chapter 8 Review   425

Chapter 9   LOGIC

    9.1   Statements; Conjunction and Disjunction   428

    9.2   Negations and Quantifiers   435

    9.3   Conditional Statements   444

    9.4   Other Forms of the Conditional   454

    9.5   Drawing Conclusions   458

    9.6   Invalid Arguments   462

    9.7   Proofs   468

    Chapter 9 Review   473

Cumulative Review   476

Appendix A   TRIGONOMETRIC TABLES   492

Appendix B   GREEK ALPHABET   495

Glossary   496

Answers to Selected Problems   501

INDEX   555

Symbols and Abbreviations   Inside front cover

Formulas and Relationships   Inside back cover

CHAPTER 1
POINTS, LINES, AND ANGLES

1.1 POINTS, CURVES, AND PLANES

As we begin the study of geometry, we need to understand the vocabulary used. As we learn new definitions, we will use certain words repeatedly. For the most part, these are words that cannot be precisely defined because any definition would require using other words that must be defined, and so on, indefinitely. While we cannot define precisely the basic terms point, line, and plane, we can develop a common intuitive understanding of them.

A POINT is an exact location in space. We use a capital letter and/or a dot to represent a point. A point has no dimension. The possible relationships between two points are: they represent the same point (COINCIDENT POINTS), or they represent different points (DISTINCT POINTS). If two points are coincident, then we can say that they INTERSECT. (Intersection means having at least one point in common.)

A connected pathway is called a CURVE. Table 1.1 illustrates the different types of curves.

A CLOSED CURVE is a curve that has no unique beginning and end. No matter where we start to trace the curve, we will end at the same point where we began.

A SIMPLE, CLOSED CURVE is a closed curve that does not intersect itself.

That is, it can be traced without lifting the pencil and the curve will not cross over itself as it is being traced. The region enclosed by a simple closed curve is called the "inside" or INTERIOR of the curve. Any point not on the curve and not in the interior of the curve lies in the "outside" or EXTERIOR of the curve.

Some simple closed curves have "indentations," or "concavities." When a simple closed curve has such an indentation, we can choose two points in the interior of the curve and connect them with a line segment that has at least one point that lies in the exterior of the simple closed curve. In the simple closed curve shown here, points A and B lie in the interior of the curve,

but point C on line segment AB ($\overline{AB}$) is in the exterior of the curve. Thus this curve has a concavity. When all possible line segments with endpoints in the interior lie entirely within or on the simple closed curve, the curve has no concavities and is called CONVEX.

When all points of a convex closed curve are the same distance from a fixed point O, then the curve is a CIRCLE, where point O is the CENTER of the circle and the fixed distance is the RADIUS of the circle.

TABLE 1.1

<u>Definitions and Illustrations of Curves.</u>

<u>Definition</u>	This set of points <u>does</u> satisfy the definition.	This set of points <u>does not</u> satisfy the definition.
A CURVE is a connected pathway (a set of points that can be traced without lifting the pencil).		

TABLE 1.1 continued.

Definition	This set of points does satisfy the definition.	This set of points does not satisfy the definition.
A CLOSED CURVE has no unique endpoints (the starting point is the ending point).		
A SIMPLE CLOSED CURVE is a closed curve that does not intersect itself.		
A CONVEX SIMPLE CLOSED CURVE is a simple closed curve where the line segment connecting any two points in the interior of the curve lies entirely in the interior of the curve.		
A CIRCLE consists of all points at a fixed distance (the radius) from a fixed point (the center).		

The term line can have two meanings: a connected pathway and a "straight line." In this book we will use CURVE when we mean a connected pathway. We will use LINE when we mean a straight line that extends indefinitely in opposite directions. Note that lines (straight lines) are a particular kind of curve. We will represent lines by giving the letters of two points contained in the line and drawing ↔ over the letters: $\overleftrightarrow{AB}$ represents the line

◄―――●―――●―――►
    A   B

. We may also refer to this line as $\overleftrightarrow{BA}$. Lines may also be represented by lower case letters: m represents the line

◄―――――――►
   m

. Note that we draw arrowheads at the ends to indicate that the line extends indefinitely. A line has no thickness or width, but it has unlimited length. Thus a line is a one-dimensional figure.

Points are said to be COLLINEAR if and only if they all lie on the same line.

```
      <----•------------------------•---->
           A                        C
                    •
                    B
```

Points A and C are contained in line $\overleftrightarrow{AC}$ and therefore are collinear. Point B does not intersect $\overleftrightarrow{AC}$ and therefore is not collinear with points A and C. We also say B does not lie on line $\overleftrightarrow{AC}$. Any two distinct points are collinear because there is a line that may be passed through them. (For any two distinct points, there is only one unique [noncoincident] line containing them.) Thus points A and B are collinear, as are points B and C. The smallest set of noncollinear points contains three points. A set of four noncollinear points could consist of the following set of points A, B, C, and D:

```
      <----•---------•-------------------•---->
           A         C                   D
                            •
                            B
```

A set of n noncollinear points could consist of n - 1 points on a line and one point not on that line.

We have several ways to refer to a part of a line. A LINE SEGMENT is that portion of a line that lies between and contains two distinct points on the line. The line segment $\overline{AB}$ is that part of $\overleftrightarrow{AB}$ that includes point A, extends to point B and includes it.  We will represent line segments by giving the letters of the two endpoints of the segment and drawing a bar (without arrows) over the letters.

A RAY is that portion of a line that lies on one side of a point P (and includes P). The ray $\overrightarrow{AB}$ is that part of the line starting at and including point A and extending through and beyond B:

```
                                              <----•---------•---->
                                                   A         B
```

Point A is called the endpoint of $\overrightarrow{AB}$. The ray $\overrightarrow{BA}$ is that part of the line starting at and including point B and extending through and beyond A:

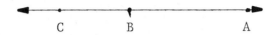

. Two rays are OPPOSITE RAYS if and only if they lie on the same line, have a common endpoint, and point in opposite directions. If A, B, and C are collinear and B is between A and C, then $\vec{BA}$ and $\vec{BC}$ are opposite rays:

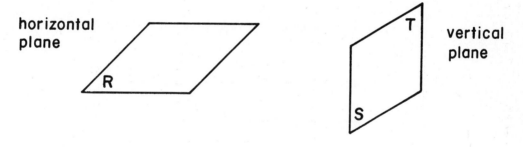

The rays $\vec{CA}$ and $\vec{BC}$, however, are not opposite rays because they do not share a common endpoint.

A PLANE is a surface such that any two points in it can be joined by a line and that line will be contained in the surface. Two examples of planes are shown in Figure 1.1. A plane can be thought of as a smooth, level surface that extends indefinitely. Either a capital letter or a pair of capital letters are used to represent a plane.

FIGURE 1.1

A plane has unlimited length and unlimited width, but has no depth. Thus a plane has two dimensions.

The possible relationships between two lines are: the two lines may represent the <u>same</u> line (are not distinct), in which case they are COINCIDENT LINES. The lines may cross each other at one point as shown in Figure 1.2. These are INTERSECTING LINES.

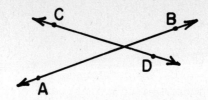

FIGURE 1.2

The two lines may be unique lines (noncoincident), lie in the same plane, and not intersect, in which case they are PARALLEL LINES. In Figure 1.3,

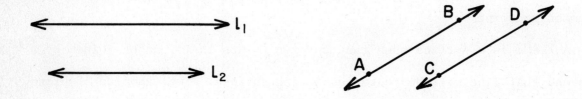

FIGURE 1.3

lines ℓ_1 and ℓ_2 are parallel, or $\ell_1 \| \ell_2$, and lines $\overleftrightarrow{AB}$ and $\overleftrightarrow{CD}$ are parallel: $\overleftrightarrow{AB} \| \overleftrightarrow{CD}$. By placing an arrowhead on each of two lines, we can indicate that the two lines are parallel, as illustrated in Figure 1.4.

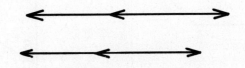

FIGURE 1.4

If the two lines are unique, do not intersect, and do not lie in the same plane (are not parallel), they are called SKEW LINES. In Figure 1.5 $\overleftrightarrow{AB}$ and $\overleftrightarrow{CD}$ are skew lines.

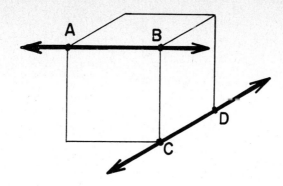

FIGURE 1.5

Lines that intersect or that are parallel lie in the same plane and are said to be COPLANAR. Note that a unique plane is determined by:

    three noncollinear points, or
    one line and one point not on the line, or
    two distinct lines that either intersect or are parallel.

The possible relationships between two planes are: they may represent the same plane (COINCIDENT PLANES). They may have a line in common (INTERSECTING PLANES). The two planes may be unique and not intersect (PARALLEL PLANES).

The possible relationships between a line and a plane are: they may not intersect, they may intersect at one point, or the line may be in the plane, that is, the plane passes through the line. When the line lies in the plane, the intersection of the line and the plane is the set of points in common: the line. These relationships are illustrated in Figures 1.6 through 1.9.

Be alert to the way relationships are phrased:

    A line will "pass through" a plane only if it does not "lie in" that plane.

    A plane will "pass through" a line only if it contains that line.

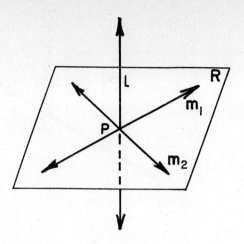

FIGURE 1.6. Line ℓ intersects plane R at point P.

Plane R contains lines m_1 and m_2.

Plane R passes through lines m_1 and m_2.

Plane R does not pass through line ℓ.

Plane R and line m_2 intersect in m_2.

The intersection of plane R and m_1 is m_1.

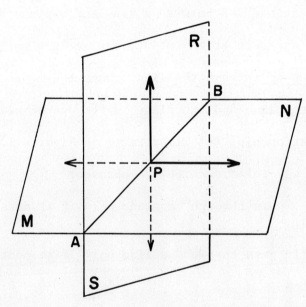

FIGURE 1.7. Plane MN and plane RS intersect in $\overleftrightarrow{AB}$.

Plane MN and plane RS pass through $\overleftrightarrow{AB}$.

$\overleftrightarrow{AB}$ lies in planes MN and RS.

$\overrightarrow{AB}$ is contained in planes MN and RS.

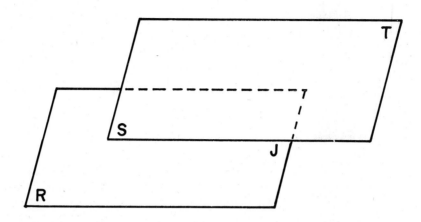

FIGURE 1.8.   Plane RJ is parallel to plane ST.

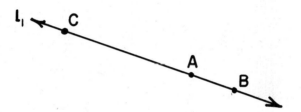

FIGURE 1.9.   The intersection of ℓ_1 and $\overrightarrow{AC}$ is $\overrightarrow{AC}$.

The intersection of $\overrightarrow{CA}$ and $\overrightarrow{AC}$ is $\overline{AC}$.

The intersection of $\overline{CB}$ and $\overline{AC}$ is $\overline{AC}$.

The intersection of C and ℓ_1 is C.

The intersection of $\overleftrightarrow{CA}$ and $\overline{AB}$ is $\overline{AB}$.

The intersection of $\overrightarrow{CB}$ and A is A.

Table 1.2 summarizes the definitions, symbols, and dimension(s) for the basic terms of this section.

TABLE 1.2

Term	Symbol(s)	Meaning	Dimension(s)
POINT	P	location in space	no dimension
CURVE	curve AB	connected pathway	either finite (fixed) or unbounded length
LINE	m, $\overleftrightarrow{AB}$	straight line that extends indefinitely in opposite directions	unbounded length, no width
LINE SEGMENT	$\overline{AB}$	portion of a line that lies between and contains two distinct points on a line	finite (fixed) length; no width
RAY	$\overrightarrow{PB}$ $\overleftarrow{BP}$	portion of a line that lies on one side of P and includes P	unbounded length; no width
PLANE	plane S, plane RT	a surface such that any two points in it can be joined by a line and that line will be contained in the surface	unbounded length; unbounded width; no thickness

EXERCISE 1.1

List the letters of all the figures that belong to each category of curve.

1. Curve
2. Convex simple closed curve
3. Closed curve
4. Simple closed curve
5. Line
6. Circle

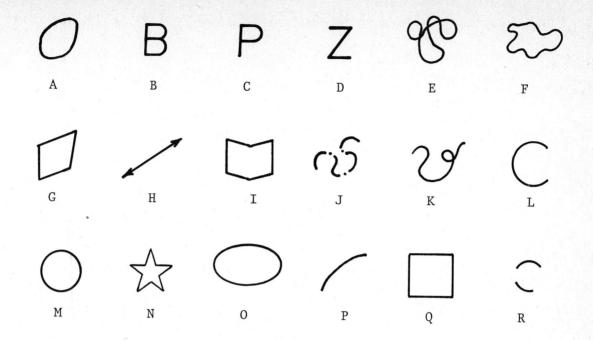

7. Make a drawing representing each of the following: a) $\overline{PQ}$ b) $\overleftrightarrow{PQ}$.

8. Make a drawing representing each of the following: a) $\overrightarrow{PQ}$ b) $\overline{RS}$ containing the point T.

9. Make a drawing representing: $\overrightarrow{AB}$ containing the point S where S is not contained in $\overline{AB}$.

10. Make a drawing representing: $\overleftrightarrow{PQ}$ containing the point T where T is contained in $\overrightarrow{PQ}$ but not in $\overrightarrow{QP}$.

11. a) Plot 4 distinct points. Draw all possible lines connecting pairs of these points. How many such segments are there? Label them.

    b) How can you place the 4 distinct points to make the number of distinct (noncoincident) segments a minimum? Illustrate. A maximum? Illustrate.

12. a) Plot 5 distinct points. Draw all possible lines connecting pairs of these points. How many such segments are there? Label them.

    b) How can you place the 5 distinct points to make the number of distinct (noncoincident) segments a minimum? Illustrate. A maximum? Illustrate.

13. Can a line always be passed through any three distinct points?

14. Can a plane always be passed through any three distinct points?

15. Can two planes ever intersect at a single point?

16. Can three distinct planes intersect in the same straight line? Can 10 planes?

Illustrate Problems 17 through 22 with drawings that satisfy the conditions given.

17. P, Q, R, and S are four noncollinear points, Q belongs to $\overleftrightarrow{PR}$ but not to $\overrightarrow{RP}$. S and P are collinear.

18. P, Q, R, S, and T are five noncollinear points, Q lies on $\overleftrightarrow{PR}$, and Q lies on $\overleftrightarrow{ST}$.

19. A, B, and C are three noncollinear points, A, B, and D are three collinear points, and A, C, and E are three collinear points.

20. A, B, and C are three collinear points, C, D, and E are three noncollinear points, and E lies on $\overrightarrow{AB}$.

21. $\overline{PR}$ contains Q, $\overline{RT}$ contains S, $\overline{QS}$ contains U.

22. $\overline{AB}$ contains D and E, $\overline{EB}$ contains C, C does not lie on $\overline{DB}$.

Refer to Figure 1.10 and indicate which of the statements in Problems 23 through 28 are true and which are false.

23. Plane AB intersects plane CD in line ℓ.

24. Plane AB passes through line ℓ.

25. Plane AB passes through $\overline{EF}$.

26. Plane CD passes through $\overline{EF}$.

27. P lines in plane CD.

28. ℓ intersects $\overline{EF}$ at G.

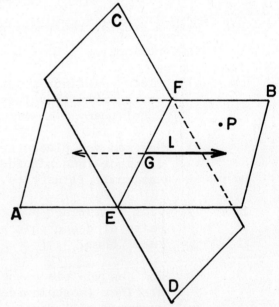

FIGURE 1.10

Draw a picture to illustrate the conditions given in each of Problems 29 through 40.

29. P, Q, R, and S are noncollinear points; R, S, and T are collinear points; and Q, R, and J are collinear points.

12

30. P, Q, and R are noncollinear points; R, J, and Q are collinear points; and P, Q, and T are collinear points.

31. P, Q, R, and S are points with R and S collinear; P, R, and T are collinear; T, P, and S are noncollinear.

32. A, B, and C are collinear; B, C, and D are noncollinear.

33. Plane AB intersects plane CD in line ℓ. Line m lies in plane CD. Line ℓ and line m do not intersect.

34. Plane RS intersects plane MN in $\overleftrightarrow{AB}$.

35. Plane AB and plane CD do not intersect.

36. Line ℓ lies in plane AB. Line m lies in plane CD. ℓ intersects m at point P.

37. A, B, C, and D are four collinear points; A is between C and D; and D is between A and B.

38. B is between A and C, and C is between A and D; A, B, C, and D are four collinear points.

39. R lies on $\overleftrightarrow{ST}$ and R does not lie on $\overrightarrow{ST}$.

40. ℓ, m, and n are three distinct lines; ℓ and m do not intersect, m and n do not intersect, and ℓ and n do not intersect.

41. For each block in Table 1.3, determine if the intersection of the figures in the row could be <u>exactly</u> the figure given in the column. Answer "yes" or "no". Some of the blanks have been filled in. Remember the intersection is the set of <u>all points in common between the two sets</u>. For each answer "yes," provide a drawing to illustrate. (One set of points can be drawn in yellow high-lighter and the other in blue high-lighter. The intersection will be the set of green points.)

TABLE 1.3

These two sets of points in the <u>same</u> plane	Point?	Segment?	Ray?	Line?	Plane?
Point and segment					
Point and ray	Yes				
Point and line					No
Point and plane			No		
Segment and segment					No
Segment and ray					
Segment and line					
Segment and plane					
Ray and ray		AB Yes			
Ray and line				No	
Ray and plane					
Line and line					
Line and plane					No

1.2 MEASUREMENT OF LINE SEGMENTS

We may measure the length of a line segment once a unit of length is established. The most common units of length are the inch, foot, yard, mile, millimeter, centimeter, meter, and kilometer. We can only <u>approximate</u> a true measurement. For instance, we may come to the following different lengths for $\overline{AC}$, depending upon how accurately we measure:

```
A                                                    C
•————————————————————————————————————————————————————•
```

To the nearest in., AC = 5 in.

To the nearest $\frac{1}{2}$ in., AC = $4\frac{1}{2}$ in.

To the nearest $\frac{1}{4}$ in., AC = $4\frac{3}{4}$ in.

To the nearest $\frac{1}{8}$ in., AC = $4\frac{5}{8}$ in.

To the nearest $\frac{1}{16}$ in., AC = $4\frac{11}{16}$ in.

To the nearest $\frac{1}{32}$ in., AC = $4\frac{21}{32}$ in.

Obviously we could continue measuring using smaller and smaller units. There is no smallest unit. (Note that when we give the length of $\overline{AC}$, we write AC without the bar.)

While it is impossible to make an exact measurement of length, we can talk about the (ideal) true measure of the length of two line segments. If two line segments $\overline{AC}$ and $\overline{FG}$ are the same length, then we say AC = FG. Two figures are CONGRUENT if and only if they are the same size and shape. Therefore if $\overline{AC}$ and $\overline{FG}$ have the same length, we can say $\overline{AC}$ is congruent to $\overline{FG}$, or $\overline{AC} \cong \overline{FG}$. Thus for line segments, $\overline{AC} \cong \overline{FG}$ and AC = FG are equivalent statements, and we may use the notation interchangeably.

Point P is the MIDPOINT of a line segment $\overline{AB}$ when AP = PB.

```
        •————————•————•
        A        P    B
```

Line segment $\overline{AB}$ BISECTS another line segment $\overline{CD}$ at E when CE = ED.

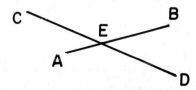

This may also be stated as $\overline{CD}$ is bisected by $\overline{AB}$. It does not mean that $\overline{AE} \cong \overline{EB}$. The statement that $\overline{FG}$ and $\overline{HI}$ bisect each other at K, means that FK = KG and KH = KI.

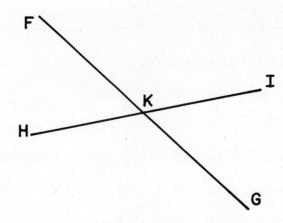

We can then say that K is the midpoint of $\overline{HI}$ and that K is also the midpoint of $\overline{FG}$.

You should be aware that the drawings in this text are not always drawn to scale. This is done so that we do not come to rely on facts that appear to be true in the drawing. Lengths of lines and sizes of angles must be given in order for us to make an accurate statement. Just because one line segment appears to bisect another does not necessarily mean that it does. The information must be given that the two segments of one line are equal, or that the line segment is bisected by the other segment, or we may construct one line segment so that it bisects the other. We can use the drawing to assume the order of points on a segment or to see that two line segments intersect. Thus we cannot use the drawings to tell us about size, but they can tell us about relationships among the parts of the drawing.

We can use a compass and a straightedge (not a ruler) to construct line segments of equal length or to construct a line segment that bisects another. Both instruments are shown in Figure 1.11. A compass is an instrument that allows us to draw circles with a given radius. A straightedge enables us to draw a line through two given points; it is NOT a ruler in that we ignore any markings of length that it may have.

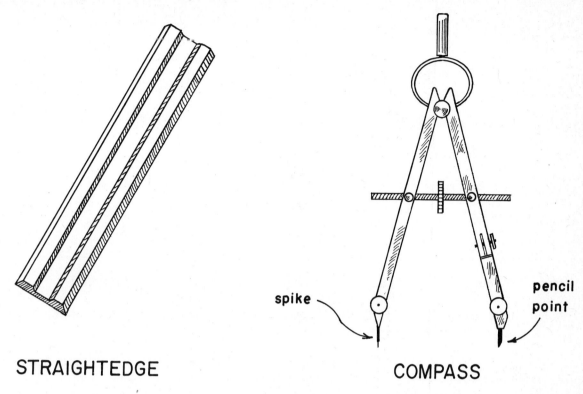

FIGURE 1.11

To draw a line through two points A and B,   place the straightedge

along the two points and draw the line through them. To draw the set of points a given distance r from a fixed point P, we set the distance between the two

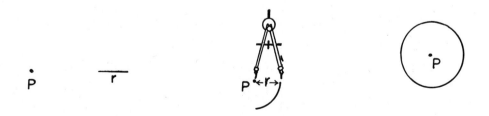

points of the compass equal to r, place the spike of the compass at P, and then draw with the pencil point of the compass the set of points at the distance r from P.

Of course when we construct a line through two points or draw a circle about the point P, we are constructing neither a perfect straight line nor a perfect circle. This is because no real object is _perfectly_ straight--a pencil line has width; the distance between the points of the compass will vary slightly as we trace the circle. Here we will ignore these limitations and will _assume_ that what we have constructed is indeed accurate.

To construct a line segment equal to a given line segment:

1. Draw with the straightedge a line segment that is longer than the given segment.

2. Mark a point near one end of the segment you have just drawn.

3. Put one point of the compass at one endpoint of the given segment and the other compass point at the other endpoint.

4. Without changing the setting of the compass, place the spike at the point you marked according to Step 2.

5. Use the pencil point of the compass to mark a point on the line segment that you have drawn.

6. The segment between the two points is equal to the given segment.

Steps 4 and 5 above can be repeated to produce line segments that are integer multiples of the given segment, as shown in Examples 1 and 2.

EXAMPLE 1   Construct a line segment three times the length of $\overline{AB}$.

```
            _____
            A                 B
```

Solution   Step 1:   Using the straightedge draw a line segment.

Step 2: Mark a point near one end of the segment.

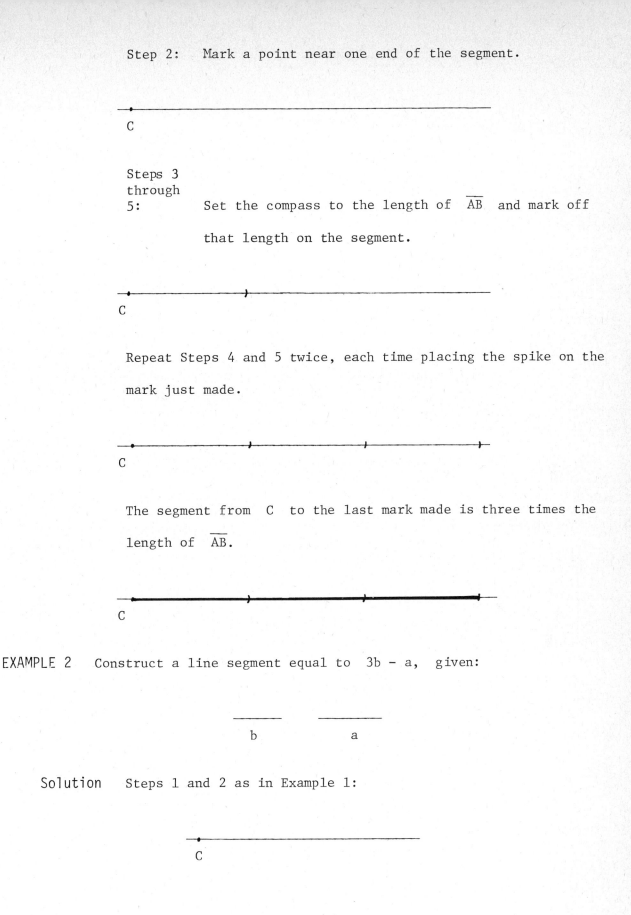

Steps 3 through 5: Set the compass to the length of $\overline{AB}$ and mark off that length on the segment.

Repeat Steps 4 and 5 twice, each time placing the spike on the mark just made.

The segment from C to the last mark made is three times the length of $\overline{AB}$.

EXAMPLE 2   Construct a line segment equal to 3b − a, given:

Solution   Steps 1 and 2 as in Example 1:

Step 3: Set the compass to the length of b.

Steps 4
and 5: Mark off three segments of length b.

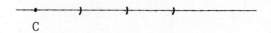
C

Repeat Steps 3 through 5 for segment a, placing the spike at the mark farthest to the right.

C

The segment from C to the last mark represents the segment 3b − a.

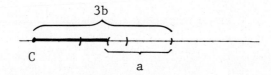

To bisect a given line segment we can use a straightedge and a compass.

To bisect a given line segment:

1. Set the compass points at a distance greater than one-half the length of the segment to be bisected but not longer than the line segment itself.

2. Place the spike of the compass at one endpoint of the segment and draw a circle.

3. Without changing the setting of the compass, place the spike of the compass at the other endpoint of the segment and draw another circle.

4. Draw a line segment connecting the points of intersection of the two circles.

5. The given line segment is bisected at the point where the line segment constructed in Step 4 intersects it.

If the circles do not intersect in Step 3, repeat the steps with the compass points farther apart.

EXAMPLE 3   Bisect the given line segment.

R   T

Solution   Steps 1 and 2:   Set the points of the compass and draw a circle with R as the center.

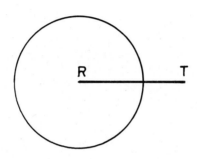

Step 3:   Draw another circle, with the same radius using T as the center.

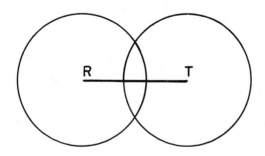

Step 4:   Draw the line segment connecting the points of intersection of the two circles.

21

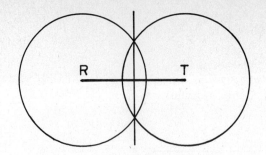

Step 5: The points of intersection of the constructed line segment and $\overline{RT}$ is the midpoint of $\overline{RT}$.

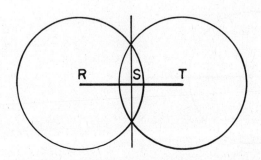

Thus $RS = ST$, and S is the midpoint of $\overline{RT}$.

We will learn in Chapter 2 why this construction enables us to bisect a line segment.

Frequently we encounter word problems involving length.

EXAMPLE 4   The lengths of two congruent line segments are $3x - 4$ and $\dfrac{5x - 3}{2}$ cm. Find the length of the line segments.

Solution   Because the line segments are congruent, we know that they are the same length, and so we have the equation:

$$3x - 4 = \frac{5x - 3}{2}$$

$$2(3x - 4) = \frac{5x - 3}{2} \cdot 2$$

$$6x - 8 = 5x - 3$$
$$x = 5$$

When $x = 5$, $3x - 4 = 11$ and $\frac{5x - 3}{2} = 11$, so the solution is:

Length of line segments = 11 cm.

*EXAMPLE 5   Find the length of two tracks having the same length if their lengths are $5x^2 - 7x + 4$ and $3x^2 + 4x + 25$ m.

Solution
$$5x^2 - 7x + 4 = 3x^2 + 4x + 25$$
$$2x^2 - 11x - 21 = 0$$
$$(x - 7)(2x + 3) = 0$$
$$x = 7 \text{ or } x = -1\tfrac{1}{2}$$

One may be tempted to discard the $-1\tfrac{1}{2}$ as a possible solution because it is negative. However, <u>a negative value for the variable may not necessarily result in a negative length</u>; therefore, each value must be substituted in the original equation or equations to determine whether or not it is a valid solution.

When $x = -1\tfrac{1}{2}$, $5x^2 - 7x + 4 = 25\tfrac{3}{4} = 3x^2 + 4x + 25$.
When $x = 7$, $5x^2 - 7x + 4 = 200 = 3x^2 + 4x + 25$.

Because each value results in a positive number, this problem has two solutions:

    Each length of track = 200 m

    <u>or</u>

    Each length of track = $25\tfrac{3}{4}$ m.

EXAMPLE 6   A board 26 ft long needs to be cut into three pieces of one length for shelves and two pieces of another length for supports. The supports must be 3 ft longer than the shelves. Find the lengths of the shelves and supports.

Solution

$$x = \text{shelf length}$$
$$x + 3 = \text{support length}$$
$$3x + 2(x + 3) = 26$$
$$3x + 2x + 6 = 26$$
$$5x = 20$$
$$x = 4$$
$$\text{Shelf length} = 4 \text{ ft}$$
$$\text{Support length} = 7 \text{ ft}$$

Students in calculus frequently are required to determine the appropriate equation for a word problem involving geometry. This book includes examples typical of calculus problems so that the student can practice setting up such equations.

**EXAMPLE 7 A piece of wire, L in. in length, is to be cut into five pieces: two of one length and three of another. If x represents the length of one of the two pieces, give the expression for the length of one of the three pieces in terms of x and L.

Solution

$$x = \text{length of one of two pieces}$$
$$2x = \text{length of both}$$
$$L - 2x = \text{total length of three pieces}$$
$$\frac{L - 2x}{3} = \text{length of one of the three pieces}$$

EXERCISE 1.2

For Problems 1 through 8, construct line segments representing each expression, given the following segments.

```
_____        _____        _____        _____
    a                b                c                d
```

1. $3a$ 2. $4b$ 3. $2a + 3c$ 4. $2b + 2d$

5. $4c - b$ 6. $3d - a$ 7. $2a - 2d$ 8. $6c - 2b$

9. Find the midpoint of $\overline{AB}$. _____

   A   B

10. Find the midpoint of $\overline{MN}$.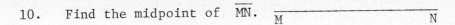

11. Bisect the segment $\overline{RS}$.

12. Bisect the segment $\overline{PQ}$.

13. Bisect the segment constructed in Problem 7.

14. Find the midpoint of the segment constructed in Problem 8.

15. Given $AC = 8$ cm and $\overline{BD}$ bisects $\overline{AC}$ in Figure 1.12, find the length of $\overline{AE}$.

16. Given $EB = 7$ in. and $\overline{AC}$ bisects $\overline{BD}$ in Figure 1.12, find the length of $\overline{BD}$.

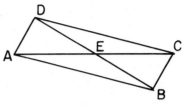

FIGURE 1.12

17. a) Make a drawing satisfying the following conditions: point A is to the left of C and to the right of B; points A, B, and C are collinear; $\overline{BC}$ is twice the length of $\overline{AB}$.

   b) Determine (if possible) the midpoint of $\overline{BC}$.

18. a) Make a drawing satisfying the following conditions: point D bisects $\overline{FG}$; $\overline{FC}$ is greater than $\overline{CG}$; C is between D and G.

   b) Which segment is longer, $\overline{DF}$ or $\overline{DC}$?

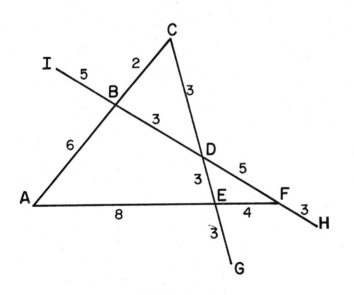

FIGURE 1.13

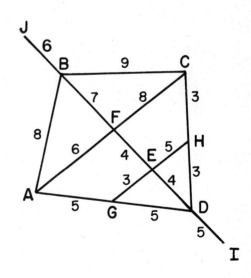

FIGURE 1.14

19. List all of the line segments that intersect $\overline{IH}$ in a single point in Figure 1.13.

20. List <u>all</u> of the line segments that intersect $\overline{IJ}$ in a single point in Figure 1.14.

21. For Figure 1.13, list each line segment that is bisected by another line segment. Give the letter for the corresponding midpoint.

22. For Figure 1.14, list each line segment that is bisected by another line segment. Give the letter for the corresponding midpoint.

23. Two line segments are the same length. The length of one is $2x + 1$ in. and the other is $3x - 6$ in. Find the length of the segments.

24. Two congruent line segments have a length expressed as $6(5x - 1) - (5x + 2) + 2$ ft and $2(9x + 1) + 4x + 1$ ft. Find the length of the segments.

25. Two steel reinforcing rods that are the same length measure $2(3x - 4) + x - 1$ ft and $6(x + 3) - (x + 16)$ ft. Find the length of the rods.

26. $3(4 - x)$ in. and $5 - x + 6(2 - x)$ in. represent the lengths of two identical pipes. Find the length of the pipes.

*27. Two identical shelves measure $7x^2 + 8x - 11$ in. and $5x^2 + 9x + 4$ in. Find the length of the shelves.

*28. Two congruent line segments measure $2x^2 - 10x + 10$ in. and $18 + 8x - 3x^2$ in. Find the length of the line segments.

*29. Two rooms of the same length measure $3x^2 - 4x - 25$ cm and $2x^2 - 4x + 144$ cm. Find the length of the rooms.

*30. Two identical control boxes measure $2x^2 + 23 - 11x$ mm and $3x^2 - 11x - 121$ mm. Find the length of the boxes.

31. One board is three feet shorter than twice the length of the other board. If the two boards together are 15 ft long, how long is each of the boards?

32. Five times the length of a rod is equal to the sum of 14 cm and three times the length of the rod. Find the length of the rod.

33. When one third of the sum of 10 in. and the length of a segment is added to the length of the segment, the result is equal to twice the length of the segment. Find the length of the segment.

34. When two times the sum of 5 mm and the length of a segment is subtracted from seven times the length of the segment, the result is 16 mm less than eleven times the length of the segment. Find the length of the segment.

**35. A board of length L ft is cut into two pieces where three times the length of one piece is 2 ft more than seven times the other. Represent this in an algebraic equation. Solve the equation for L in terms of x.

**36. A wire of length W cm is cut into two pieces where four times the shorter piece is 13 cm shorter than twice the sum of 3 and the longer piece. Express this as an algebraic equation. Solve for W in terms of x, the length of the shorter piece.

**37. A piece of fencing, L ft in length, is to be divided into four sections: two of one length and two of another. If x is the length of one longer piece, give the expression for one shorter piece in terms of x and L.

**38. A piece of fencing, L ft in length, is to be divided into three sections: two of one length and one longer piece. If x is the length of the longer piece, give the expression for one shorter piece in terms of x and L.

1.3 ANGLES AND THEIR MEASUREMENT

In this section, we will be concerned with angles and their properties. Two rays that have a common endpoint form an ANGLE. Thus $\overrightarrow{BA}$ ($\overleftrightarrow{AB}$) and $\overrightarrow{BC}$ form an angle ABC ($\angle$ABC).

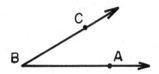

The common endpoint of the two rays is called the VERTEX of the angle. The rays $\overrightarrow{BA}$ and $\overrightarrow{BC}$ are called the SIDES of the angle. When we write $\angle$ACB or $\angle$DBA, the middle letter _always_ represents the vertex of the angle.

The order of the other letters does not matter. That is, $\angle$DBA is the same angle as $\angle$ABD: the angle formed by $\overrightarrow{BA}$ and $\overrightarrow{BD}$ with vertex B.

Sometimes we will refer to ∠B or ∠1 if there is no chance for ambiguity:

For the drawings above, we can refer to ∠B and to ∠1, ∠2, ∠3, and ∠4. For the drawing below, however, we cannot refer to ∠A without being ambiguous:

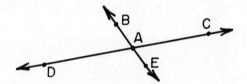

We must be specific and use ∠CAB, ∠DAC, and so on to indicate exactly which angle we mean.

Is the notation ∠PQR completely unambiguous? Could it possibly represent two different angles? Let us look at two rays $\overrightarrow{RQ}$ and $\overrightarrow{PQ}$ that intersect at Q:

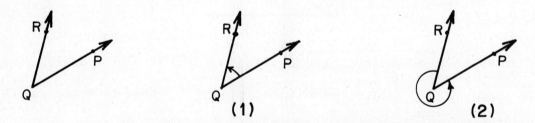

We see that we <u>could</u> mean two different angles. However, in elementary geometry we usually mean the smaller angle. If we mean the larger angle, we indicate that as shown on the right above. Below, if we wish to refer to the larger angle ∠RTS, we draw the circle segment to indicate the larger angle:

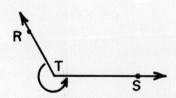

Angles are measured in degrees (°) or in radians (rad). We can think of an angle with a measure of 0 degrees (0°), or 0 rad, as one formed by two coincident rays:

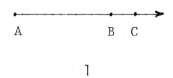

1

Rays $\vec{AB}$ and $\vec{AC}$ form an angle of 0° or 0 rad. As the rays rotate apart, they form larger and larger angles:

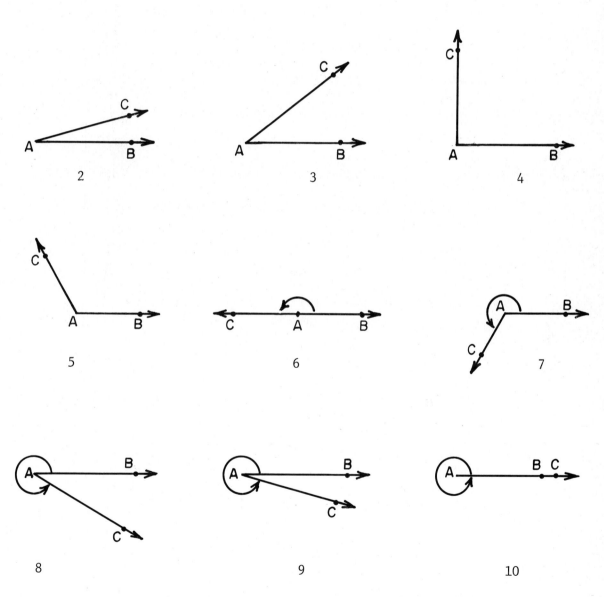

As the rays continue to rotate apart, they become opposite rays as shown in 6. They then come closer together until they coincide again in 10. The angle formed in 10 is defined to be an angle of 360° or 2π rad. Thus 1° is $\frac{1}{360}$ of a full rotation. (In Chapter 6, when we have the necessary definitions, we will discuss the meaning of 1 rad.)

When the rays forming an angle are opposite rays, as in 6, the angle is one half of a rotation and contains 180°, or π rad; it is called a STRAIGHT ANGLE. When the rays form an angle that is one fourth of a rotation, as in 4, the angle contains 90° or $\frac{\pi}{2}$ rad and is a RIGHT ANGLE and the rays are PERPENDICULAR. A right angle is indicated by putting a "box" at the vertex: ⌐ . If two line segments, rays, or lines are perpendicular, we use the notation ⊥. That is, $\overline{AB} \perp \overline{BC}$ means $\overline{AB}$ is perpendicular to $\overline{BC}$. Angles that contain less than 90° (or $\frac{\pi}{2}$ rad) are ACUTE ANGLES, as in 2 and 3. An angle that contains less than 180° (π rad) but more than 90° ($\frac{\pi}{2}$ rad) is an OBTUSE ANGLE (5). An angle that contains more than 180° (π rad) is a REFLEX ANGLE (7, 8, 9, and 10).

The instrument used to measure the number of degrees in an angle is a PROTRACTOR. It is pictured in Figure 1.15.

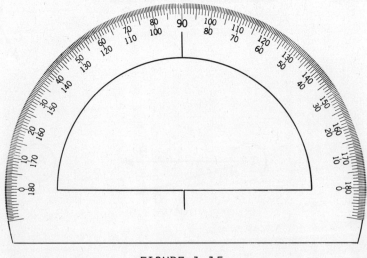

FIGURE 1.15

When measuring an angle, place the protractor so that the base line of it coincides with one ray and the center of the protractor's base coincides with the vertex, as shown in Figure 1.16.

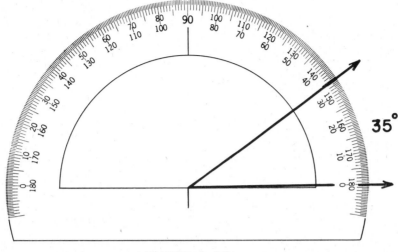

FIGURE 1.16

If the ray coincides with the bottom <u>right</u> side of the protractor, use the inner set of numbers to read where the other ray crosses the protractor to determine the number of degrees in the angle. The angle in Figure 1.16 measures 35°. If the ray coincides with the bottom <u>left</u> side of the protractor, use the outer set of numbers to measure the angle. The angle in Figure 1.17 is 115°.

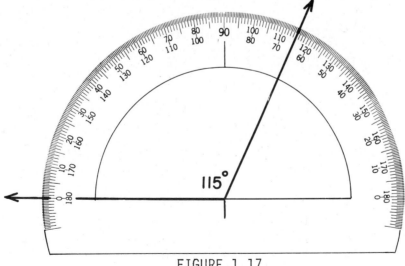

FIGURE 1.17

The protractor we have shown here enables you to measure angles to the nearest degree. Larger protractors are more accurate--you can measure to parts of a degree. One degree can be divided into 10ths and 100ths, yielding measurements such as 13.75°. One degree can also be divided into 60ths, each of which is called a MINUTE, so 60′ (60 minutes) = 1°.

> To convert from decimal notation to minutes we multiply the decimal by the unit fraction $\frac{60 \text{ min}}{1 \text{ degree}} = \frac{60'}{1°}$.

EXAMPLE 1   Convert 13.75° to degrees and minutes.

Solution $\frac{0.75 \; \cancel{\text{degree}}}{1} \cdot \frac{60 \text{ min}}{1 \; \cancel{\text{degree}}} = 45.00 \text{ min} = 45'.$

The answer is 13°45′.

Most frequently, however we need to convert minutes to decimal notation so that we may use a calculator.

> To convert minutes to decimal notation, we multiply the minutes by the unit fraction $\frac{1 \text{ degree}}{60 \text{ minutes}} = \frac{1°}{60'}$.

EXAMPLE 2   Convert 171° 16′ to decimal notation.

Solution $\frac{16 \; \cancel{\text{minutes}}}{1} \cdot \frac{1 \text{ degree}}{60 \; \cancel{\text{minutes}}} = 0.266\overline{6} \text{ degrees} = 0.266\overline{6}°$

The answer is 171.27° (to two decimal places).

In both of the examples above, the multiplication was done with the unit fraction, which equals one. This is helpful because we can cancel like units from the numerator and denominator, leaving the desired unit. If the unit we wish to use in the answer is <u>not</u> left, then we know the problem is set up incorrectly. It is important to check to be sure that the units are correct.

Because we have been given that 0° = 0 rad, 180° = π rad, and 360° = 2π rad, we can determine that $1 = \frac{360°}{360°} = \frac{2\pi \text{ rad}}{360°}$. Therefore $\frac{2\pi \text{ rad}}{360°} = 1$.

> To convert from degrees to radians, we multiply the number of degrees by the unit fraction $\frac{2\pi \text{ radians}}{360 \text{ degrees}} = \frac{2\pi \text{ rad}}{360°}$.

EXAMPLE 3 Convert 30° to radians.

Solution

$$\frac{30 \text{ degrees}}{1} \cdot \frac{2\pi \text{ rad}}{360 \text{ degrees}}$$

$$\frac{\cancel{3}^1}{1} \cdot \frac{2\pi \text{ rad}}{\cancel{36}_{12}}$$

$$1 \cdot \frac{\cancel{2}^1 \pi \text{ rad}}{\cancel{12}_6}$$

$$\frac{\pi \text{ rad}}{6}$$

Thus $30° = \frac{\pi}{6}$ rad.

We can also use a unit fraction to convert from radians to degrees because we know $1 = \frac{2\pi \text{ rad}}{2\pi \text{ rad}} = \frac{360°}{2\pi \text{ rad}}$. Therefore $\frac{360°}{2\pi \text{ rad}} = 1$.

> To convert from radians to degrees, we multiply the number of radians by the unit fraction $\frac{360 \text{ degrees}}{2\pi \text{ radians}} = \frac{360°}{2\pi \text{ rad}}$.

EXAMPLE 4 Convert $\frac{2\pi}{3}$ rad to degrees.

Solution $\frac{2\pi}{3} \text{ radians} \cdot \frac{360 \text{ degrees}}{2\pi \text{ radians}} = 120 \text{ degrees} = 120°$

Thus $\frac{2\pi}{3}$ rad $= 120°$.

Many times we must express radians as an exact answer using π, as in $\frac{3\pi}{4}$ rad or 2π rad. There are other times when it is more convenient to approximate. This occurs most frequently when using a calculator to evaluate the trigonometric functions of the angle as we will do in Chapter 7.

EXAMPLE 5   Convert 90° to radians, correct to four decimal places.

Solution $90° = \dfrac{90 \text{ \sout{degrees}}}{1} \cdot \dfrac{2\pi \text{ radians}}{360 \text{ \sout{degrees}}}$

$= \dfrac{\pi}{2}$ rad

$= 1.57079633$ rad   (to 8 decimal places)

or $90° = 1.5708$ rad   (to 4 decimal places)

EXAMPLE 6   Convert 1.76 rad to degrees in decimal notation.

Solution $1.76 \text{ rad} = \dfrac{1.76 \text{ \sout{rad}}}{1} \cdot \dfrac{360°}{2\pi \text{ \sout{rad}}}$

$= 100.84057°$

$1.76 \text{ rad} = 100.8406°$ (to 4 decimal places)

Two angles that contain the same number of degrees or radians are CONGRUENT ANGLES; thus congruent angles are EQUAL ANGLES. We symbolize ∠A congruent to ∠B by ∠A ≅ ∠B.

Two angles that have a common vertex and a common side and that lie on opposite sides of their common side are called ADJACENT ANGLES. ∠ABC and ∠CBD, below, are adjacent angles.

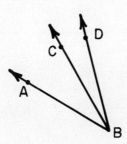

Two angles with measures that total 180° or π rad are SUPPLEMENTARY ANGLES.

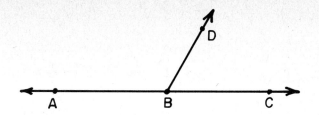

Given that ∠ABC above is a straight angle, we may conclude that ∠ABD and ∠DBC are supplementary angles because a straight angle contain 180° or π rad.

Two angles with measures that total 90° or $\frac{\pi}{2}$ radians are COMPLEMENTARY ANGLES.

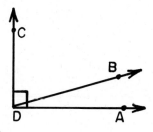

Since we are given that ∠CDA is a right angle, we know here that ∠CDB and ∠BDA are complementary angles.

When two lines intersect, they form four angles, and the nonadjacent angles, called VERTICAL ANGLES, are equal.

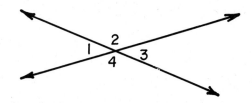

In the drawing above, ∠1 and ∠3 are vertical angles and thus are equal; ∠2 is equal to ∠4 because they also are vertical angles; ∠1 and ∠2 are supplementary as are ∠3 and ∠4, ∠3 and ∠2, and ∠1 and ∠4.

Remember that we can tell relationships from drawings, but we cannot determine size. Thus we can tell that one line segment intersects another, but we cannot assume that they are perpendicular just because they "look" perpendicular in the drawing. Likewise, we cannot judge two angles as equal unless we are given facts to justify that conclusion: that they are vertical angles or that they are the same size.

EXAMPLE 7   Find the measure of ∠CBD if ∠ABD is a straight angle and ∠ABC = 108°.

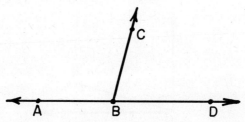

Solution   The two adjacent angles ∠ABC and ∠CBD form ∠ABD, which is a straight angle. Therefore the smaller angles are supplementary. ∠ABC + ∠CBD = ∠ABD. Let ∠CBD = x:

$$108° + x = 180°$$
$$x = 72°$$

Therefore   ∠CBD = 72°

EXAMPLE 8   Find the measure of ∠1 and ∠2 if ∠1 = 3x + 2 degrees, ∠2 = 7x - 12 degrees, and ∠AOB is a straight angle.

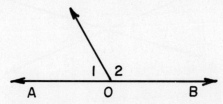

Solution   ∠1 and ∠2 are supplementary angles so their sum is 180°.
Writing this as an equation, we get:

36

$$(3x + 2) + (7x - 12) = 180°$$
$$10x - 10 = 180°$$
$$10x = 190°$$
$$x = 19°$$

$\angle 1 = 3x + 2$	$\angle 2 = 7x - 12$
$= 3(19°) + 2$	$= 7(19°) - 12$
$\angle 1 = 59°$	$\angle 2 = 121°$

Check $\angle 1 + \angle 2 = 180°$

$$59° + 121° \stackrel{?}{=} 180°$$
$$180° \stackrel{\checkmark}{=} 180°$$

Therefore the solution is $\angle 1 = 59°$ and $\angle 2 = 121°$.

EXAMPLE 9 Two angles are complementary. If one angle is $\frac{\pi}{12}$ rad more than the other, determine the size of each.

Solution Let x = smaller angle

$x + \frac{\pi}{12}$ = larger angle

Because the angles are complementary, their sum is $\frac{\pi}{2}$ rad.

This can be written:
$$x + x + \frac{\pi}{12} = \frac{\pi}{2}$$
$$2x = \frac{5\pi}{12}$$
$$x = \frac{5\pi}{24}$$

smaller angle = $\frac{5\pi}{24}$ rad   larger angle = $\frac{5\pi}{24} + \frac{\pi}{12}$

$= \frac{7\pi}{24}$ rad

Check $\frac{5\pi}{24} + \frac{7\pi}{24} = \frac{12\pi}{24} = \frac{\pi}{2}$ rad

The angles are $\frac{5\pi}{24}$ rad and $\frac{7\pi}{24}$ rad.

EXAMPLE 10 Two angles are complementary. Doubled, the smaller angle is 18° more than the larger angle. Find the number of degrees in the angles.

Solution   Let x = smaller angle

2x - 18° = larger angle

Because the two angles are complementary, their sum is 90°.

Writing this as an equation:

$$x + (2x - 18°) = 90°$$
$$3x = 108°$$
$$x = 36°$$

smaller angle = 36°   larger angle = 2x - 18°
                                         = 2(36) - 18°
                                         = 72 - 18°
                                         = 54°

Check   36° + 54° = 90°

The angles are 36° and 54°.

EXAMPLE 11   Find the number of degrees in $\angle 1, \angle 2,$ and $\angle 4$ formed by the intersection of two lines if $\angle 3 = 49°$.

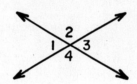

Solution $\angle 1$ and $\angle 3$ are vertical angles and are equal, so $\angle 1 = 49°$.

$\angle 2$ and $\angle 3$ are supplementary so $\angle 2 = 180° - \angle 3$
                                                       $= 180° - 49°$
                                               $\angle 2 = 131°$

$\angle 2$ and $\angle 4$ are vertical angles so $\angle 4 = 131°$.

Therefore $\angle 1 = 49°$, $\angle 2 = 131°$, and $\angle 4 = 131°$.

EXERCISE 1.3

1. Measure the following angles to the nearest whole degree.

a) 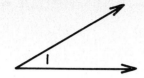   b)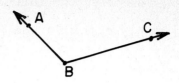

2. Measure the following angles to the nearest whole degree.

   a) 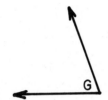   b)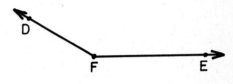

3. Draw an angle of:

   a)   15°. Label it ∠1.

   b)   122°. Label it ∠ABC.

   c) $\frac{\pi}{4}$ rad. Label it ∠P.

   d) $\frac{3\pi}{2}$ rad. Label it ∠DEF.

4. Draw an angle of:

   a)   58°. Label it ∠F.

   b)   180°. Label it ∠DBC.

   c) $\frac{5\pi}{4}$ rad. Label it ∠2.

   d) $\frac{\pi}{6}$ rad. Label it ∠RST.

5. Convert the following to decimal notation (to the nearest hundredth).

   a)   127°18′   b)   24°21′   c)   87°59′

6. Convert the following to decimal notation (to the nearest hundredth).

   a)   61°6′   b)   15°33′   c)   174°40′

7. Convert the following to degrees and minutes (to the nearest minute).

   a)   86.43°   b)   183.793°   c)   34.22°

8. Convert the following to degrees and minutes (to the nearest minute).

   a)   104.87°   b)   57.023°   c)   79.95°

9. Complete the following tables. Answers should be exact.

Degrees		30°	45°			120°		150°	
Radians	0			$\frac{\pi}{3}$	$\frac{\pi}{2}$		$\frac{3\pi}{4}$		π

10. Complete the following table. Answers should be exact.

Degrees	180°		225°	240°		300°			360°
Radians		$\frac{7\pi}{6}$			$\frac{3\pi}{2}$		$\frac{7\pi}{4}$	$\frac{11\pi}{6}$	

C 11. Convert the following to radians (to the nearest ten-thousandth).

   a) 41°   b) 306°   c) 169°   d) 80°

C 12. Convert the following to radians (to the nearest ten-thousandth).

   a) 200°   b) 93°   c) 9°   d) 65°

C 13. Convert the following to degrees in decimal notation (to the nearest hundredth).

   a) $\frac{\pi}{7}$ rad   b) 1.37 rad   c) $\frac{5\pi}{16}$ rad   d) 0.3 rad

C 14. Convert the following to degrees in decimal notation (to the nearest hundredth).

   a) 0.2 rad   b) $\frac{7\pi}{16}$ rad   c) 0.834 rad   d) $\frac{2\pi}{5}$ rad

15. Which of the angles in Problems 1 and 3 are acute angles?

16. Which of the angles in Problems 2 and 4 are obtuse angles?

17. Find the complements to the following angles:

   a) 14°   b) $\frac{3\pi}{7}$ c) 32°   d) y°   e) 2x°

18. Find the supplements to the following angles:

   a) 19°   b) $\frac{5\pi}{6}$ c) 90°   d) x°   e) 3x − 2°

19. If two lines intersect so that $\angle 3 = 22°$, find the number of degrees in $\angle 1$, $\angle 2$, and $\angle 4$.

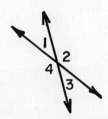

20. If two lines intersect so that $\angle 6 = \dfrac{2\pi}{5}$ rad, find the measure of $\angle 5$, $\angle 7$, and $\angle 8$.

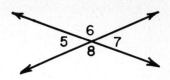

21. If two lines intersect as given in the drawing to the right, find the number of degrees in each angle.

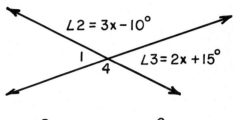

22. If two lines intersect as given in the drawing to the right, find the number of degrees in each angle.

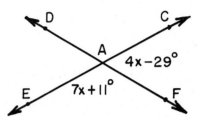

23. Two angles are supplementary. Three times the smaller is five less than twice the larger. Find the number of degrees in each angle.

24. Two angles are supplementary. Four times the smaller is $\dfrac{3\pi}{8}$ rad more than twice the larger. Find the measure of each angle.

25. Two angles are complementary. If five times the sum of the smaller and 8 is 49° greater than two times the larger, find the number of degrees in each angle.

C 26. Two angles are complementary. The sum of the smaller and 0.834 when divided by 4 is 0.437 rad less than the larger. Find the number of radians in each angle.

27. The complement of an angle is 6° more than six times the angle. Find the angle and its complement.

28. The supplement of an angle is 12° less than seven times the angle. Find the angle and its supplement.

1.4 CONSTRUCTING AND BISECTING ANGLES

Given an angle, we can construct an angle congruent to it using a compass and a straightedge.

To construct an angle congruent to a given angle:

1. Draw a line segment and mark one endpoint P to be the vertex of the new angle.

2. Place the spike of the compass at the vertex of the given angle and draw a circle.

3. Without changing the setting of the compass, place the spike at P on the line segment and draw another circle.

4. Place the spike of the compass at one point where the circle drawn intersects the given angle, then adjust the compass so that the other tip coincides with the other point of intersection with the given angle.

5. Without changing the setting of the compass, place the spike of the compass at the point on the line segment where the circle intersects it. Now draw a circle.

6. Draw the line segment from P to one of the two points of intersection of the two circles.

7. The angle you have constructed is congruent to the given angle because the two angles have the same measure.

EXAMPLE 1   Construct an angle congruent to ∠A.

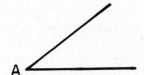

Solution   Step 1:   Draw a line segment and mark an endpoint to be the vertex.

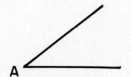

Steps 2
and 3:   Draw a circle about A. Then draw a circle about P without changing the setting of the compass.

42

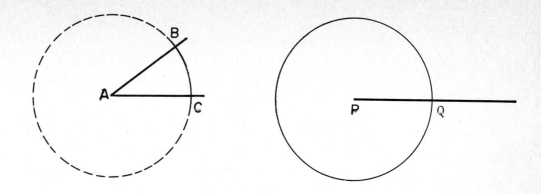

Steps 4 and 5: Adjust the compass so that one tip coincides with B and the other with C. Without changing the setting, place the spike of the compass at Q and draw a circle.

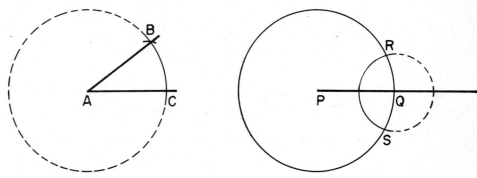

Steps 6 and 7: Draw the line segment from P to R (or to S). Then ∠QPR is congruent to ∠CAB or ∠QPS ≅ ∠CAB.

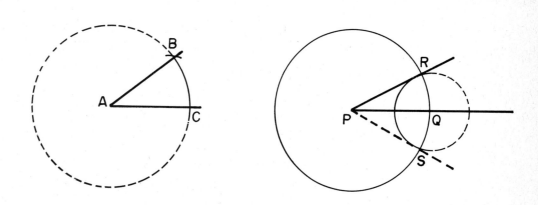

> To construct an angle twice the size of a given angle:
>
> 1. Follow Steps 1 through 5 for the construction of a congruent angle.
>
> 2. Without changing the compass setting, place the spike on the intersection of the two circles and draw a circle.
>
> 3. Draw a line segment from P to the point of intersection of this circle and the one with center at P.

The construction that bisects a line segment also gives us a line segment perpendicular to the given segment. Therefore we can use that same construction to construct a right angle (90° = $\frac{\pi}{2}$ rad = 1.5708 rad).

> To construct a right angle:
>
> 1. Draw a line segment $\overline{AB}$.
>
> 2. Bisect $\overline{AB}$ using the procedure given on p. 20.
>
> 3. The line segment that bisects $\overline{AB}$ is perpendicular to $\overline{AB}$ and therefore forms a right angle.

EXAMPLE 2   Construct a $\frac{\pi}{2}$ rad angle.

Solution   Step 1:   Draw a line segment $\overline{AB}$.

A   B

Step 2:   Bisect $\overline{AB}$.

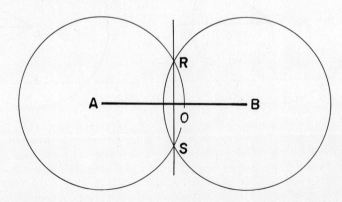

Step 3: The line segment $\overline{RS}$ is perpendicular to $\overline{AB}$ so

$$\angle ROB = \angle ROA = \angle AOS = \angle SOB = \frac{\pi}{2} \text{ rad}.$$

Given a line, we can construct a perpendicular through a given point not on that line.

To construct a perpendicular to a given line through a given point not on the line:

1. Set the tips of the compass to a distance <u>greater</u> <u>than</u> the distance from the point to the line segment.

2. Set the spike of the compass at the point and draw a circle; label the points of intersection of the circle and the line segment points A and B.

3. Adjust the tips of the compass to be greater than half the distance between A and B.

4. Without changing the setting, place the spike at A and draw a circle, then at B and draw another circle.

5. Draw a line segment connecting the points of intersection of the two circles just drawn, extending it if necessary to pass through the given point; if the given point does not lie on the constructed line segment, check the accuracy of the construction.

6. The line segment constructed is perpendicular to the given line.

EXAMPLE 3   Construct a perpendicular to segment c containing the point P.

P
•

———————
c

Solution   Steps 1 and 2:   Adjust the compass and draw a circle about P.

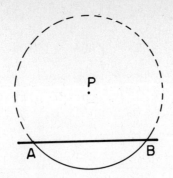

Steps 3
and 4: Adjust the compass to a distance greater than half the distance from A to B. Draw a circle about A, then draw a circle about B.

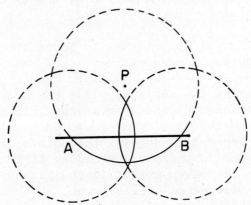

Step 5: Draw a line segment through the points of intersection of the two circles having A and B as centers. If necessary, extend it to pass through P.

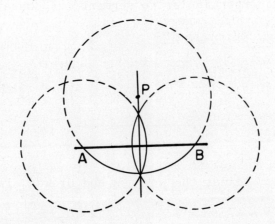

Step 6: The line segment constructed is the desired segment because it contains point P and is perpendicular to the given line segment.

To construct a perpendicular to a given line passing through a given point, P, on the line:

1. Set the spike of the compass at P and draw a circle; label the points of the intersection of the circle and the line points A and B.

2. Follow Steps 3 through 6 of the construction for a perpendicular to a given line through a given point not on the line.

We can also bisect an angle using a compass and straightedge.

To bisect an angle:

1. Place the spike of the compass on the vertex of the angle and draw a circle. Call the points of intersection of the circle with the sides of the angle points A and B.

2. Adjust the compass so the tips are apart less than the distance between A and B, but more than half that distance.

3. Place the spike at A and draw a circle.

4. Without changing the setting, place the spike at B and draw another circle.

5. Draw a line segment through the points of intersection of the circles drawn in Steps 3 and 4 and the vertex of the angle. If the line segment does not contain the vertex, then you have made an error. Check the accuracy of the circles drawn in Steps 3 and 4.

6. The line segment constructed bisects the given angle.

EXAMPLE 4   Bisect ∠C.

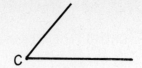

Solution   Step 1:   Draw a circle with the vertex of ∠C at the center.

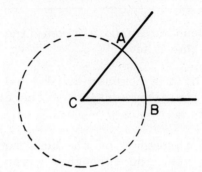

Steps 2
to 4:   Adjust the compass and draw equal circles with points A and B as centers.

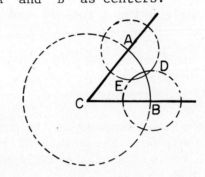

Step 5:   Draw the line segment $\overline{ED}$ containing C.

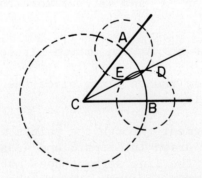

Step 6:   ∠DCA ≅ ∠DCB.

Now that we can bisect angles, we can construct many different angles.

EXAMPLE 5   Construct an angle containing 135°.

Solution   Because 180° - 135° = 45°, we can first construct a 90° angle.

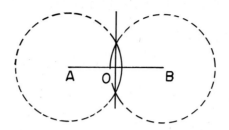

We can then bisect the 90° angle to get one of 45° and the supplementary angle of 135°.

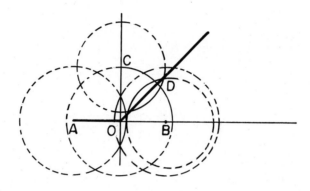

Since ∠DOB measures 45°, ∠AOD measures 135°. Thus ∠AOD is the desired angle.

We can also construct an angle of 60° (equal to $\frac{\pi}{3}$ rad). This construction is based on the facts that the sum of the angles in a triangle is 180° (π rad) and that the angles in an equilateral triangle are equal. The reasons justifying this construction are covered in Chapter 2.

> To construct a 60° ($\frac{\pi}{3}$ rad) angle:
>
> 1. Draw a line segment $\overline{AB}$.
>
> 2. Fix the compass at a distance less than or equal to the length of $\overline{AB}$.
>
> 3. Insert the spike at A and draw a circle.
>
> 4. Without changing the setting, place the spike at C, the point of intersection of the circle just drawn, and $\overline{AB}$. Draw another circle.
>
> 5. Draw the line segment from one point of intersection of the circles to A. The angle so formed measures 60° or $\frac{\pi}{3}$ rad.

EXAMPLE 6   Construct a $\frac{\pi}{3}$ rad angle.

Solution   Step 1:   Draw a line segment $\overline{AB}$.

```
  A              B
```

Steps 2
and 3:   Set the compass to a distance less than or equal to $\overline{AB}$ and draw a circle about A.

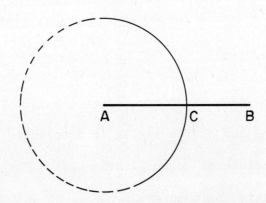

Step 4:   Without changing the setting, draw another circle about C.

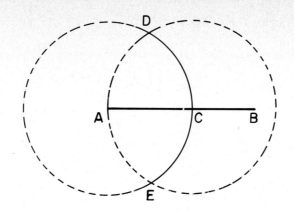

Step 5: Draw the line segment from D to A (or E to A, D to C, or E to C).

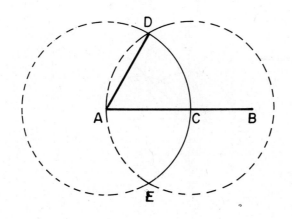

Then $\angle DAC = \frac{\pi}{3}$ rad (or $\angle EAC$, $\angle DCA$, $\angle ECA = \frac{\pi}{3}$ rad).

EXAMPLE 7 Construct an angle of 75°.

Solution Because 75° = 45° + 30°, 60° + 15°, or 90° - 15°, we have several constructions that would give us the desired result. Let us use the fact that 45° + 30° = 75°, and construct one angle of 45° and another of 30°. To construct a 45° angle, we first form a 90° angle.

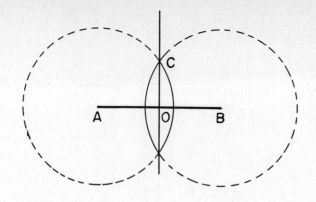

Now we bisect ∠BOC to obtain a 45° angle - ∠BOD.

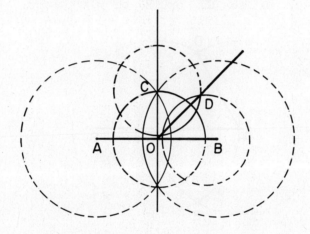

We next construct a 60° angle ∠TRS.

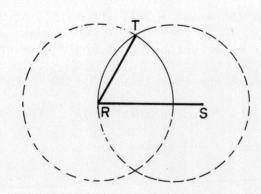

Now we bisect ∠TRS to obtain an angle of 30° - ∠URS.

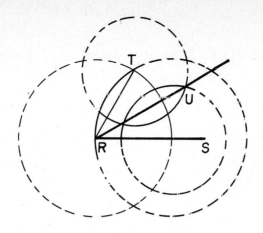

To make the drawings easier to follow, ∠BOD and ∠URS are shown without the circles and line segments that we used to construct them.

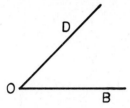
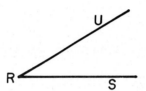

For the final step, we need to construct an angle congruent to ∠BOD as an adjacent angle to ∠URS, using $\overline{UR}$ as the common side.

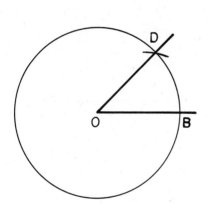
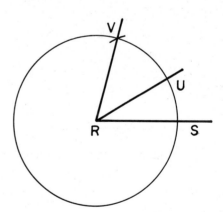

Because ∠BOD = ∠URV = 45° and ∠SRU = 30°, ∠SRV = 75°, and thus is the desired angle.

EXERCISE 1.4

For Problems 1 through 4, construct an angle congruent to the given angle.

1.

2.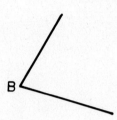

3.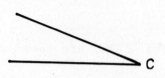

4.

5. Construct an angle equal to ∠A + ∠C above.

6. Construct an angle equal to ∠B + ∠D above.

7. Construct an angle of 90°.

8. Construct an angle of $\frac{3\pi}{2}$ rad.

9. Construct a line segment perpendicular to segment a and passing through P.

   • P   • Q

10. Construct a line segment perpendicular to segment b and passing through Q.

11. Construct a line segment perpendicular to $\overline{AC}$ passing through B.

12. Construct a line segment perpendicular to $\overline{RT}$ passing through S.

13. Bisect ∠A in Problem 1.

14. Bisect ∠D in Problem 4.

15. Bisect ∠φ below.

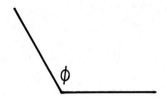
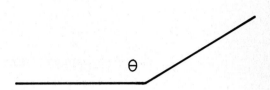

16. Bisect ∠θ above.

For Problems 17 through 24, construct angles of the given size.

17. 22.5°   18. 15°   19. $\frac{5\pi}{4}$ rad   20. $\frac{7\pi}{6}$ rad

21. 105°   22. 52.5° 23. $\frac{5\pi}{12}$ rad 24. $\frac{7\pi}{12}$ rad

CHAPTER 1 REVIEW

1. For each figure, give the term(s) appropriate to it from the following list:

       curve   convex simple closed curve
       circle   simple closed curve
       line   closed curve

   If none is appropriate, write "none."

   a)  b)

   c)  d)

   e)  f)

   g)  h)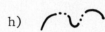

2. Match each item in Column I with the appropriate item from Column II. Items in Column II can be used once, more than once, or not at all.

   <u>Column I</u>

   1. has no dimension
   2. connected pathway
   3. intersection of a point and a plane
   4. line segment
   5. line
   6. ·P
      ←•————•→
        c   d
   7. •————→
      P
   8. non-intersecting lines in a plane

   <u>Column II</u>

   A. skew
   B. $\overleftrightarrow{AB}$
   C. parallel
   D. collinear
   E. ray
   F. $\overline{QR}$
   G. plane
   H. coplanar
   I. point
   J. noncollinear
   K. curve

Fill in the blank for Problems 3 through 7.

3. Points that lie on the same line are called _____.

4. Two lines not intersecting, not parallel and not in the same plane are called _____.

5. Two planes representing the same plane are called _____.

6. _____ distinct points determine a line.

7. _____ distinct points determine a plane.

8. Make one drawing to represent: $\overleftrightarrow{PQ}$ containing R; $\overline{RP}$ less than $\overline{RQ}$; R not contained in $\overrightarrow{PQ}$.

9. Make one drawing to represent: R, S, T, and Q are four non-collinear distinct points; Q lies on $\overleftrightarrow{RS}$; R does not lie on $\overrightarrow{SQ}$.

10. Make one drawing to represent: A, B, and C are noncollinear points; B, S, and C are collinear points; S, A, and T are collinear points.

11. $\overline{AB}$ contains C; $\overline{BD}$ contains E; F lies on $\overrightarrow{CB}$. Must F lie on $\overrightarrow{CA}$? (Support your answer with a drawing.)

12. Make one drawing to represent: plane FG intersects plane BC in line m; line 1 intersects line m and passes through FG at T.

13. Make one drawing to represent: plane CD contains line g; plane FG contains line m; the lines g and m intersect at T.

14. Plane PQ contains m_1; plane RS contains m_2. If the two planes are parallel, must the two lines be parallel?

Answer Problems 15 through 19 "yes" or "no."

15. Can the intersection of a point and a ray in the same plane be exactly one segment?

16. Can the intersection of a ray and a ray in the same plane be exactly one point?

17. Can the intersection of a segment and a line in the same plane be exactly one ray?

18. Can the intersection of a ray and a ray in the same plane be exactly one line?

19. Can the intersection of a plane and a line in the same plane be exactly one point?

20. List <u>all</u> of the line segments in Figure 1.18 that intersect $\overline{AD}$ in a single point.

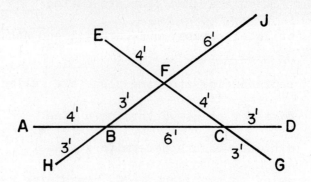

FIGURE 1.18

21. List all of the segments in Figure 1.18 that are bisected by another segment.

22. Two line segments are congruent and have length $5 + x + 3(9 - x)$ and $7(x + 2)$ ft. Find the length of the segments.

*23. Two boards of the same length are $x^2 + 2x + 2$ in. and $2x^2 - 3x + 2$ in. long. Find the length of the boards.

**24. A board of L feet is cut into two pieces. One foot less than the first piece is equal to twice the second piece. Express this as an algebraic equation. Solve for L in terms of x, the length of the first piece.

25. Give all possible types of relationships three distinct planes in space may have. Illustrate with drawings.

26. Convert 121.35° to degrees and minutes.

C 27. Convert 74°39′ to decimal notation (correct to four places).

28. Convert 200° to radians. Give an exact answer.

29. Convert $\frac{4\pi}{9}$ rad to degrees.

C 30. Convert 1.143 rad to degrees in decimal notation (correct to four places).

31. For each of the angles in Problems 26 through 30, state whether it is acute, obtuse, or reflex.

32. Find the complement of an angle of 19°.

33. If two lines intersect so that $\angle 4 = \dfrac{3\pi}{7}$ rad, find the measure of the other angles.

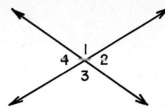

34. If two lines intersect so that $\angle 1 = 6x + 5°$ and $\angle 2 = 2x + 1°$, find the number of degrees in each angle.

35. Two angles are supplementary. One is 9° less than six times the other. Find the angles.

36. Find the number of degrees in $\angle A$.

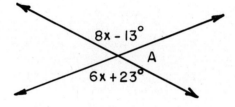

37. Find the number of degrees in each angle.

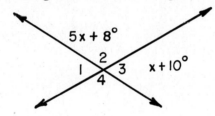

Use only a straightedge, a compass and a pencil for Problems 38 through 44.

38. Construct a line segment equal to $3a + c - 2b$ given

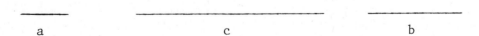

        a   c   b

39. Construct the midpoint of $\overline{AB}$.

       A   B

40. Bisect $\angle A$.

41. Construct the perpendicular bisector of $\overline{MN}$.

   M   N

42. Construct a line segment perpendicular to $\overline{AB}$ through Q.

   • Q

   A   B

43. Construct an angle of 157.5°.

44. Construct an angle of $\frac{7\pi}{24}$ rad.

CHAPTER 2
TRIANGLES

2.1 BASIC DEFINITIONS

We have talked about lines and angles and simple closed curves. Let us recall that a SIMPLE, CLOSED CURVE is a closed curve that does not intersect itself. When a simple closed curve consists entirely of line segments, the figure is called a POLYGON, where the line segments are SIDES of the polygon. The endpoints of the line segments are VERTICES of the polygon. The (interior) ANGLES of a polygon are the angles formed by two adjacent sides. A polygon that is a convex simple closed curve is a CONVEX POLYGON. An EQUILATERAL POLYGON has all sides of equal length; all sides are congruent. An EQUIANGULAR POLYGON has all angles of equal measure; that is, all angles are congruent.

The polygon with the fewest number of sides is a TRIANGLE, which has three sides. (Why can we not have a two-sided polygon?) A polygon with four sides is a QUADRILATERAL. There are many more categories of polygons, and we will study many of them in later sections.

Definitions and illustrations of polygons are shown in Table 2.1.

TABLE 2.1

Definitions and Illustrations of Polygons

Definition	This set of points *does* satisfy the definition.	This set of points *does not* satisfy the definition.
A POLYGON is a simple, closed curve composed entirely of line segments.		
A CONVEX POLYGON is a polygon that is also a convex simple closed curve.		
A QUADRILATERAL is a polygon with four sides.		
A TRIANGLE is a polygon with three sides.		

EXERCISE 2.1

List the letters of all the figures that belong to each category of curve.

1. Curve
2. Convex simple closed curve
3. Closed curve
4. Simple closed curve
5. Quadrilateral
6. Polygon
7. Triangle
8. Convex polygon

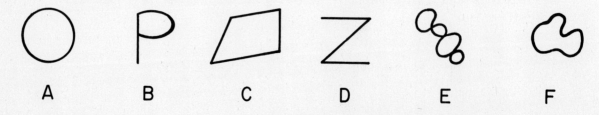

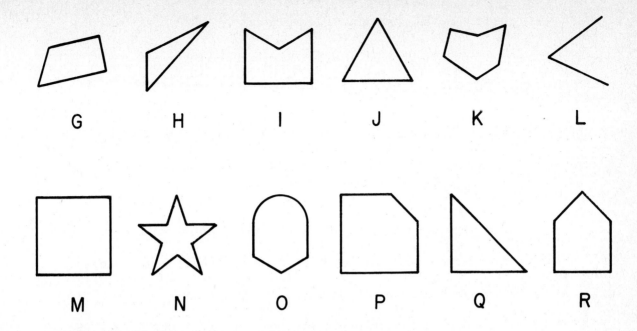

2.2 TRIANGLES AND THEIR CLASSIFICATION

The polygon with the fewest sides, the TRIANGLE, has three angles, as its name implies, and therefore, has three VERTICES. When we wish to refer to a triangle, we usually list the names of the vertices: is symbolized by "triangle ABC" or △ABC. is symbolized triangle DFG or △DFG. The order in which the letters are given does not matter: △DFG, △FDG, △DGF, △FGD, △GFD, and △GDF all represent the same triangle. Triangle DFG has vertices D, F, and G; angles ∠D, ∠F, and ∠G; and sides $\overline{GF}$, $\overline{FD}$, and $\overline{DG}$. When the vertices are labeled with capital letters, the same lower case letter refers to the side opposite the angle. Thus for △DFG, side $\overline{DF}$ may also be labeled side g; side $\overline{FG}$, side d; and side $\overline{DG}$, side f.

Triangles can be classified by the number of sides that are congruent. A triangle with all three sides congruent is an EQUILATERAL TRIANGLE. A triangle with at least two sides congruent is an ISOSCELES TRIANGLE. A triangle with no sides congruent is a SCALENE TRIANGLE.

The angles opposite the equal sides in an isosceles triangle are congruent. Therefore in △ABC, if $\overline{AB} \cong \overline{BC}$, then ∠A ≅ ∠C

In a scalene triangle no two angles are congruent.

Because the angles opposite equal sides are congruent, the angles in an equilateral triangle are all equal. If △RST is an equilateral triangle, then the lengths of the sides are the same: t = r = s. Because t = r, we have ∠T ≅ ∠R. Similarly, because r = s, ∠R ≅ ∠S. Thus ∠T ≅ ∠R ≅ ∠S. An equilateral triangle is also equiangular.

Triangles can also be classified by the type of angle in the triangle. An ACUTE TRIANGLE has three acute angles. An OBTUSE TRIANGLE has one obtuse angle. A RIGHT TRIANGLE has one right angle.

Using these two classification systems, we can label triangles. Some examples are:

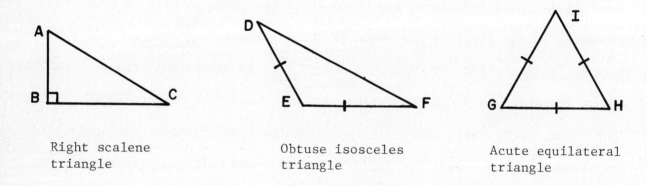

Right scalene triangle   Obtuse isosceles triangle   Acute equilateral triangle

Drawings can be marked to give us information about the figure. In △ABC above, ∠ABC is marked with ⌐ to indicate it is a right angle. In △DEF, sides $\overline{DE}$ and $\overline{EF}$ are marked with —|— to indicate that these sides are congruent. Similar marks in △GHI indicate that sides $\overline{GH}$, $\overline{HI}$, and $\overline{IG}$ are all congruent.

Because we know that angles opposite congruent sides in a triangle are congruent, we may indicate that these angles are congruent by adding marks to the appropriate figures:

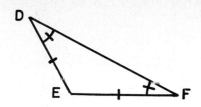
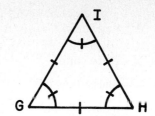

> The sum of the angles in a triangle is 180° or π rad.

The fact that the sum of the angles in a triangle is equal to a straight angle is proved in Chapter 3.

We can use the fact that the sum of the angles in a triangle is 180° to solve several types of problems.

EXAMPLE 1   Given a triangle in which ∠1 = 25° and ∠3 = 76°, find the number of degrees in ∠2.

Solution   Let x = number of degrees in ∠2

$$x + 25° + 76° = 180°$$

$$x = 79°$$

Thus,   ∠2 = 79°.

EXAMPLE 2   Given a triangle in which two angles are 47°19′ and 89°49′, find the number of degrees in the third angle.

Solution   47°19′

+ 89°49′

136°68′ = 137°8′   (Remember 60′ = 1°.)

Because the sum of the angles in a triangle must equal 180°, the measurement of the third angle can be obtained as follows:

180°   = 179°60′

− 137°8′ = 137° 8′

42°52′

The third angle contains 42°52′.

EXAMPLE 3   The angles of a triangle are in the ratio of 3:4:2. Find the number of radians in each angle.

Solution   3x = angle one
4x = angle two
2x = angle three

$$3x + 4x + 2x = \pi \text{ rad}$$
$$9x = \pi$$
$$x = \frac{\pi}{9}$$

angle one = $\frac{\pi}{3}$ rad

angle two = $\frac{4\pi}{9}$ rad

angle three = $\frac{2\pi}{9}$ rad

EXAMPLE 4   Given a triangle in which $\angle 1 = 4x + 1°$, $\angle 2 = 3x - 3°$, and $\angle 3 = 12x + 11°$, find the number of degrees in each angle.

Solution
$$4x + 1 + 3x - 3 + 12x + 11 = 180$$
$$19x + 9 = 180$$
$$x = 9°$$

$\angle 1 = 4x + 1°$	$\angle 2 = 3x - 3°$	$\angle 3 = 12x + 11°$
$= 4(9) + 1$	$= 3(9) - 3$	$= 12(9) + 11$
$\angle 1 = 37°$	$\angle 2 = 24°$	$\angle 3 = 119°$

Check $\angle 1 + \angle 2 + \angle 3 = 180°$
$$37° + 24° + 119° = 180°$$
$$180° = 180°$$

Thus $\angle 1 = 37°$, $\angle 2 = 24°$, and $\angle 3 = 119°$.

EXAMPLE 5   In a triangle, the second angle is one more than twice the first, and the third is 19° larger than the first. Find the number of degrees in each angle.

Solution
x = first angle
2x + 1 = second angle
x + 19 = third angle

$$x + 2x + 1 + x + 19 = 180$$
$$4x + 20 = 180$$
$$x = 40$$

Thus, the first angle = 40°, the second angle = 81°, and the third angle = 59°.

EXERCISE 2.2

1. For triangle PQR, name each vertex and the side opposite it.

2. In the drawing below add the appropriate markings to indicate that $\overline{AB} \cong \overline{CD}$ and that $\angle DBC \cong \angle BCA$.

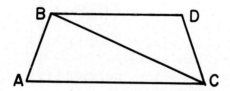

3. Are all isosceles triangles equilateral?

4. Are all equilateral triangles isosceles?

5. For each triangle below, determine whether it is acute, right, or obtuse.

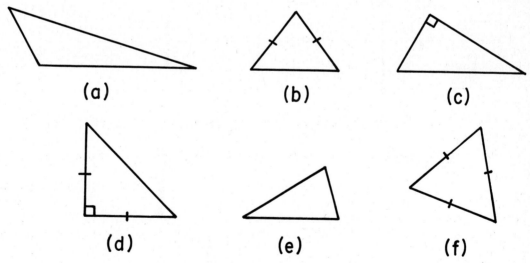

6. For each triangle above, determine whether it is scalene, isosceles, or equilateral.

7. List all the triangles in the
   drawing to the right and
   state whether each is right, obtuse,
   or acute.

8. List all of the triangles in the
   drawing to the right and state
   whether each is scalene, isosceles,
   obtuse scalene, or right isosceles.

9. Determine the number of triangles
   in the figure below.

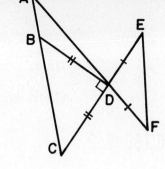

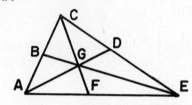

10. In the table below, write "possible" if it is possible to have such a triangle; write "not possible" if it is not. For example, it is not possible to have an obtuse triangle that is equilateral (why?) so "not possible" has been entered in that blank.

	Scalene	Isosceles and not equilateral	Equilateral
Acute			
Right			
Obtuse			Not possible

11. What is the maximum number of acute angles a triangle may have?

12. a) What is the maximum number of right angles a triangle may have?

    b) What is the maximum number of obtuse angles a triangle may have?

For each part in Exercises 13 and 14 you are given the number of degrees in two angles of a triangle. Find the number of degrees in the third angle.

13. a) 40° & 90°   b) 24.7° & 95.34°   c) 35°17′ & 72°48′

14. a) 33° & 75°   b) 14 1/2° & 39 3/5°   c) 106°37′ & 22°41′

For each part in Exercises 15 and 16 you are given the number of radians in two angles of a triangle. Find the number of radians in the third angle.

15. a) $\dfrac{\pi}{8}$ rad & $\dfrac{5\pi}{16}$ rad   b) $\dfrac{5\pi}{9}$ rad & $\dfrac{\pi}{6}$ rad   c) 0.973 rad & 1.021 rad

16. a) $\dfrac{3\pi}{8}$ rad & $\dfrac{3\pi}{16}$ rad   b) $\dfrac{7\pi}{12}$ rad & $\dfrac{\pi}{9}$ rad   c) 2.043 rad & 0.883 rad

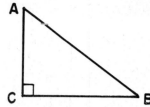

17. In a right triangle ABC, $\angle A = 66°31'$.
    Find the measure of $\angle B$.

18. In a right triangle ABC, $\angle B = \dfrac{3\pi}{8}$ rad.
    Find the measure of $\angle A$.

Give the number of degrees in each angle in Problems 19 and 20.

19. 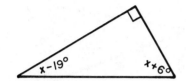   20.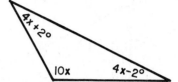

21. A triangle has two angles with equal measures. The third angle is twice as large as either one of the others. Give the number of radians in each angle and name the type of triangle.

22. In triangle ABC, $\angle A$ is ten times as large as $\angle B$, but $\angle B$ is only one-seventh the size of $\angle C$. Give the number of degrees in each angle.

23. The angles of a triangle are in the ratio 5:12:13. Give the number of degrees in each angle.

24. In triangle DEF, angle D is one-half the size of $\angle E$, and $\angle F$ is 30° more than $\angle E$. Give the measures of angles D, E, and F.

25. What is the measure in degrees of an angle in an equilateral triangle?

26. A triangle has $\angle A$ that is five more than four times $\angle B$. If $\angle A$ is 17°, how many degrees are in $\angle C$?

27. One angle is one-third the size of the second angle. The third angle is five more than the second. Find the number of degrees in each angle of this triangle.

28. A triangle has two equal angles. Find the number of degrees in the equal angles if the third angle is 50°.

29. $\angle A$ is three more than twice $\angle B$. $\angle C$ is three less than one-third $\angle B$. How many degrees are in each angle of this triangle?

30. In a given triangle, one angle is one-fifth the largest angle. The third angle is $\dfrac{\pi}{5}$ rad larger than the smallest angle. Find the number of radians in each angle.

2.3 RIGHT TRIANGLES AND THE PYTHAGOREAN THEOREM

The right triangle is especially important to study because of its special properties. Because it is so important we have special terms for its sides. The sides that form the right angle are called LEGS of the right triangle. The side opposite the right angle is called the HYPOTENUSE. The hypotenuse is always the longest side in the right triangle.

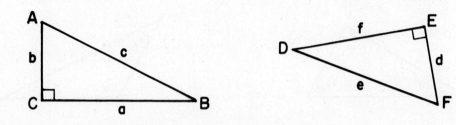

In the drawings above recall that ⌐ indicates a right angle, so $\overline{AC}$ and $\overline{CB}$ are legs of $\triangle ABC$, and $\overline{AB}$ is the hypotenuse. For $\triangle DEF$, d and f are the legs and e is the hypotenuse.

The sides of the right triangle have a special relationship to each other that is given by the Pythagorean theorem.

The Pythagorean Theorem

Given a right triangle with legs of length a and b and hypotenuse of length c, then $a^2 + b^2 = c^2$.

If the length of the sides of a triangle satisfy $a^2 + b^2 = c^2$, then the triangle is a right triangle with hypotenuse c.

EXAMPLE 1   A right triangle has legs of length 5 cm and 12 cm. Find the length of the hypotenuse.

Solution
$$a = 5 \text{ cm}$$
$$b = 12 \text{ cm}$$
$$c = x \text{ cm}$$

$$a^2 + b^2 = c^2$$
$$5^2 + 12^2 = x^2$$
$$169 = x^2$$
$$\pm 13 = x$$

Because we wish to measure length in positive units, we discard the solution -13. The hypotenuse is 13 cm in length.

EXAMPLE 2 A right triangle has a 6 in. leg and a 9 in. hypotenuse. Find the length of the other leg.

Solution
$$a = 6 \text{ in.}$$
$$b = x \text{ in.}$$
$$c = 9 \text{ in.}$$
$$a^2 + b^2 = c^2$$
$$6^2 + x^2 = 9^2$$
$$36 + x^2 = 54$$
$$x^2 = 18$$
$$x = \pm \sqrt{18} = \pm 3\sqrt{2}$$

The other leg is $3\sqrt{2}$ in. long.

Using a calculator, we get 4.243 or ∼4 inches. (The sign "∼" means "approximately.")

Your instructor will indicate whether you should leave your answers exact in radical form $(3\sqrt{2})$ or whether you should use a calculator to approximate the answer. You should become familiar with both forms of answers.

There are three right triangles that occur very frequently. The first is the 3:4:5 right triangle, on page 72. To verify that it is a right

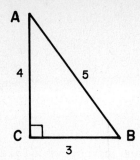 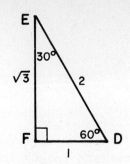 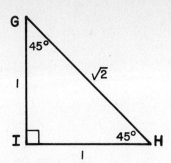

triangle, let us use the Pythagorean theorem:

$$a^2 + b^2 = c^2$$
$$3^2 + 4^2 \stackrel{?}{=} 5^2$$
$$9 + 16 \stackrel{?}{=} 25$$
$$25 \stackrel{\checkmark}{=} 25$$

Therefore, a triangle with sides in the ratio 3:4:5 is a right triangle.

The second important triangle has angles of 30°, 60°, and 90°, and sides in the ratio $1:\sqrt{3}:2$ (in the middle, above). Let us verify that this ratio satisfies the Pythagorean theorem.

$$a^2 + b^2 = c^2$$
$$1^2 + (\sqrt{3})^2 \stackrel{?}{=} 2^2$$
$$1 + 3 \stackrel{?}{=} 4$$
$$4 \stackrel{\checkmark}{=} 4$$

Thus this ratio does satisfy the theorem. Note we did not indicate any unit of measurement for the sides. The units do not matter, as long as the <u>same</u> unit is used for all three sides.

The third important triangle is the isosceles right triangle, which has two 45° angles and a 90° angle (on the right, above). Because the triangle is isosceles (the two legs are equal) we can give each leg the unit length "one." Let us then determine the length of the hypotenuse:

$$a^2 + b^2 = c^2$$
$$1^2 + 1^2 = x^2$$
$$2 = x^2$$
$$\pm \sqrt{2} = x$$
$$\sqrt{2} = x$$

We choose the positive root because we are measuring length. Thus for the 45° right triangle the sides are in the ratio $1:1:\sqrt{2}$.

You should memorize the ratios for each of these three triangles because they occur frequently. Because the 30°-60°-90° triangle and the 45°-45°-90° are especially important in trigonometry and calculus, <u>you should memorize them in both degrees and radians</u>. (See Problems 1 and 2 of the Exercises at the end of this section for the radian measure for these triangles.)

The Pythagorean theorem and these three basic triangles allow us to work a wide variety of problems.

EXAMPLE 3   Determine whether a triangle with sides of length 12 cm, 4 cm, and 13 cm is a right triangle.

  Solution   Because 13 cm is the longest side of the triangle, it must be the hypotenuse if the triangle is indeed a right triangle. Therefore, we can let:

$$a = 4 \text{ cm}$$
$$b = 12 \text{ cm}$$
$$c = 13 \text{ cm}$$

$$a^2 + b^2 = c^2$$
$$4^2 + 12^2 \stackrel{?}{=} 13^2$$
$$16 + 144 \stackrel{?}{=} 169$$
$$160 \neq 169$$

Because $160 \neq 169$, the triangle is <u>not</u> a right triangle.

EXAMPLE 4   Given a 30°-60°-90° triangle with hypotenuse of 12 in., find the length of each leg.

Solution   In a 30°-60°-90° triangle, we know the sides are in the ratio $1:\sqrt{3}:2$. We draw two 30°-60°-90° triangles oriented the same so we may easily compare them. The sides of the first are labeled 1, $\sqrt{3}$, and 2. In the second triangle we label the hypotenuse 12 in., the shorter leg y and the longer leg x. We can use the drawings to guide us when we set up the proportions. We know that 12 in. is to 2 as the longer leg is to $\sqrt{3}$:

$$\frac{12 \text{ in.}}{2} = \frac{x}{\sqrt{3}}$$

$$x = 6\sqrt{3} \text{ in.}$$

We have $x = 6\sqrt{3}$ in. or 10.3923 in. (to 4 places). To find the length of the shorter leg y, we use the drawings to help set up the proportion:

$$\frac{12 \text{ in.}}{2} = \frac{y}{1}$$

$$y = 6 \text{ in.}$$

Thus y = 6 in. The two legs thus are 6 in. and $6\sqrt{3}$ in. (10.3923 in.).

Note that in setting up the second proportion in the example, we could have used the result from the first proportion: $6\sqrt{3}$ in. to $\sqrt{3}$ instead of 12 to 2. We chose, however, to use the information given in order to avoid introducing in the second answer any error we might have made in solving the first proportion.

EXAMPLE 5   Given an isosceles right triangle with hypotenuse of 10 ft, find the length of the legs.

Solution Because the right triangle is isosceles, we know that it must be a $\frac{\pi}{4}$ rad - $\frac{\pi}{4}$ rad - $\frac{\pi}{2}$ rad triangle and the sides are in the ratio $1:1:\sqrt{2}$. We draw two triangles oriented the same so we may easily compare them. The sides of the first are labeled 1, 1, and $\sqrt{2}$. In the second triangle, we label the hypotenuse 10 ft and each of the legs x. We can use the drawings to guide us when we set up the proportions. The proportion we need is

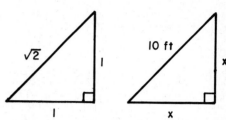

$$\frac{10 \text{ ft}}{\sqrt{2}} = \frac{x}{1}$$

$$\sqrt{2}x = 10$$

$$x = \frac{10}{\sqrt{2}}$$

$$= \frac{10}{\sqrt{2}} \cdot \frac{\sqrt{2}}{\sqrt{2}}$$

$$= 5\sqrt{2} \text{ ft or } 7.0711 \text{ ft (to 4 places)}$$

The legs are $5\sqrt{2}$ ft (7.0711 ft).

EXAMPLE 6 Given the triangle at the right with meters as the unit, find the length of each side.

Solution Because it is a right triangle, we can use the Pythagorean theorem where:

$$a = 2x + 2$$
$$b = 2x - 1$$
$$c = 3x$$

$$a^2 + b^2 = c^2$$

$$(2x+2)^2 + (2x-1)^2 = (3x)^2$$

$$4x^2 + 8x + 4 + 4x^2 - 4x + 1 = 9x^2$$

$$8x^2 + 4x + 5 = 9x^2$$

$$0 = 9x^2 - 8x^2 - 4x - 5$$

$$0 = x^2 - 4x - 5$$

$$0 = (x-5)(x+1)$$

Either $x - 5 = 0$ or $x + 1 = 0$

$\qquad\qquad x = 5 \qquad\qquad x = -1$

If $x = 5$, then $\qquad\qquad$ If $x = -1$, then

$a = 2x + 2 = 12$ m $\qquad\qquad a = 2x + 2 = 0$ m

$b = 2x - 1 = 9$ m $\qquad\qquad b = 2x - 1 = -3$ m

$c = 3x = 15$ m $\qquad\qquad c = 3x = -3$ m

Because we get negative and zero values when $x = -1$, we must discard that root. The solution to the problem then is:

$a = 12$ m

$b = 9$ m

$c = 15$ m

When working word problems using geometry, you should draw a picture representing the physical relationships and enter any information that you have. Frequently making the drawing gives enough clues so that you will then know how to proceed with the problem.

EXAMPLE 7   A guy wire attached to an upright pole 18 ft above level ground is 24 ft long and has a stake attached to the other end. How far from the base of the pole should the stake be driven so that the wire will be taut? (The stake will be driven far enough into the ground so that the wire will touch the ground.)

Solution   First a drawing is made. The drawing is of a right triangle so we may use the Pythagorean theorem to solve for x.

$$a = 18 \text{ ft}$$
$$b = x \text{ ft}$$
$$c = 24 \text{ ft}$$
$$a^2 + b^2 = c^2$$
$$18^2 + x^2 = 24^2$$
$$x^2 = 576 - 324$$
$$x^2 = 252$$
$$x = \pm \sqrt{252} = \pm 6\sqrt{7}$$
$$= 6\sqrt{7} \text{ ft or } \sim 15.8745 \text{ ft}$$

The stake should be driven in the ground ~15.87 ft from the pole.

EXAMPLE 8   One airplane leaves an airport and flies due north at 300 knots (nautical miles [n mi] per hour). At the same time, another airplane leaves the same airport and flies due west at 320 knots. After $1\frac{1}{2}$ hours, how far apart are the planes?

Solution   First make a drawing. After $1\frac{1}{2}$ hours the first airplane has traveled 450 n mi and the second, 480 n mi, so we enter that information on the drawing. Now we can use the Pythagorean theorem.

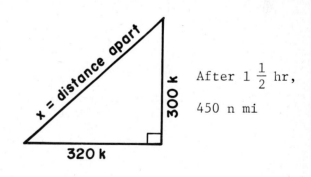

After $1\frac{1}{2}$ hr, 450 n mi

After $1\frac{1}{2}$ hr, 480 n mi

77

$$a = 450 \text{ n mi}$$
$$b = 480 \text{ n mi}$$
$$c = x \text{ n mi}$$

$$a^2 + b^2 = c^2$$
$$(450)^2 + (480)^2 = x^2$$
$$202{,}500 + 230{,}400 = x^2$$
$$432{,}900 = x^2$$
$$\pm 30\sqrt{481} = x$$
$$\sim 657.9514 \text{ n mi} = 30\sqrt{481} = x$$

The airplanes are about 657.95 n mi apart after $1\frac{1}{2}$ hours.

EXAMPLE 9 Given the same problem as presented in Example 8, determine how far apart the airplanes are after t hours.

Solution The drawing is the same as for Example 8. Now, however, we must determine how far each has traveled after t hours. Because the distance traveled is equal to the rate times the time ($D = r \cdot t$), the first airplane has traveled 300t n mi and the second 320t n mi. Using the Pythagorean theorem, we get:

$$a = 300t \text{ n mi}$$
$$b = 320t \text{ n mi}$$
$$c = x \text{ n mi}$$

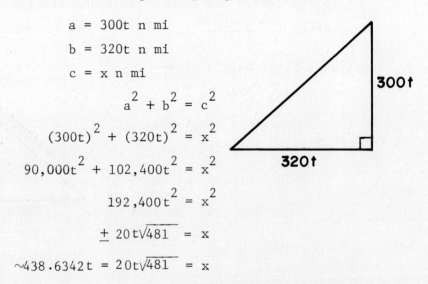

$$a^2 + b^2 = c^2$$
$$(300t)^2 + (320t)^2 = x^2$$
$$90{,}000t^2 + 102{,}400t^2 = x^2$$
$$192{,}400t^2 = x^2$$
$$\pm 20t\sqrt{481} = x$$
$$\sim 438.6342t = 20t\sqrt{481} = x$$

The airplanes are $20t\sqrt{481}$ n mi or approximately 438.63t n mi apart after t hours.

While both the radical form of the answer and its decimal approximation have been given, the exact answer (radical form) is the one that will be most useful in a calculus course. Only after performing all of the operations necessary, would one calculate the decimal approximation for the radical form of the answer.

EXAMPLE 10   A woman is building a rectangular box and wishes to reinforce the box by placing a taut wire from one top corner diagonally across the box to the opposite lower corner. If the dimensions of the box are 3 × 4 × 8 ft, find the final length of the wire. (Ignore waste.)

Solution   A drawing is made to represent the problem. While we have a diagonal $\overline{AG}$ representing the length of the wire, we do not yet have a right triangle. However, if we draw the line segment from E to G, we will have a right triangle AEG with $\overline{AG}$ as the hypotenuse. (Note that there are two other possible right triangles we could have chosen.) Because we have the length of $\overline{FG}$ (3 ft) and $\overline{EF}$ (8 ft), we can use Pythagorean theorem to calculate the length of $\overline{EG}$.

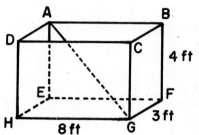

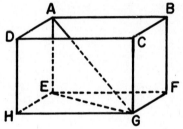

$$a = 3 \text{ ft}$$
$$b = 8 \text{ ft}$$
$$c = x \text{ ft}$$
$$a^2 + b^2 = c^2$$
$$3^2 + 8^2 = x^2$$
$$9 + 64 = x^2$$

$$73 = x^2$$
$$\pm\sqrt{73} = x$$
$$\sqrt{73} \text{ ft} = \text{length of } \overline{EG}$$

Note that because we must use the length of $\overline{EG}$ to calculate $\overline{AG}$, and we want our result to be more accurate, we use the exact value (radical form) rather than approximate it with a calculator.

Now we have enough information to calculate the length of $\overline{AG}$, the hypotenuse of the right triangle AEG.

$$a = AE = 4 \text{ ft}$$
$$b = EG = \sqrt{73} \text{ ft}$$
$$c = x \text{ ft}$$
$$a^2 + b^2 = c^2$$
$$4^2 + (\sqrt{73})^2 = x^2$$
$$16 + 73 = x^2$$
$$89 = x^2$$
$$\pm\sqrt{89} = x$$
$$9.434 \text{ ft} = \sqrt{89} \text{ ft} = x$$

The final length of the wire is $\sqrt{89}$ ft or approximately 9.43 ft.

EXAMPLE 11 A woman is fly fishing in a stream. The tip of her fishing rod is 6 ft above the surface of the water and her fly rests on the water 12 ft away (8 ft from a point directly under the tip of the pole). Assuming that her line is taut, how much line does she have out (from the tip of her rod to the fly)? If she does not move, keeps her rod steady, and reels in the line at a constant rate of 4 in. per second, how far away from her is the fly after 9 seconds?

Solution First we need a drawing to represent the situation. Once we have the drawing labeled, we see the line, x, is the hypotenuse of a 3-4-5 right triangle, with each side multiplied by 2. Thus x = 10 ft.

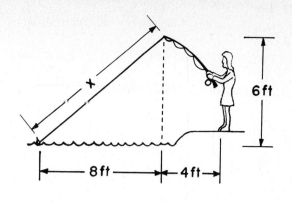

She has 10 ft of line out.

We must change 4 in. per second to ft per second so our units will be the same. The rate of 4 in. per second is equivalent to $\frac{4}{12} = \frac{1}{3}$ ft per second.

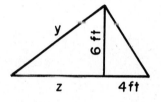

To calculate the distance z after 9 seconds:

Let y = Line out after 9 seconds

$$y = 10 \text{ ft} - 9(\tfrac{1}{3}) \text{ ft}$$
$$= 10 - 3$$
$$= 7 \text{ ft}$$

If y = 7 ft, we can use the Pythagorean theorem to find z.

$$a^2 + b^2 = c^2$$
$$6^2 + z^2 = 7^2$$
$$z^2 = 49 - 36$$
$$z^2 = 13$$
$$z = \pm\sqrt{13} \text{ or } 3.61$$

Because z represents a distance, we use only the positive root: $z = \sqrt{13}$ ft or 3.61 ft (to the nearest hundredth).

To find the distance the fly is away from her, we must add 4 ft to z, giving 7.61 ft. The fly is 7.61 ft away from her after 9 seconds.

EXAMPLE 12 Given the same problem as presented in Example 11, determine how far the fly is from her after t seconds (t < 12).

Solution We must first determine the expression for y, the line out after t seconds.

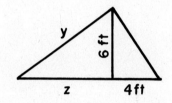

$$y = 10 \text{ ft} - t(\tfrac{1}{3}) \text{ ft}$$
$$y = 10 - \tfrac{t}{3}$$

We can use this expression in the Pythagorean theorem to obtain an expression for z.

$$a^2 + b^2 = c^2$$
$$6^2 + z^2 = (10 - \tfrac{t}{3})^2$$
$$36 + z^2 = 100 - \tfrac{20t}{3} + \tfrac{t^2}{9}$$
$$z^2 = \tfrac{t^2}{9} - \tfrac{20t}{3} + 64$$
$$z = \sqrt{\tfrac{t^2}{9} - \tfrac{20t}{3} + 64}$$

We use only the positive root because z measures distance. The distance the fly is from the woman after t seconds is z + 4 ft, or:

$$4 + \sqrt{\tfrac{t^2}{9} - \tfrac{20t}{3} + 64}$$

EXERCISE 2.3

1. Determine what the angles for the 45°-45°-90° triangle are when measured in radians.

2. Determine what the angles for the 30°-60°-90° triangles are when measured in radians.

For Problems 3 and 4 identify the legs and the hypotenuse for each right triangle given.

3.

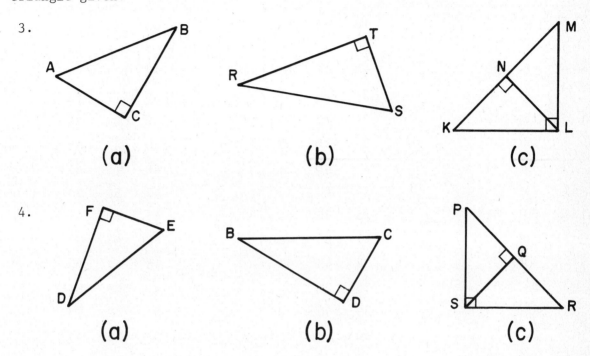

    (a)   (b)   (c)

4.

    (a)   (b)   (c)

5. Listed below are the sets of three whole numbers (Pythagorean triples) that satisfy the relationship $a^2 + b^2 = c^2$, where $c \leq 100$. One such set is 3, 4, 5. If each number is multiplied by a constant h, then the set 3h, 4h, 5h is another Pythagorean triple. Determine how many BASIC patterns there are in the list below. That is, count as one pattern 3, 4, 5; 6, 8, 10; and all other multiples of the 3, 4, 5 set. The labeling of the patterns has been begun for you.

A	B	C	Pattern	A	B	C	Pattern	A	B	C	Pattern
3	4	5	1	18	80	82	___	35	84	91	___
5	12	13	2	20	21	29	___	36	48	60	___
6	8	10	1	20	48	52	___	36	77	85	___
7	24	25	3	21	28	35	___	39	52	65	___
8	15	17	___	21	72	75	___	39	80	89	___
9	12	15	___	24	32	40	___	40	42	58	___
9	40	41	___	24	45	51	___	40	75	85	___
10	24	26	___	24	70	74	___	42	56	70	___
11	60	61	___	25	60	65	___	45	60	75	___
12	16	20	___	27	36	45	___	48	55	73	___
12	35	37	___	28	45	53	___	48	64	80	___
13	84	85	___	28	96	100	___	51	68	85	___
14	48	50	___	30	40	50	___	54	72	90	___
15	20	25	___	30	72	78	___	57	76	95	___
15	36	39	___	32	60	68	___	60	63	87	___
16	30	34	___	33	44	55	___	60	80	100	___
16	63	65	___	33	56	65	___	65	72	97	___
18	24	30	___								

6. Which were the three most frequently used patterns in Problem 5?

7. Find the value of x for each triangle. Answers should be exact.

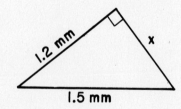
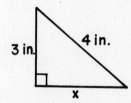
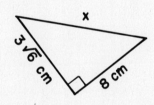

For Problems marked with a "C", compute the answer correct to two decimal places unless instructed to do otherwise.

8. Find the value of x for each triangle. Answers should be exact.

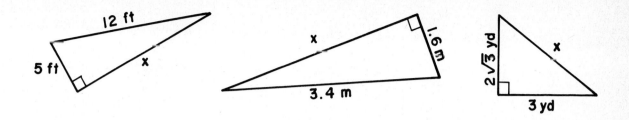

C 9. Find the value of x for each triangle. Answers should be accurate to four decimal places.

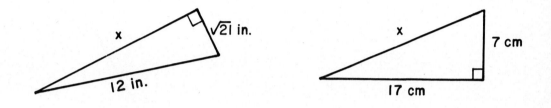

C 10. Find the value of x for each triangle. Answers should be accurate to four decimal places.

11. Is the triangle with sides of length 2.8 cm, 5.3 cm, and 4.5 cm a right triangle?

12. Is the triangle with sides of length 2.4 ft, 4 ft, and 3 ft a right triangle?

Refer to Figure 2.1 for Problems 13 and 14. In each case, find the lengths of the missing segments. Express answers in simplest radical form.

		AC	AB	BC
13.	a)	6 cm		
	b)		12 in.	
14.	a)	$\sqrt{3}$ m		
	b)		7 ft	

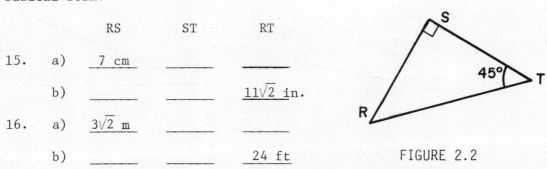

FIGURE 2.1

Refer to Figure 2.2 for Problems 15 and 16. Find the lengths of the missing segments. Answers should be in simplest radical form.

		RS	ST	RT
15.	a)	7 cm		
	b)			$11\sqrt{2}$ in.
16.	a)	$3\sqrt{2}$ m		
	b)			24 ft

FIGURE 2.2

Refer to Figure 2.3 for Problems 17 and 18. In each case, find the lengths of the missing segments. Give answers correct to four decimal places.

		DE	EF	DF
C 17.	a)	5.08 m		
	b)		11.3 in.	
C 18.	a)			1.27 cm
	b)		15.2 in.	

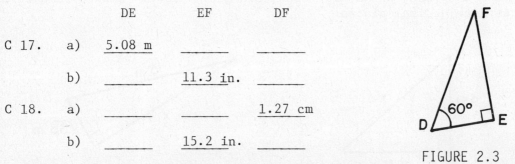

FIGURE 2.3

Refer to Figure 2.4 for Problems 19 and 20. In each case, find the lengths of the missing segments. Give answers correct to four decimal places.

		WX	XY	WY
C 19.	a)	14.09 ft		
	b)			7.03 mm
C 20.	a)		4.21 in.	
	b)			25.07 cm

FIGURE 2.4

86

For Problems 21 and 22 refer to the figure at the right: AB = BC = DC = AD and ∠DAE = 45°. Find the lengths of the missing segments. Leave answers in simplest radical form.

	AB	AE	DE	DC	AD	EC	BC
21.	___	7 mm	___	___	___	___	___
22.	___	___	___	8 ft	___	___	___

**23. For Problems 23 and 24, refer to Figure 2.3. If DE = 4x, give the expressions for EF and DF.

**24. If EF = 2x, give the expression for DE and DF.

For Problems 25 and 26 refer to the Figure 2.4.

**25. If WX = 5x, give the expressions for WY and XY.

**26. If WY = 3x, give the expressions for WX and XY.

*27. Find the length of each side in the given right triangle.

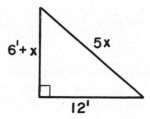

*28. Find the length of each side of the triangle.

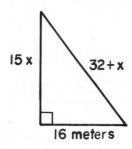

29. The foot of a 25-ft ladder leaning against the wall of a building is 15 ft from the base of the building. How high up the building does the ladder reach?

C 30. A football field measures 100 yd by 33 yd. How far must a player run from one corner to the other on the diagonal to make a touchdown?

C 31. A local telephone company erected a 50-ft high telephone pole. To this pole they attached a cable, which they then secured to the ground 50 ft from the base of the pole. How long was the tension cable?

32. If a flag pole 45 ft high casts a shadow 24 ft long, what is the distance from the top of the pole to the far end of the shadow?

C 33. Two airplanes leave at a 90° angle from the same airport at the same time. How far apart will they be after 2 hours of flying if they each travel at 120 mi per hour?

C 34. An airplane travels at 600 mi per hour, while a jet travels at 1125 mi per hour. If the airplane is flying from Pittsburgh to Miami and the jet is flying from Pittsburgh to Los Angeles, after 2 hours and 20 minutes how far away are the two aircraft from each other? (Assume that the paths form a 90° angle.)

C 35. A box of length 6 in. and width 4 in. is 2 in. high. Find the diagonal of the box.

36. Find the diagonal of a box 9 in. wide, 12 in. long, and 8 in. high.

*37. One leg of a right triangle is three in. longer than another. The hypotenuse is $\sqrt{17}$ in. Find the length of the legs in inches.

*38. To drive from Greenville to Cooperstown, you must take Route 41 to Thomastown, then Route 981 into Cooperstown. However, the highway department has announced plans to build a new four-lane freeway directly connecting Greenville with Cooperstown. From the information given in Figure 2.5, calculate how many miles shorter the proposed highway will be than the present route of travel.

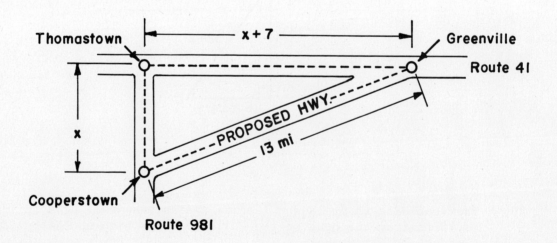

FIGURE 2.5

*39. One leg of a right triangle is three in. more than three times the other side. The hypotenuse is 25 in. long. Find the length of each of the legs.

C 40. A 7 ft pole is placed on a 45° angle and attached to the top of another pole placed at a 135° angle. If they were both on level ground, what is the distance from the bottom of one pole to the bottom of the other pole?

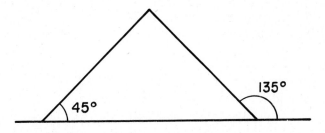

41. Figure 2.6 represents an L-shaped building. On the line segment $\overline{AD}$, a stake is driven at C, 8 ft from the building and 10 ft from A. If $\angle ABC$ is a right angle, find the length of $\overline{AB}$ and $\overline{AD}$.

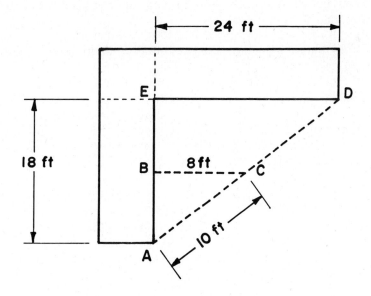

FIGURE 2.6

42. The competition for the coveted DY Cup World Championship speedboat race is about to begin. Two prominent competitors in this fast and often dangerous sport vie for the trophy. They are Mrs. Hudspeth (last year's winner) and two-time Indianapolis 500 winner, Mrs. Douthit. The Douthit boat, dubbed "Ifpeethencue" is capable of 30 mph while the more powerful Hudspeth craft, "Screamin' Scalene" turns a hair-raising 36 mph.
The race committee has finished marking the course. The boats will travel up the lake in a straight line, for 3 mi after which they veer 90° to the right, and continue across the lake for 4 mi. The end of this stretch is marked by an orange buoy. The boats must round the buoy and head straight back to the starting point to commence another lap of the course. Three laps must be completed; the first to complete all of the laps is the winner.
At the sound of the starter's pistol, the Douthit craft rips away from the starting line, leaving the stalled Hudspeth boat behind. Her lead widens. Thirteen minutes pass before the Hudspeth boat is restarted and continues the race. Who wins the prestigious DY cup? (Assume all boats travel at their top speed from start to finish.)

C 43. A 24-ft ladder is leaning against a wall with the foot of the ladder 4 ft from the base of the wall. The base of the ladder is pulled away from the wall at a rate of 4 ft per second. How far is the top of the ladder from the ground after 3 seconds?

C 44. A girl is flying a kite at noon; the sun is directly overhead. She holds the string 3 ft above the ground and the wind holds the kite at a constant altitude of 53 ft. How much string does she have out when the shadow of the kite is 100 ft away from her? If she then lets more string out at a rate of 6 ft per second, how far is the shadow from her after 3 seconds?

**45. In Problem 43 how far is the top of the ladder from the ground after t seconds ($t < 5$)?

**46. In Problem 44 how far is the shadow from her after t seconds?

C 47. A 1-ft cube is hooked to the end of a 25-ft long rope. The rope passes around a pulley mounted 10 ft above the floor. The loose end of the rope is held 4 ft above the floor by a man who begins to walk away from the cube at a rate of 2 ft per second. How long does he walk before all slack is out of the rope? How far off the floor is the cube after 10 seconds?

C 48. A boat is pulled toward a dock by a rope passed through a ring on the dock that is 3 ft higher than the bow of the boat (where the other end of the rope is attached). How far is the boat from the base of the dock when 12 ft of rope is out? If the rope is then pulled in at a rate of 2 ft per second, how far out is the boat after 2 seconds?

**49. For Problem 47, how far off the floor is the cube after t seconds?

**50. For Problem 48 how far out is the boat after t seconds ($t \leq 4.5$)?

2.4 CONSTRUCTION OF TRIANGLES

We have learned previously how to construct a line segment congruent to a given segment and an angle congruent to a given angle. Because line segments and angles are the elements of triangles, we can apply our knowledge of constructions to triangles.

First let us construct a triangle given the lengths of the three sides.

To construct a triangle given three line segments:

1. Construct a line segment congruent to one of the given line segments and call that segment a.

2. Set the compass to the length of another of the given line segments.

3. Without changing the setting of the compass, place the spike at one endpoint of segment a and draw a circle.

4. Set the compass to the length of the remaining line segment given.

5. Without changing the setting of the compass, place the spike at the <u>other</u> endpoint of segment a and draw a circle.

6. Draw a line segment from one point of intersection of the circles to one endpoint of segment a.

7. Then draw the segment from the same point of intersection of the circles to the other endpoint of a.

8. The three line segments so constructed form the desired triangle.

EXAMPLE 1   Construct a triangle given segments a, b, and c.

```
———————    ———————    ———————
    a          b          c
```

Solution   Step 1:   Construct a line segment congruent to a.

```
C———————————B
      a
```

Steps 2
and 3:   Set the compass to the length of b, place the spike at one endpoint of a, and draw a circle.

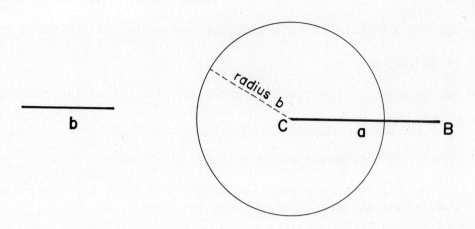

Steps 4
and 5:   Set the compass to the length of c, place the spike at the other endpoint of a, and draw a circle.

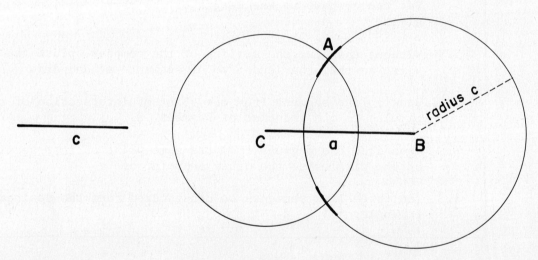

Steps 6
and 7:   Draw a line segment from A, one point of intersection of the two circles, to C. Then draw a segment from A to B.

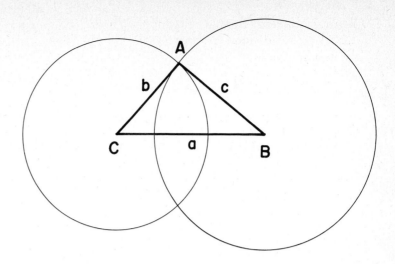

Step 8: The △ABC has sides congruent to segments a, b, and c. Therefore △ABC is the desired triangle.

In order to form such a triangle, what must be the relative lengths of the sides? The sum of the lengths of any two sides must be longer than the third side. This is demonstrated in Problems 5 and 6 in the Exercises at the end of this section.

We can also construct a triangle given two angles and the included side (the common side of the given angles).

To construct a triangle given two angles and the included side:

1. Call the angles ∠A and ∠B, and the included side, segment c.

2. Construct an angle congruent to ∠A with vertex at A and segment c as one side.

3. Construct an angle congruent to ∠B with vertex at B and segment c as one side. ∠B must be constructed on the same side of c as was ∠A.

4. Extend the sides of ∠A and ∠B so that they intersect. Call the point of intersection C.

5. The triangle so formed, △ABC, is the desired triangle.

EXAMPLE 2 Construct a triangle given ∠A, ∠B, and segment c, the side included between the two angles.

Solution   Step 1:   Construct a segment congruent to c; name the endpoints A and B.

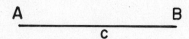

Step 2:   Construct ∠A with vertex at A and with c as one side. (Refer to Section 1.4 if necessary.)

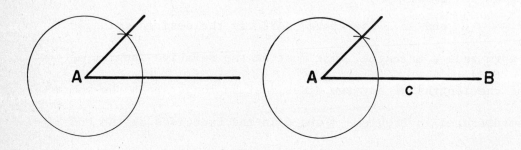

Step 3:   Construct ∠B with vertex at B and with c as one side. ∠B must be on the same side of c as ∠A is.

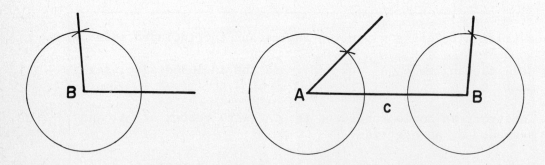

Step 4:   Extend the sides of ∠A and ∠B so that they intersect. Call the point of intersection C.

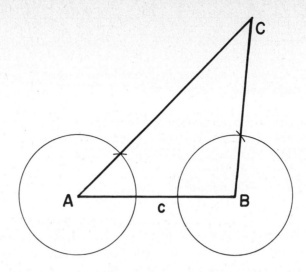

Step 5: △ABC is the desired triangle because it contains ∠A and ∠B with c as the side included between them.

In order to form a triangle given two angles and the include side, what must be the relative sizes of the angles? The sum of the two angles given must be less than π rad (180°). This is demonstrated in Problems 11 and 12 in the Exercises at the end of this section.

We can construct a triangle given two sides and the included angle (the angle formed by them).

> To construct a triangle given two sides and the included angle:
>
> 1. Construct a line segment congruent to one of the given segments.
>
> 2. Construct an angle congruent to the given angle with the vertex as one endpoint of the constructed segment and with the constructed segment as one side.
>
> 3. Set the compass to the length of the other given segment.
>
> 4. Without changing the setting, place the spike at the vertex of the angle constructed and draw a circle. Label as point A the intersection of this circle with the second side of the angle just constructed.
>
> 5. Draw the line segment from A to the other endpoint of the line segment first constructed.
>
> 6. The triangle thus constructed is the desired triangle.

EXAMPLE 3   Construct a triangle given sides a and b with ∠C as the included angle.

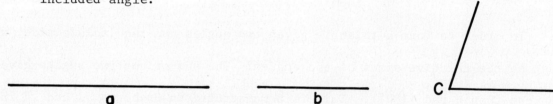

Solution   Step 1:   Construct a segment congruent to a. Name one endpoint C and the other B.

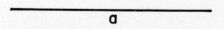

Step 2:   Construct ∠C with C as the vertex and a as one side.

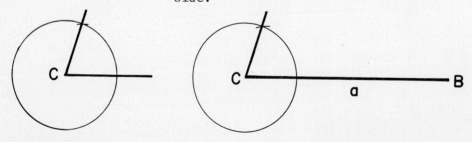

Steps 3
and 4: Set the compass to the length of segment b, then mark off that length on the side of ∠C just drawn.

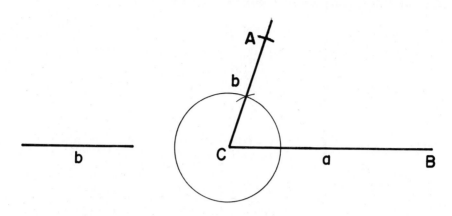

Step 5: Draw the line segment from A to B.

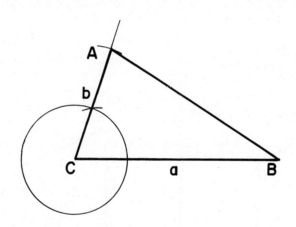

Step 6: The △ABC just constructed has as sides the segments a and b with ∠C as the angle formed by them. Therefore, △ABC is the desired triangle.

In order to construct a unique triangle, does the given angle have to be the angle included between the two given sides? That is, is there only one triangle that can be constructed if we are given two sides and an angle not the included angle? Problems 17 and 18 in the Exercises at the end of this section show the answer depends on the relative sizes of the sides and the given angle.

EXAMPLE 4   Construct a right triangle with legs of lengths 7 cm and 3 cm.

Solution   First construct a right angle. (Refer to Section 1.4 if necessary.)

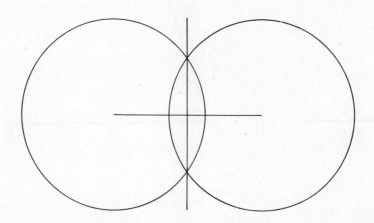

Extend the legs if necessary, then mark off 7 cm on one leg and 3 cm on the other. Draw the line segment between the two marks just made.

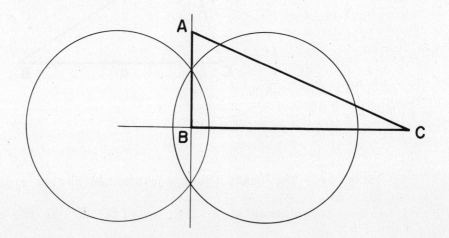

The desired triangle is △ABC because ∠ABC is a right angle and AB = 3 cm and BC = 7 cm.

Is there only one triangle we can construct if we are given the three angles? Problems 31 and 32 in the Exercises at the end of this section demonstrate that there are many such triangles possible.

EXERCISE 2.4

Use a compass and straightedge only for Problems 1 through 18.

1. Construct a triangle given the segments a, b, and c.

   _____a_____   _____b_____   _____c_____

2. Construct a triangle given the segments d, e, and f.

   _____d_____   _____e_____   _____f_____

3. Construct a triangle given the segments g, h, and i.

   _____g_____   _____h_____   _____i_____

4. Construct a triangle given the segments j, k, and l.

   ____j____   _____k_____   _____l_____

5. Construct a triangle, if possible, given the segments a, b, and c.

   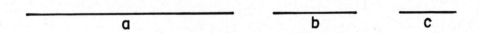

6. Explain the result obtained in Problem 5.

7. Construct a triangle given ∠A, ∠B, and segment c, the side included between the angles.

   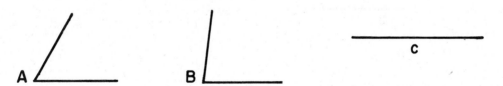

8. Construct a triangle given ∠D, ∠E, and segment f, the side included between the angles.

   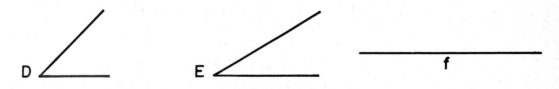

9. Construct a triangle given ∠P, ∠Q, and segment r, the side included between the angles.

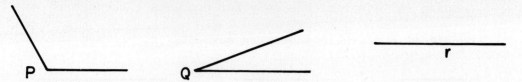

10. Construct a triangle given ∠S, ∠T, and segment r, the side included between the angles.

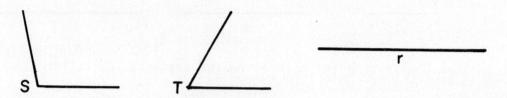

11. Construct, if possible, a triangle given ∠X, ∠Y, and segment z, the side included between the angles.

12. Explain the result obtained in Problem 11.

13. Construct a triangle given segments b and c and ∠A, the included angle.

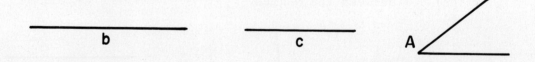

14. Construct a triangle given segments r and s and ∠T, the included angle.

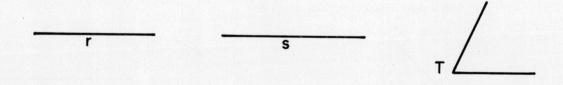

15. Construct a triangle given segments f and g with ∠H, the included angle.

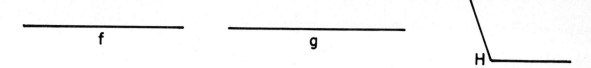

16. Construct a triangle given segments j and k with ∠L, the included angle.

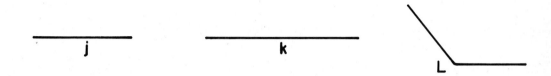

17. Construct, if possible, a triangle given segments a and b with ∠B not the included angle. (Hint: Construct ∠B first, then mark off segment a on one side of B. Set the compass to the length of segment b and place the spike at the end of segment a, which is not the vertex B. Draw a circle.) Is there only one such triangle possible?

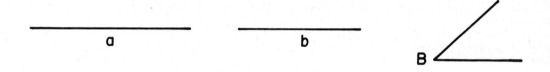

18. Explain the result obtained in Problem 17.

Use a protractor and ruler for Problems 19 through 31.

19. Construct a triangle with sides of lengths 7 cm, 10 cm, and 15 cm.

20. Construct a triangle with sides of lengths 12 cm, 8 cm, and 6 cm.

21. Construct a triangle with one angle of 60° and one of 45°, where the included side is 2.5 in.

22. Construct a triangle with one angle of 30° and one of 120°, where the included side is 1.5 in.

23. Construct a triangle with sides of 6 cm and 4 cm, where the included angle is $\frac{2\pi}{3}$ rad.

24. Construct a triangle with sides of 10 cm and 4.5 cm, where the included angle is $\frac{\pi}{6}$ rad.

25. Construct a right triangle with legs of lengths 3 in. and 4 in.

26. Construct a right triangle with legs of lengths 5 cm and 12 cm.

27. Construct an isosceles right triangle with legs of length 5 cm.

28. Construct an isosceles right triangle with legs of length 2 in.

29. Construct an equilateral triangle with 2.5 in. sides.

30. Construct an equilateral triangle with 4.6 cm sides.

31. Construct a triangle with angles of $\frac{\pi}{4}$ rad, $\frac{\pi}{4}$ rad, and $\frac{\pi}{2}$ rad.

32. Is the triangle constructed in Problem 31 unique? (Is there only one such triangle that can be drawn?) Are all such triangles the same size?

*33. Using as a model the instructions for constructions in this section, write out a similar procedure to construct a triangle given one leg and the hypotenuse.

*34. Using as a model the instructions for constructions in this section, write out a similar procedure to construct a triangle given an acute angle and the hypotenuse.

*35. Using as a model the instructions for constructions in this section, write out a similar procedure to construct a triangle given two angles and the side opposite one of them.

2.5 CONGRUENCE

We have previously discussed congruent line segments and congruent angles. We now need a common understanding of what we mean when a pair of two-dimensional figures is congruent. Two figures are CONGRUENT if and only if they can be made to coincide by placing one on top of the other or by turning one over and placing it on top of the other. That is, the figures are congruent if by moving one (a _translation_), turning one (a _rotation_), and/or turning one over (a _reflection_), it will coincide with the other. Thus, A and B in Figure 2.7 are congruent because we can rotate A and move (translate) it, and the two figures will coincide. C and D are congruent because we can "turn over" (reflect) D, and move (translate) it so that it will coincide with C. A and C, however, are not congruent. While they _are_ the _same shape_, they are _not_ the _same size_.

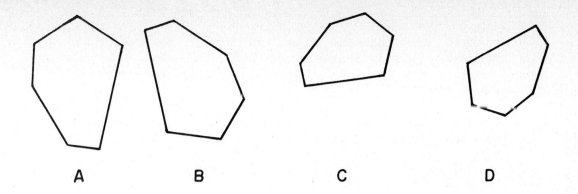

FIGURE 2.7

Two triangles therefore will be congruent if and only if <u>each</u> of the following is true:

1. Each side of the first triangle can be paired with a side of the second (called the corresponding side).

2. Each angle of the first triangle can be paired with an angle of the second (called the corresponding angle).

3. Each side and angle of the second triangle is paired with a side and angle of the first.

4. Each of the sides of the first triangle is congruent to its corresponding side in the second.

5. Each of the angles of the first triangle is congruent to its corresponding angle in the second.

Fortunately, in order to know that two triangles are congruent we do not need to know that each of these facts is true. We can have less information and still be assured that the two triangles are congruent. We saw in Section 4 that we could construct a triangle if we were only given the three sides. All triangles constructed with the given sides would be congruent. Thus we only need to know that corresponding sides of two triangles are congruent in order to know that the triangles are congruent.

> Two triangles are congruent if each pair of corresponding sides is congruent. This is abbreviated: SSS correspondence is a congruence.

EXAMPLE 1   Determine whether △ABC ≅ △FED.

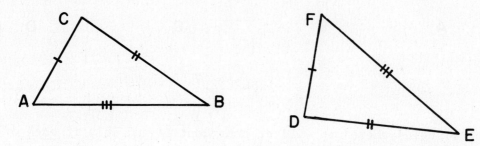

Solution   From the markings entered on the drawings, we know that $\overline{AC} \cong \overline{FD}$, $\overline{CB} \cong \overline{DE}$, and $\overline{AB} \cong \overline{FE}$. Therefore each corresponding pair of sides is congruent and the triangles have an SSS correspondence, so △ABC ≅ △FED.

Note that the sides of △ABC are always on the left and the sides of △FED are always on the right. Always keeping the references to one triangle first and those to the other second makes the presentation easier to understand.

Because the two triangles △ABC and △FED are congruent, the corresponding angles are congruent. We must determine which angle in △FED corresponds to ∠A. Now ∠A is the angle formed by the side with —|— and the side with —|||—, therefore the angle corresponding to it must be formed by the corresponding sides in △FED, or ∠F. Because ∠C is formed by the side with —|— and —||—, the angle corresponding to it must be formed by the corresponding sides in △FED, or ∠D. Now we know that ∠A ≅ ∠F and ∠C ≅ ∠D, so ∠B ≅ ∠E.

EXAMPLE 2   Given △DEF ≅ △RST, and DE = 4 cm, EF = 7 cm, DF = 9 cm, RS = 7 cm, ST = 4 cm. Determine the length of RT.

Solution   First we must draw each triangle and enter the given information.

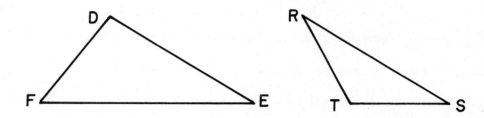

As we make the drawing we do not know which side of the first triangle corresponds to each side of the second triangle. (Here the triangles <u>deliberately have not been drawn as congruent</u> to emphasize that we do <u>not</u> know the correspondence between the two triangles.) Now that we have a drawing we can enter the given measurements.

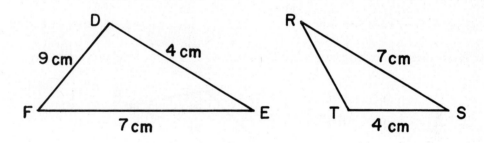

We can now tell that $\overline{DE} \cong \overline{TS}$ and $\overline{EF} \cong \overline{RS}$. We can enter this on the drawings by adding the ─┼─ marks:

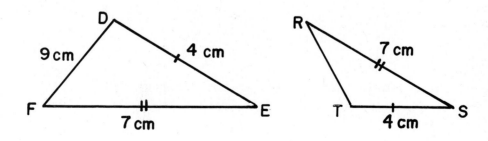

Because the triangles are congruent, $\overline{DF} \cong \overline{RT}$. Therefore $RT = 9$ cm.

(We also know that $\angle E$ corresponds to $\angle S$ and they are congruent; $\angle F$ corresponds to $\angle R$ and $\angle F \cong \angle R$; and $\angle D \cong \angle T$.)

In the previous section we saw that we could construct a unique triangle if we were given a side and the angles at its endpoints. All triangles constructed with the given side and angles would be congruent. Therefore we know two triangles are congruent if we know two angles in one triangle are congruent to the corresponding angles in a second triangle and that the included side of the first is congruent to the corresponding side of the second.

Two triangles are congruent if two pairs of corresponding angles are congruent and the pair of corresponding sides between those angles is congruent. This is abbreviated: ASA correspondence is a congruence.

Special attention should be paid to the order of the letters ASA. That order indicates an angle, an adjacent side, and the next angle in that order.

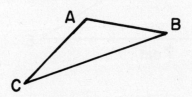

For △ABC, any of the following sets satisfies the sequence ASA:

$\angle C$, $\overline{CB}$, $\angle B$
$\angle B$, $\overline{AB}$, $\angle A$
$\angle C$, $\overline{CA}$, $\angle A$

<u>None</u> of the following sets satisfies the sequence ASA:

∠C, $\overline{CB}$, ∠A
∠C, $\overline{AC}$, ∠B
∠B, $\overline{CB}$, ∠A
∠B, $\overline{AB}$, ∠C
∠A, $\overline{AB}$, ∠C
∠A, $\overline{AC}$, ∠B

All of the above pairs can be represented by the sequence AAS (or SAA); that is, the side is not the side common to the two angles. Once we know the measure of two angles in a triangle, we know the size of the third. (Why?) Therefore if two pairs of angles in two triangles are congruent, then the third pair is also congruent. Thus if we know we have AAS correspondence for a pair of triangles, then we must also have an ASA correspondence, therefore the triangles are congruent.

Many times the drawing will give information useful to the problem.

EXAMPLE 3   Show that △ADC ≅ △BDC.

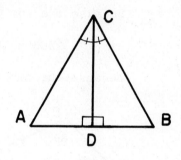

Solution   From the drawing we know that ∠ADC = 90° = ∠CDB, so these two angles are congruent. We also are given that ∠ACD ≅ ∠BCD. We now know that two pairs of angles are congruent, so we could use ASA correspondence if we know that the corresponding sides between these angles were congruent. From the drawing we see that the desired side for each triangle is $\overline{CD}$, and $\overline{CD} \cong \overline{CD}$. Therefore, we have ASA correspondence for the two triangles, so they are congruent.

EXAMPLE 4   If $\triangle ECB \cong \triangle DBC$, $\overline{EB} \cong \overline{DC}$, $\angle DBC = 37°$, and $\angle CDB = 88°$, find the number of degrees in $\angle EBD$ and $\angle BEC$.

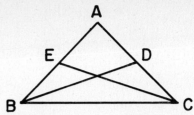

Solution   Because the two congruent triangles overlap, first redraw them as distinct triangles and mark the given information on the drawings. (Remember $\overline{BC} \cong \overline{CB}$.) Because we know that two pairs of sides are congruent and we know that the triangles are congruent, $\overline{EC} \cong \overline{DB}$. We also have the following corresponding

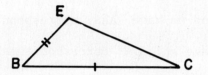 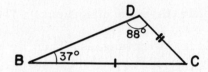

angles: $\angle BEC \cong \angle CDB$, $\angle ECB \cong \angle DBC$, and $\angle EBC \cong \angle DCB$.

Because the sum of the angles in a triangle is 180°,

$$\angle DCB = 180° - (37° + 88°) = 55°$$

Because the two triangles are congruent we now have:

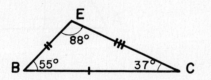 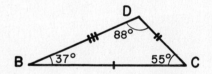

Going back to the original drawing, we have:

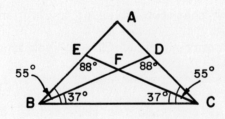

Therefore: $\angle EBD = 55° - 37° = 18°$
             $\angle BFC = 180° - 2(37°) = 106°$

The answer is: $\angle EBD = 18°$
                 $\angle BFC = 106°$

108

In Section 4 we constructed triangles given two sides and the included angle (the angle formed by the sides). Triangles constructed with the given sides and included angle are congruent. We thus can show two triangles are congruent if we know that two sides of one triangle are congruent to the corresponding sides in the second triangle and that the included angles are congruent.

> Two triangles are congruent if two pair of corresponding sides are congruent and the pair of angles formed by these sides is congruent. This is abbreviated: SAS correspondence is a congruence.

EXAMPLE 5 Show that $\triangle ABC \cong \triangle EDC$ if $\overline{AE}$ intersects $\overline{BD}$ at C.

Solution From the drawing we have $\overline{AC} \cong \overline{EC}$ and $\overline{BC} \cong \overline{DC}$. Now we need to know that $\overline{AB} \cong \overline{DE}$ to use SSS correspondence or that $\angle ACB \cong \angle ECD$ to use SAS correspondence. We

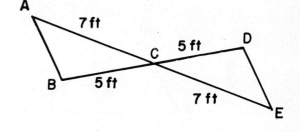

do not know anything about the length of $\overline{AB}$ or $\overline{DE}$ so we cannot use the SSS correspondence. We do know that $\angle ACB \cong \angle ECD$ because they are vertical angles formed by the intersection of $\overline{AE}$ and $\overline{BD}$. Now we have SAS correspondence and $\triangle ABC \cong \triangle EDC$.

We have seen that if we have an AAS correspondence between two triangles that we can also show an ASA correspondence between them. We might think that the same would be the case for SSA correspondence—that it would guarantee a congruence between the triangles. This is <u>not</u> the case, however, as we saw in Problems 17 and 18 in the Exercise for Section 4.

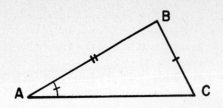
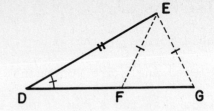

One example of such a correspondence is △ABC and △DEF or △DEG. Because both triangles △DEF and △DEG have SSA correspondence with △ABC, we see that a SSA correspondence between two triangles does NOT guarantee congruence.

An AAA correspondence also does not guarantee congruence. (See Problems 31 and 32 in the Exercise for Section 4.) One triangle may be much larger than the other as is the case with △RST and △XYZ below.

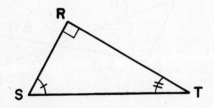
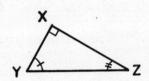

EXERCISE 2.5

Problems 1 and 2 have pairs of congruent triangles. For each pair list the corresponding pairs of sides and angles.

1.

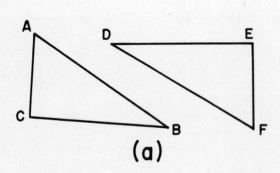
(a)

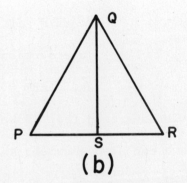
(b)

110

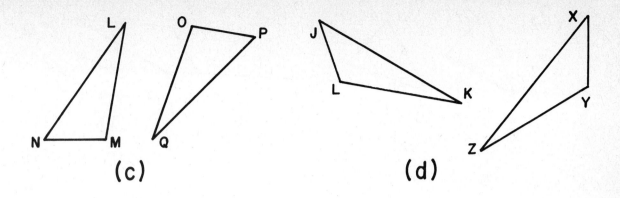

2.

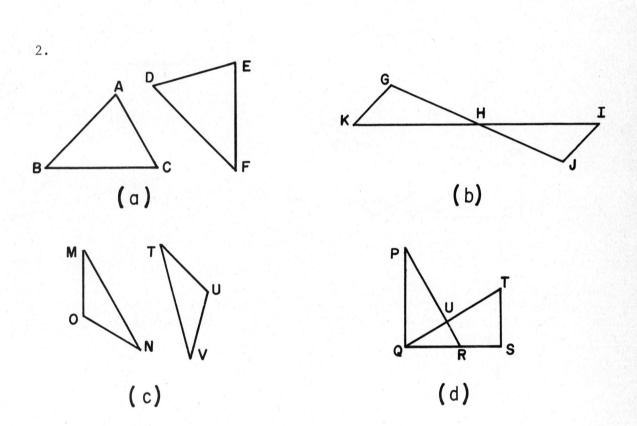

For Problems 3 and 4, determine if the pairs of triangles are congruent with the information given. If they are congruent, state which correspondence applies: SSS, SAS, or ASA.

3.

111

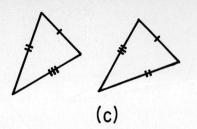
(c)

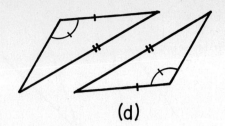
(d)

4.

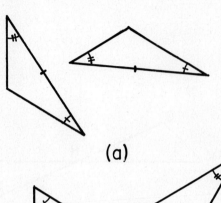
(a)

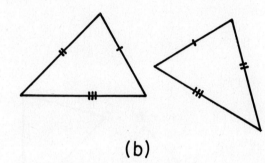
(b)

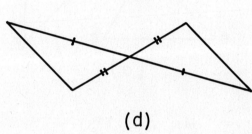

(c)

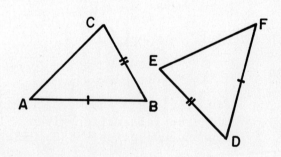
(d)

For Problems 5 through 10, determine what additional information is necessary in order to show the given triangles congruent.

5.

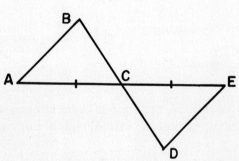

△ABC ≅ △EDC
   by SAS? _____

   by ASA? _____

6.

△ABC ≅ △FDE
   by SSS? _____

   by SAS? _____

112

7.

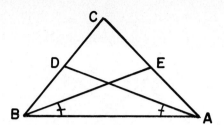

8.

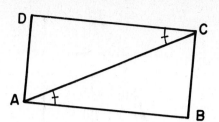

△DBA ≅ △EAB
   by SAS? _____

   by ASA? _____

△ABC ≅ △CDA
   by SAS? _____

   by ASA? _____

9.

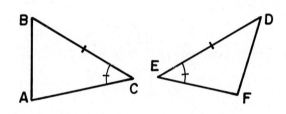

10.

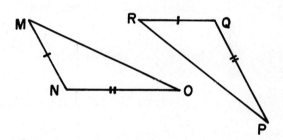

△BAC ≅ △FDE
   by ASA? _____

   by SAS? _____

△MNO ≅ △RQP
   by SAS? _____

   by SSS? _____

11. a) $\overline{AB}$ would be congruent to $\overline{DC}$ if △_____ ≅ △_____ .

    b) $\overline{CR}$ would be congruent to $\overline{BR}$ if △_____ ≅ △_____ .

    c) ∠1 would be congruent to ∠2 if △_____ ≅ △_____ .

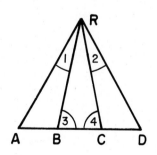

12. a) $\overline{PR}$ would be congruent to $\overline{SQ}$ if $\triangle\underline{\quad} \cong \triangle\underline{\quad}$.

b) $\overline{PQ}$ would be congruent to $\overline{SR}$ if $\triangle\underline{\quad} \cong \triangle\underline{\quad}$ or if $\triangle\underline{\quad} \cong \triangle\underline{\quad}$.

c) $\overline{PT}$ would be congruent to $\overline{ST}$ if $\triangle\underline{\quad} \cong \triangle\underline{\quad}$.

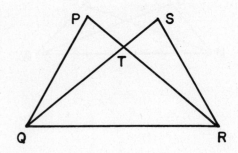

13. Are two triangles congruent if each angle of one is congruent to the corresponding angle of the other?

14. If you know that two triangles are equilateral, what is the minimum additional information you would need to conclude that they are congruent?

Use Figure 2.8 for Problems 15 through 18. Assume in $\triangle ABC$, $\overline{EC}$ intersects $\overline{BD}$ at F.

15. If $\triangle ADB \cong \triangle AEC$, $\overline{EA} \cong \overline{AD}$, $\angle EBD \cong 29°$, $\angle A \cong 87°$, find the number of degrees in $\angle AEC$ and $\angle DCE$.

16. If $\triangle ADB \cong \triangle AEC$, $\overline{EA} \cong \overline{AD}$, $\angle EBD = 42°$, and $\angle ADB = 62°$, find the number of degrees in $\angle CDB$ and $\angle EFB$.

17. If $\triangle EFB \cong \triangle DCF$, $\overline{EB} \cong \overline{DC}$, $\angle EBC = 73°$, and $\angle DCF = 19°$, find the number of degrees in $\angle BFC$ and $\angle FBC$.

18. If $\triangle EFB \cong \triangle DFC$, $\angle EFD = 1.38$ rad, $\angle EBF = 0.66$ rad, find the number of radians in $\angle FBC$, $\angle DCB$, and $\angle A$.

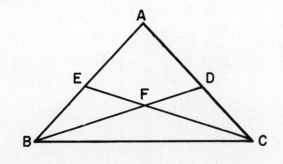

FIGURE 2.8

Use Figure 2.9 for Problems 19 through 22. Assume $\overline{AC}$ intersects $\overline{BD}$ at E.

19. If △ABE ≅ △DCE, $\overline{AE} \cong \overline{CE}$, ∠A = 43°, and ∠DEC = 68°, find the number of degrees in ∠ABE and ∠AED.

20. If △ABE ≅ △DCE, $\overline{BE} \cong \overline{DE}$, ∠ABE = 39°, ∠AED = 123°, ∠DCB = 97°, find the number of degrees in ∠A and ∠EBC.

21. If △ABC ≅ △DCB, $\overline{AB} \cong \overline{DC}$, ∠D = 0.97, ∠DBC = 0.74, find the number of radians in ∠DCB and ∠EBA.

22. If △ABC ≅ △DCB, $\overline{AB} \cong \overline{DC}$, ∠ABD = 62°, ∠DCB = 108°, find the number of degrees in ∠D and ∠AEB.

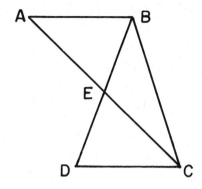

FIGURE 2.9

2.6 PROOFS OF CONGRUENCE

When we are given a geometry problem (or any problem, for that matter), we wish to go about solving it in a careful, logical manner. For many problems we do not have to justify the steps taken. Now, however, we will work with problems where the focus will be on carefully organized and justified reasoning. This is an important process because it will help us to think through problems more carefully.

> We must be cautious about what we assume as we work. Just because two figures "look" as though they are congruent does not mean that they are. Just because one line "looks" perpendicular to another does not mean it is so. Therefore, be careful about what you assume from drawings.

We MAY ASSUME the general relationships given in a drawing. For instance, in the drawing below, we can assume that three line segments intersect at F, and that figures ACF and DAF are triangles.

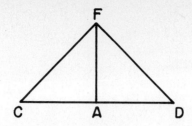

We MAY NOT ASSUME that A lies halfway between C and D (that is, that $\overline{AC} \cong \overline{DA}$). Nor may we assume that $\overline{FA} \perp \overline{CD}$. We may not be sure that $\overline{CD}$ is a line segment; we could, however, if we were given that figure FCD was a triangle. (Why?)

Some of the other assumptions we <u>may</u> make are:

1. Things equal (congruent) to the same or to equal (congruent) things are equal (congruent) to each other (TRANSITIVE PROPERTY).

2. An object (quantity) is congruent (equal) to itself (REFLEXIVE PROPERTY).

3. Equal quantities added to two equal quantities results in equal quantities (ADDITION OF EQUALS).

4. Equal quantities subtracted from two equal quantities results in equal quantities (SUBTRACTION OF EQUALS).

5. Two equal quantities multiplied by equal quantities results in equal quantities (MULTIPLICATION OF EQUALS).

6. Two equal quantities divided by equal (non-zero) quantities results in equal quantities (DIVISION OF EQUALS).

7. One quantity may be substituted for its equal in any equation or expression (SUBSTITUTION OF EQUALS).

8. The whole of a quantity is equal to the sum of its parts.

The eight statements above are axioms, which are statements that are assumed to be true in mathematics.

The following geometrical statements have been observed to be true in previous work. Here we will assume this list is true as a basis for our logical presentation that follows.

1. Two distinct straight lines intersect in at most one point.

2. A line segment has only one midpoint.

3. All right angles are congruent.

4. All straight angles are congruent.

5. Complements of the same or of congruent angles are congruent.

6. Supplements of the same or of congruent angles are congruent.

7. Vertical angles are congruent.

8. If two triangles are congruent, then their corresponding parts are congruent (CPCTC).

9. If two sides and the included angle of one triangle are congruent to the corresponding parts of another triangle, then the two triangles are congruent (SAS).

10. If two angles and the included side in one triangle are congruent to the corresponding parts of another triangle, then the two triangles are congruent (ASA).

11. If three sides of one triangle are congruent to the corresponding sides of another triangle, then the two triangles are congruent (SSS).

12. If two sides of a triangle are congruent, the angles opposite these sides are congruent. (BASE ANGLES OF AN ISOSCELES TRIANGLE ARE CONGRUENT.)

13. If two angles of a triangle are congruent, the sides opposite these angles are congruent. (IF THE BASE ANGLES ARE CONGRUENT, THEN THE TRIANGLE IS ISOSCELES.)

14. An equilateral triangle is equiangular.

15. An equilangular triangle is equilateral.

16. The sum of the angles in a triangle is 180°.

17. If two angles in one triangle are congruent to the corresponding angles in another triangle, then the third angle in the first is congruent to the third angle in the second.

We can use statements from either set together with facts from a given problem to justify our conclusions. Let us look at an example of a proof.

EXAMPLE 1   Given: $\overline{AB} \perp \overline{DC}$

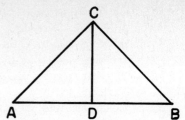

$\overline{AD} \cong \overline{BD}$

Prove: $\triangle ADC \cong \triangle BDC$

Solution   Before we begin writing the proof we first need to determine how we will prove the desired result. For this problem, we wish to show two triangles congruent. We know we have three ways to prove two triangles are congruent: ASA, SAS, or SSS. For the triangles in this problem, we are given one pair of sides congruent. Because $\overline{AB} \perp \overline{DC}$, we also know $\angle ADC$ and $\angle BDC$ are right angles and therefore are congruent. As we look more closely at $\triangle ADC$ and $\triangle BDC$, we see that they share a common side.

We enter all this information on the drawing and see that we have marked SAS on each triangle. We therefore can use SAS to justify the congruence of the two

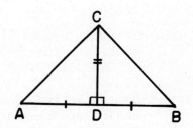

triangles. Now that we have determined a logical way to obtain the desired result, we can begin writing the proof. We first write down one given fact in the left column and that it is given, in the right column. We then write the conclusion we obtain from the given fact and our reason for it.

	Statement	Reason
1.	$\overline{AB} \perp \overline{DC}$	Given
2.	$\angle ADC = 90°$	1 & def. of $\perp$
3.	$\angle BDC = 90°$	1 & def. of $\perp$
A 4.	$\angle ADC \cong \angle BDC$	2, 3, & def. of $\cong$

For our reason justifying Statement 2, we write both the reason and the statement number of the given fact: because two segments are perpendicular (Statement 1) we know they form right angles (definition of perpendicular). We can justify Statement 4 because we know two angles that contain the same number of degrees (Statements 2 and 3) are congruent (definition of congruence). This gives us one pair of congruent angles. We note that by writing A beside Statement 4.

We can now record the other given information.

	Statement	Reason
1.	$\overline{AB} \perp \overline{DC}$	Given
2.	$\angle ADC = 90°$	1 & def. of $\perp$
3.	$\angle BDC = 90°$	1 & def. of $\perp$
A 4.	$\angle ADC \cong \angle BDC$	2, 3 & def. of $\cong$
S 5.	$\overline{AD} \cong \overline{BD}$	Given
S 6.	$\overline{CD} \cong \overline{CD}$	Common side

Note that while we are using SAS, the lines are not in that order; they are in the order ASS. Thus while it is *not* necessary to *write* the lines in the order SAS, we must have the angle included between the given sides on the *figure*.

Now that we have established congruence for the desired pairs of sides and angles, we may finish the proof by stating the two triangles are congruent, as desired, once again giving the number for the needed statements.

	Statement	Reason
1.	$\overline{AB} \perp \overline{DC}$	Given
2.	$\angle ADC = 90°$	1 & def. of $\perp$
3.	$\angle BDC = 90°$	1 & def. of $\perp$

A	4.	∠ADC ≅ ∠BDC	2,3 & Def. of ≅
S	5.	$\overline{AD} \cong \overline{BD}$	Given
S	6.	$\overline{CD} \cong \overline{CD}$	Common side
	7.	△ADC ≅ △BDC	5, 4, 6 & SAS

The proof is easiest to follow if the parts for one triangle are written on the left of the congruence sign and the parts for the other are on the right side.

These are the important procedures to follow when writing a proof:

1. Rewrite the facts that are given (the premises) under "Statements" and indicate that they are "Given."

2. Each successive line must be a "Given" or follow from what precedes it. (That is, we could not have written Statement 7 <u>before</u> we had all of the other statements written.)

3. For each statement we make, we <u>must</u> give a reason. The reason can be that the fact is given, a definition, one of the statements from pages 116 or 117, or a combination of these.

4. Before showing that some fact is true, we must have stated the appropriate premises. (To show that two triangles are congruent, we may show that two sides, two angles, and two other sides are congruent, where the angle is included between the sides. These are the premises we need in order to conclude that two triangles are congruent by SAS.)

We sometimes must show that the two triangles are congruent in order to justify the conclusion sought in the given problem.

EXAMPLE 2 Given: E is the midpoint of $\overline{AB}$

∠A ≅ ∠B

∠DEA ≅ ∠FEB

Prove: $\overline{DE} \cong \overline{FE}$

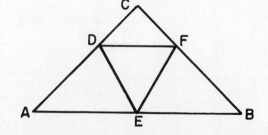

Solution   First mark the given facts on the drawing. Because E is the midpoint of $\overline{AB}$, we know $\overline{AE} \cong \overline{BE}$, and we enter that fact on the drawing. We see that we have an angle, a side, and an angle in one triangle congruent to an angle, a side, and an angle in the other, so we can use ASA to prove the two triangles, $\triangle AED$ and $\triangle BEF$, congruent. We will then have the corresponding sides $\overline{DE}$ and $\overline{FE}$ congruent. We can now write out the proof.

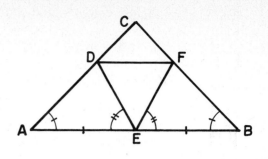

	Statement	Reason
A 1.	$\angle A \cong \angle B$	Given
2.	E is the midpoint of $\overline{AB}$	Given
S 3.	$\overline{AE} \cong \overline{BE}$	2 & Definition of midpoint
A 4.	$\angle DEA \cong \angle FEB$	Given
5.	$\triangle DEA \cong \triangle FEB$	1, 3, 4 and ASA
6.	$\overline{DE} \cong \overline{FE}$	5 & Corresponding parts of congruent triangles are congruent (CPCTC).

Sometimes adding a line segment (or segments) to a drawing makes the proof easier.

EXAMPLE 3   Given: $\overline{AD} \cong \overline{CB}$
$\overline{AB} \cong \overline{DC}$
Prove: $\angle A \cong \angle C$

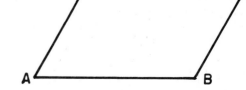

121

Solution   If we could show that two triangles, △DAB and △DCB, are congruent, we could use corresponding parts to show that ∠A ≅ ∠C. We may add the line segment $\overline{DB}$ to form the two triangles.

We also mark the given congruences on the drawing so we can see more easily what is needed for the proof. We can add one mark to $\overline{AD}$ and $\overline{CB}$ to indicate that they are a pair of congruent sides. We add two marks to each of $\overline{AB}$ and $\overline{DC}$ so we know that they are a different pair of congruent sides.

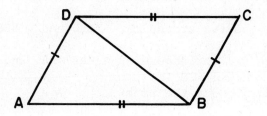

	Statement	Reason
S 1.	$\overline{AD} \cong \overline{CB}$	Given
S 2.	$\overline{AB} \cong \overline{CD}$	Given
S 3.	$\overline{DB} \cong \overline{DB}$	Common side
4.	△DAB ≅ △BCD	1, 2, 3 & SSS
5.	∠A ≅ ∠C	4 & corresponding parts of congruent triangles are congruent (CPCTC).

EXERCISE 2.6

Remember that figures may not be drawn to scale.

1. Given: △ARS with W the midpoint of $\overline{AR}$
   ∠AWS ≅ ∠RWS

   Prove: △AWS ≅ △RWS

2. Use the figure for Problem 1.

   Given: △ARS with $\overline{AS}$ ≅ $\overline{RS}$
   and ∠1 ≅ ∠2

   Prove: ∠A ≅ ∠R

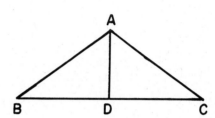

3. Given the figure with

   ∠BDA ≅ ∠CDA

   $\overline{BD}$ ≅ $\overline{CD}$

   Prove: ∠ABD ≅ ∠ACD

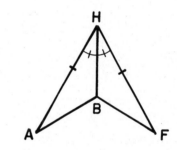

4. Given: $\overline{AH}$ ≅ $\overline{FH}$

   ∠AHB ≅ ∠FHB

   Prove: ∠A ≅ ∠F

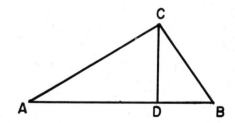

5. Given: $\overline{CD}$ ⊥ $\overline{AB}$ and

   $\overline{AD}$ ≅ $\overline{BD}$

   Prove: △ADC ≅ △BDC

123

6. Given: ∠M ≅ ∠Y

   ∠MKP ≅ ∠YXZ

   $\overline{MK} \cong \overline{XY}$

   Prove: $\overline{PK} \cong \overline{ZX}$

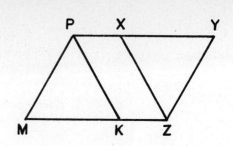

7. Given: $\overline{BD}$ and $\overline{AE}$ intersect at C

   $\overline{CE} \cong \overline{CB}$

   ∠E ≅ ∠B

   Prove: ∠D ≅ ∠A

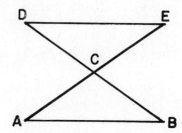

8. Given: $\overline{RV} \cong \overline{ST}$

   $\overline{RQ} \cong \overline{SP}$

   ∠VRQ ≅ ∠TSP

   Prove: $\overline{QV} \cong \overline{PT}$

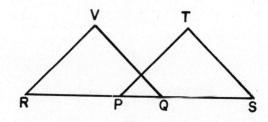

9. Given: $\overline{KG} \perp \overline{GH}$

   $\overline{LH} \perp \overline{GH}$

   ∠KHG ≅ ∠LGH

   Prove: $\overline{KH} \cong \overline{LG}$

10. Given: $\overline{AC} \cong \overline{BC}$

    ∠CAE ≅ ∠CBD

    Prove: △ACE ≅ △BCD

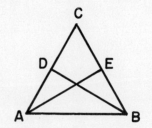

*11. Given: $\overline{QS}$ and $\overline{RT}$ bisect each other.

   Line segments $\overline{AB}$, $\overline{QR}$ and $\overline{TS}$

   Prove: $\overline{AP} \cong \overline{BP}$

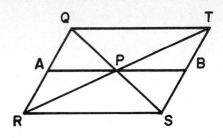

*12. Given: $\overline{AB}$ and $\overline{AC}$ are line segments

   $\overline{AD} \cong \overline{AE}$

   $\overline{BD} \cong \overline{CE}$

   $\overline{BG} \cong \overline{CH}$

   $\angle X \cong \angle Y$

   Prove: $\overline{AG} \cong \overline{AH}$

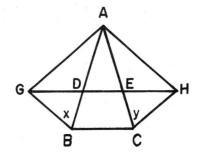

2.7 MORE PROOFS OF CONGRUENCE

Frequently developing a proof can require several intermediate steps to obtain the facts necessary to arrive at the desired conclusion. These intermediate steps may be to show that a pair of triangles is congruent so that one can justify that another pair of triangles is congruent. Sometimes we need to "add" or "subtract" two sets of congruent line segments or angles to obtain the desired congruence.

EXAMPLE 1   Given: $\overline{EF} \cong \overline{DF}$

   $\overline{AD} \cong \overline{EB}$

   Prove: $\angle EAF \cong \angle DBF$

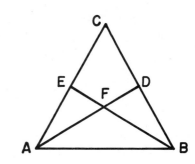

Solution First we must label the drawing with the given information and determine what we wish to prove. From the drawing we can see that if we can show that △EAF ≅ △DBF we can obtain the result that ∠EAF ≅ ∠DBF. Those triangles were chosen because we are given more information about them than the other pair of triangles having ∠EAF and ∠DBF as vertices. Let us see whether we have enough information to prove those triangles congruent. We redraw just the triangles under consideration so we may more clearly see the relationships between them. We know that one corresponding side is congruent. We also see from the new drawing that ∠EFA and ∠DFB are vertical angles, which are therefore equal. That gives us an SA correspondence. Now we need to find either a side congruence ($\overline{AF} \cong \overline{FB}$, to use SAS) or an angle congruence (∠FEA ≅ ∠FDB, to use ASA). We know nothing about the angles so we must examine our information about the sides. We do know $\overline{EF} \cong \overline{DF}$ and that $\overline{AD} \cong \overline{BE}$. We also know that if we subtract equal quantities from equal quantities the results are equal. Therefore AD - DF = BE - EF or $\overline{AF} \cong \overline{BF}$ (AF = BF). Recall that AF = BF states the lengths are equal, and $\overline{AF} \cong \overline{BF}$ states the line segments are congruent. Now we do have an SAS

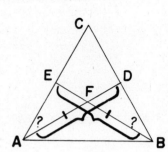

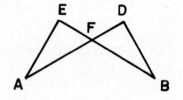

correspondence to show that the triangles are congruent, and we can write up the proof.

	Statement	Reason
S 1.	$\overline{EF} \cong \overline{DF}$	Given
A 2.	$\angle EFA \cong \angle DFB$	Vertical angles
3.	$\overline{AD} \cong \overline{BE}$	Given
4.	$AD - DF = BE - EF$	1, 3, & Subtraction of equals
S 5.	$\overline{AF} \cong \overline{BF}$	Restatement of 4
6.	$\triangle FAE \cong \triangle FBD$	1, 2, 5 & SAS
7.	$\angle EAF \cong \angle DBF$	6 & CPCTC

In Chapter 1 we learned that we could bisect a line segment using a compass and straightedge. We also learned that this construction gives us a perpendicular to the line segment. Let us prove that this is always the case.

EXAMPLE 2   Given: $\overline{ST}$

$\overline{SA} \cong \overline{SB} \cong \overline{TA} \cong \overline{TB}$
(by construction)

Prove: C is the midpoint of $\overline{ST}$

$\angle ACS = \angle ACT = 90°$

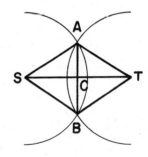

Solution

	Statement	Reason
S 1.	$\overline{SA} \cong \overline{TA}$	Given by construction (the points on a circle are all the same distance from the center of the circle)
S 2.	$\overline{SB} \cong \overline{TB}$	Given by construction (see Reason 1.)
S 3.	$\overline{AB} \cong \overline{AB}$	Common side
4.	$\triangle ABS \cong \triangle ABT$	1, 2, 3 & SSS

A	5.	∠SAC ≅ ∠TAC	4 & CPCTC
S	6.	$\overline{SA}$ ≅ $\overline{TA}$	See Reason 1.
S	7.	$\overline{AC}$ ≅ $\overline{AC}$	Common side
	8.	△SAC ≅ △TAC	5, 6, 7 & SAS
	9.	$\overline{SC}$ ≅ $\overline{TC}$	8 & CPCTC
	10.	C is the midpoint of $\overline{ST}$	9 & Definition of midpoint
	11.	∠ACS ≅ ∠ACT	8 & CPCTC
	12.	∠ACS + ∠ACT = 180°	SCT is a straight angle (given)
	13.	∠ACS = ∠ACT = 90°	11, 12 & two angles that are equal and have a sum of 180° each must equal 90°

There are frequently different ways to write the same proof. For instance, instead of Statement 13 above, the proof might have been written, starting with Statement 12:

	Statement	Reason
	.	.
	.	.
	.	.
12.	∠ACS + ∠ACT = 180°	∠SCT is a straight angle (given)
13.	∠ACS + ∠ACS = 180°	11, 12 & Substitution
14.	2∠ACS = 180°	13 & Addition
15.	∠ACS = 90°	14 & Division
16.	∠ACT = 90°	11, 15 & Substitution

EXERCISE 2.7

1. Given: $\overline{AE}$ intersects $\overline{BD}$ at C

   $\overline{AC} \cong \overline{CE}$

   $\overline{DC} \cong \overline{CB}$

   Prove: $\angle B \cong \angle D$

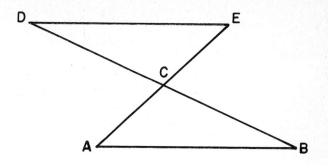

2. Given: $\overline{AC} \perp \overline{AB}$

   $\overline{DE} \perp \overline{BD}$

   B is the midpoint of $\overline{AD}$

   $\overline{AC} \cong \overline{DE}$

   Prove: $\triangle BAC \cong \triangle BDE$

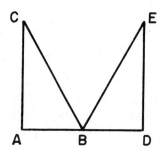

3. Given: Segments $\overline{MK}$, $\overline{SV}$, $\overline{KT}$, and $\overline{MN}$

   $\angle K \cong \angle T$

   $\overline{KR} \cong \overline{TR}$

   Prove: $\overline{SK} \cong \overline{TV}$

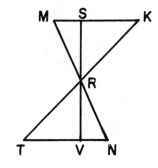

4. Given: a segment $\overline{RS}$ and points T and U on the opposite sides of $\overleftrightarrow{RS}$ so that $\overline{TR} \cong \overline{UR}$ and $\overline{TS} \cong \overline{US}$

   Prove: $\angle T \cong \angle U$ (Draw a figure.)

5. Given: $\overline{CD} \cong \overline{CB}$

   $\overline{AB} \perp \overline{CE}$

   $\overline{ED} \perp \overline{AC}$

   Prove: $\triangle ABC \cong \triangle EDC$

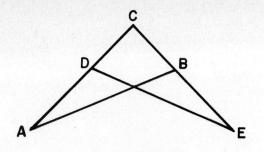

6. Given: $\overline{RT} \cong \overline{TS}$

   $\overline{RP} \cong \overline{PS}$

   Prove: $\angle TRP \cong \angle TSP$

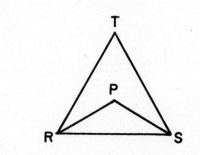

7. Given: In the 3-dimensional figure, points Q, R, and S lie in plane EF;

   $\overline{QS} \cong \overline{QR}$

   $\angle PQS \cong \angle PQR$

   Prove: $\triangle PQR \cong \triangle PQS$

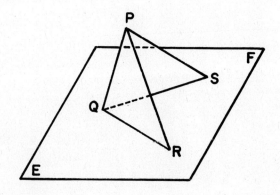

8. Given: $\overline{AF} \cong \overline{BG}$

   $\angle A \cong \angle B$

   $\overline{AE} \cong \overline{BD}$

   Prove: $\overline{DG} \cong \overline{EF}$

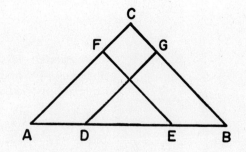

9. Given: M is the midpoint of
   $\overline{AC}$;

   $\overline{BM} \perp \overline{AC}$

   Prove: △ACB is isosceles

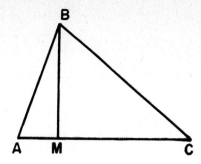

10. Chapter 1 presented a method to construct one angle congruent to another angle. Prove that the procedure given will always produce a congruent angle. (Hint: Prove two triangles are congruent.)

*11. Chapter 1 presented a method to construct a perpendicular to a line through a given point not on that line. Prove that the procedure given will always produce a right angle.

12. Chapter 1 presented a method to bisect an angle using a compass and straightedge. Prove that the procedure given will always produce congruent angles.

13. A method to construct a 60° ($\frac{\pi}{3}$ rad) angle was given in Chapter 1. Prove that this construction will produce an equilateral triangle, and therefore, a 60° angle.

14. Prove that the base angles of an isosceles triangle are congruent. (Hint: Use an angle bisector.)

15. Prove that if a leg and the hypotenuse of one triangle are congruent respectively to one leg and the hypotenuse of another triangle, the triangles are congruent. (Hint: Use the results from Problem 14 and a common side.)

CHAPTER 2 REVIEW

1. Match each item in Column I with the appropriate item(s) from Column II. Items in Column II can be used once, more than once, or not at all.

Column I		Column II	
a)	Convex polygon	A.	△
b)	equilateral triangle	B.	⬡
c)	sum of the angles in a triangle	C.	3:4:5
d)	obtuse scalene triangle	D.	π
e)	simple closed curve	E.	$1:\sqrt{3}:2$
f)	ratio of the sides in a 30°-60°-90° triangle	F.	⌂
g)	right isosceles triangle	G.	△ (equilateral marked)
h)	scalene triangle	H.	2π
i)	isosceles triangle	I.	$1:1:\sqrt{2}$
		J.	▽ (isosceles marked)
		K.	(obtuse triangle)
		L.	(right triangle)

2. List all triangles in the figure to the right and state whether each is

    a) right, obtuse, or acute

    b) scalene, isosceles, or equilateral.

    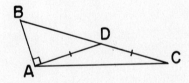

3. List all triangles in the figure to the right.

    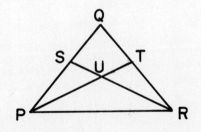

In Exercises 4 through 7 find the number of degrees or radians in the third angle of a triangle if the first two angles are as given.

4. 72°15' & 24°51'

5. $\frac{4\pi}{9}$ rad & $\frac{\pi}{6}$ rad

6. 13.43° & 61.21°

7. 1.63 rad & 0.48 rad

8. If $\angle A = 2x - 1°$, $\angle B = 3x - 5°$, and $\angle C = 4x + 15°$, find the number of degrees in each angle of the triangle.

9. The angles in a triangle are in the ratio 2:4:9. Find the number of degrees in each angle.

10. In a right triangle, the smallest angle is half the next largest angle. Find the number of radians in each angle of the triangle. (Give an exact answer.)

11. Determine the value of x for each triangle. (Give an exact answer.)

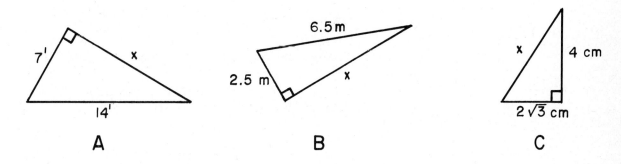

A   B   C

C 12. Determine the value of x for each triangle, correct to two decimal places.

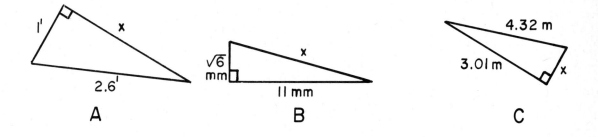

A   B   C

13. Determine the value of x and y for each triangle. Give an exact answer.

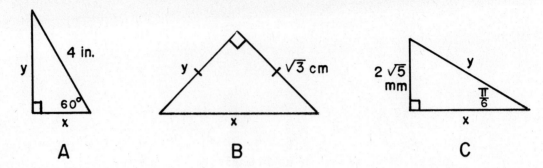

*14. Find the length of each side of the given triangle.

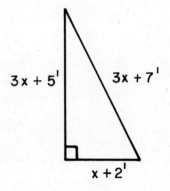

For all "C" problems, give answers to nearest hundredth.

*15. One leg of a right triangle is 1 cm longer than the other leg. The hypotenuse is 9 cm longer than the shorter leg. Find the dimensions of the triangle.

C 16. Two planes leave an airport at the same time and at a 90° angle from each other. The first plane flies at a speed of 120 miles per hour, and the second one flies at 200 miles per hour. If each flies a straight course, how far apart are they after 45 minutes?

C 17. Will a 21 inch umbrella fit in an attaché case 18" × 12" × 3"?

C 18. An oceanographic ship is conducting tests to determine the salinity of the ocean at specific positions in the Pacific. You are in charge of seeing that the Nansen Sample Bottle (used to obtain water samples) goes to the reporting depth of 1500 meters. The Captain is anxious to get to port to see his girlfriend, however, and he refuses to stop the ship completely. The ship is slowly moving forward as you winch down the cable with the Nansen Bottle attached to it. Using the angle gauge you find that your cable is going into the water at 45 degrees. The winch has an automatic meter counter that measures cable to the nearest tenth of a meter. This counter is located 3 meters above the ocean surface, and it is on a boom directly over the ocean. How much cable do you want to let out to get the bottle at the desired depth of 1500 meters?

C 19. If the cable is let out at a constant rate of 10 m per second in Problem 18, how far down is the Nansen Sample Bottle after 1 minute?

**20. For Problem 19, find the expression for the depth of the Nansen Sample Bottle after t minutes.

C 21. Two cycles are going to race down a straight track in parallel lanes that are 50 feet apart. If the first cyclist travels at a (fixed) rate of 880 ft per minute and the second at 900 ft per minute, find the distance separating the two after two minutes.

**22. For Problem 21, find the expression for the distance separating the two cyclists after t minutes.

For Problems 23 through 27, use only a compass and straightedge.

23. Construct a 30°-60°-90° triangle with segment r as the shortest side.

        ————————
            r

24. Construct a triangle given ∠A, ∠B and segment c as the included side.

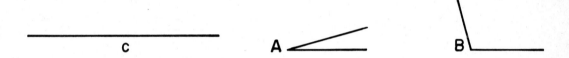

25. Construct a right triangle given segments r and s as the two legs.

   ————————   ————
       r   s

26. Construct a triangle given segments p, q, and r as the sides.

  ——————   ————————   ——————————
    p   q   r

27. Construct a $\frac{\pi}{4} - \frac{\pi}{4} - \frac{\pi}{2}$ rad triangle with segment t as the longer leg.

        ————————————
              t

28. Construct an equilateral triangle with sides of 6.5 cm using a compass, straightedge, and ruler.

For Problems 29 and 30, determine what additional information is necessary in order to show the given triangles congruent.

29.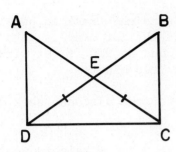

30.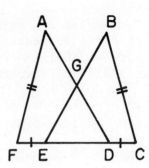

$\triangle AED \cong \triangle BEC$

 by SAS? _____

 by ASA? _____

$\triangle ADF \cong \triangle BEC$

 by SSS? _____

 by SAS? _____

31. Given: $\overline{AE} \perp \overline{AD}$

 $\overline{BF} \perp \overline{AD}$

 $\overline{AB} \cong \overline{CD}$

 $\angle BDF \cong \angle ACE$

 Prove: $\triangle AEC \cong \triangle BFD$

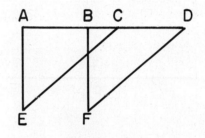

32. Given: $\overline{AB} \cong \overline{AD}$

 $\overline{BC} \cong \overline{DE}$

 Prove: $\angle C = \angle E$

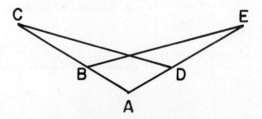

33. Given: Figure ABCD with $\overline{BX} \cong \overline{CY}$

 $\overline{AB} \perp \overline{BC}$

 $\overline{DC} \perp \overline{BC}$

 Prove: $\overline{XC} \cong \overline{YB}$

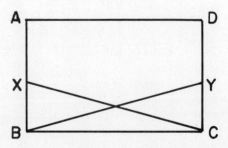

136

For Problems 34 through 36, use Figure 2.10.

34. Given: $\overline{BD}$ is the perpendicular bisector of $\overline{AC}$

    Prove: $\angle A \cong \angle C$

35. Given: $\overline{AG}$ and $\overline{CE}$ bisect each other at D

    Prove: $\angle DEG \cong \angle DCA$

36. Given: D is the midpoint of $\overline{BF}$
    Segment $\overline{AG}$
    $\overline{AC} \perp \overline{BF} \perp \overline{EG}$

    Prove: $\overline{DG} \cong \overline{AD}$

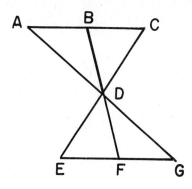

FIGURE 2.10

37. Given: $\triangle ADE$ with $\overline{AE} \cong \overline{ED}$
    $\overline{AC} \cong \overline{BD}$

    Prove: $\overline{BE} \cong \overline{CE}$

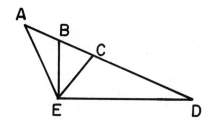

38. Given: $\angle AEB \cong \angle DEC$ in $\triangle AED$
    $\overline{AB} \cong \overline{CD}$
    $\angle ACE \cong \angle DBE$

    Prove: $\overline{AE} \cong \overline{ED}$

137

CHAPTER 3
PARALLEL LINES AND SIMILAR TRIANGLES

In Chapter 1, we defined two parallel lines as two lines that lie in the same plane and do not intersect. We also defined perpendicular lines as intersecting lines that form a right angle. Now we will study parallel and perpendicular lines and their use in triangles.

3.1 PARALLEL LINES

A line that crosses two lines is called a TRANSVERSAL. In Figure 3.1 $\overleftrightarrow{CD}$ is a transversal to $\overleftrightarrow{AB}$ and $\overleftrightarrow{EF}$.

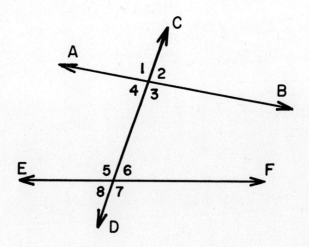

FIGURE 3.1

The transversal forms a total of eight angles. The angles lying between two lines and on opposite sides of the transversal are called ALTERNATE INTERIOR ANGLES. These angles share a segment of the transversal as a common side. Thus we will always have two pairs of alternate interior angles. For instance, $\angle 4$ and $\angle 6$, and $\angle 3$ and $\angle 5$ are the two pairs in Figure 3.1.

The angles lying outside of the two lines and on opposite sides of the transversal are called ALTERNATE EXTERIOR ANGLES. These angles each have a side contained in one of the lines and the other side contained in the transversal. In Figure 3.1 $\angle 1$ and $\angle 7$, and $\angle 2$ and $\angle 8$ are the two pair of alternate exterior angles.

CORRESPONDING ANGLES are angles on the same side of the transversal and which are either both above or both below the two lines. The pairs of corresponding angles in Figure 3.1 are: $\angle 1$ and $\angle 5$, $\angle 2$ and $\angle 6$, $\angle 3$ and $\angle 7$, and $\angle 4$ and $\angle 8$.

When a transversal crosses two parallel lines, then certain pairs of angles are congruent.

If two parallel lines are cut by a transversal then:

1. Alternate interior angles are congruent.
2. Alternate exterior angles are congruent.
3. Corresponding angles are congruent.

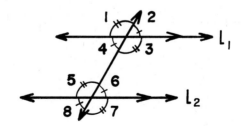

Thus, when $\ell_1 \| \ell_2$ we have two sets of congruent angles:

$$\angle 1 \cong \angle 3 \cong \angle 5 \cong \angle 7$$

$$\angle 2 \cong \angle 4 \cong \angle 6 \cong \angle 8$$

Further, any angle from the first set is supplementary to any angle in the second set. For instance, $\angle 1$ is supplementary to $\angle 4$.

EXAMPLE 1   Determine the number of degrees in each angle if $\angle 3 = 38°$.

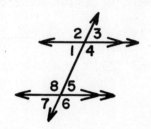

Solution   Because the two lines are each marked with an arrowhead, we know they are parallel. Knowing this, we can use the facts stated above.

$\angle 3$ and $\angle 5$ are corresponding angles and therefore are equal, so $\angle 5 = 38°$.

$\angle 5$ and $\angle 1$ are alternate interior angles and therefore are equal, so $\angle 1 = 38°$.

$\angle 1$ and $\angle 7$ are corresponding angles and therefore are equal, so $\angle 7 = 38°$.

$\angle 3$ and $\angle 2$ are supplementary angles, so $\angle 2 = 180° - 38° = 142°$.

$\angle 2$ and $\angle 8$ are corresponding angles and therefore are equal, so $\angle 8 = 142°$.

$\angle 8$ and $\angle 4$ are alternate interior angles and therefore are equal, so $\angle 4 = 142°$.

$\angle 2$ and $\angle 6$ are alternate exterior angles and therefore are equal, so $\angle 6 = 142°$.

The drawing can now be labeled as shown below.

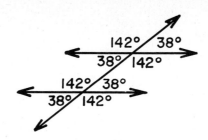

Thus we see that ∠1 = ∠3 = ∠5 = ∠7 = 38° and
∠2 = ∠4 = ∠6 = ∠8 = 142°.

Note that interior angles on the same side of the transversal are supplementary.

If two lines are cut by a transversal and

1. alternate interior angles are congruent, or
2. the alternate exterior angles are congruent, or
3. corresponding angles are congruent,
   then the lines are parallel.

Thus alternate interior angles are equal if and only if the lines are parallel. We can use this to determine the degrees of angles.

EXAMPLE 2   In the drawing shown, determine the value of x so that the lines will be parallel.

Solution   First we number each angle for easy reference. Because they are vertical angles, ∠4 = ∠2. If the lines are to be parallel, ∠6 = ∠4 = 143°21'. Now the desired angle, ∠7, and ∠6 are supplementary so we have:

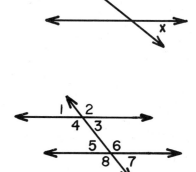

141

$$\angle 7 = 180° - 143°21'$$
$$= 36°39'$$

If the lines are to be parallel, $x = 36°39'$.

> Two lines are parallel if and only if the interior angles on the same side of the transversal are supplementary.

EXAMPLE 3   Determine the number of degrees in $\angle 3$ and $\angle 8$ if $\angle 1 = 3x + 2°$ and $\angle 7 = 4x - 20°$.

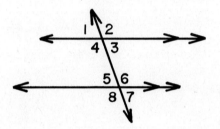

Solution   Because the lines are parallel,

$$\angle 1 = \angle 3 = \angle 5 = \angle 7 \text{ so}$$

$$3x + 2 = 4x - 20$$

$$22 = x$$

$\angle 1 = 3x + 2$	$\angle 7 = 4x - 20$
$= 3(22) + 2$	$= 4(22) - 20$
$= 68°$	$= 68°$

Because both values are 68° the solution is correct.

We know that $\angle 1 = \angle 3$, so $\angle 3 = 68°$. $\angle 3$ and $\angle 6$ are supplementary, so $\angle 6 = 180° - 68° = 112°$.

$\angle 8$ and $\angle 6$ are vertical angles and are therefore equal, so $\angle 8 = 112°$.

The solution is $\angle 3 = 68°$ and $\angle 8 = 112°$.

EXAMPLE 4 Given $\ell_1 \| \ell_2$, determine the number of degrees in each angle.

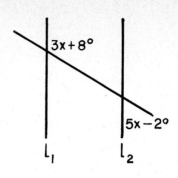

Solution First, number each angle.

$\angle 2 = \angle 8 = \angle 4 = \angle 6$ and

$\angle 1 = \angle 3 = \angle 5 = \angle 7$, so

$\angle 7$ and $\angle 8$ are supplementary.

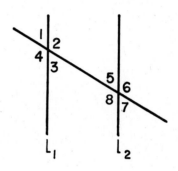

$$\angle 7 + \angle 8 = 180°$$
$$5x - 2 + 3x + 8 = 180°$$
$$8x = 174°$$
$$x = 21.75°$$

$\angle 7 = 5x - 2$ $\angle 8 = 3x + 8$

$ = 5(21.75) - 2$ $ = 3(21.75) + 8$

$ = 108.75 - 2$ $ = 65.25 + 8$

$ = 106.75°$ $ = 73.25°$

$\angle 7 = \angle 5 = \angle 3 = \angle 1 = 106.75°$ or $106°45'$

$\angle 8 = \angle 6 = \angle 4 = \angle 2 = 73.25°$ or $73°15'$

EXAMPLE 5   For the given drawing determine the number of degrees in ∠4, ∠7, and ∠8 if ∠5 = 33° and ∠1 = 78°.

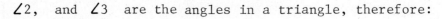

Solution   Because ∠5 and ∠4 are supplementary,

∠4 = 180° - 33° = 147°.

∠3 and ∠5 are vertical angles so ∠3 = 33°. ∠1, ∠2, and ∠3 are the angles in a triangle, therefore:

∠1 + ∠2 + ∠3 = 180°

78° + ∠2 + 33° = 180°

∠2 = 180° - 111°

∠2 = 69°

∠7 and ∠5 are corresponding angles formed when a transversal cuts two parallel lines, so they are equal. Therefore ∠7 = 33°. ∠2, ∠7, and ∠8 together form a straight angle. Therefore:

∠2 + ∠7 + ∠8 = 180°

69° + 33° + ∠8 = 180°

∠8 = 180° - 102°

= 78°

The solution is: ∠4 = 147°, ∠7 = 33°, and ∠8 = 78°.

EXAMPLE 6 Given the drawing to the
right, is △ABE ≅ △BDE?

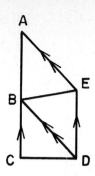

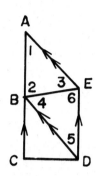

Solution We know ∠3 ≅ ∠4 because they are alternate interior angles formed by a transversal cutting two parallel segments ($\overline{AE}$ and $\overline{BD}$). Likewise ∠2 ≅ ∠6 because they are alternate interior angles formed by a transversal cutting $\overline{AC}$ and $\overline{DE}$. We also have $\overline{BE}$ as a common side of the two triangles so we have an ASA correspondence: ∠3 ≅ ∠4, $\overline{BE} \cong \overline{BE}$, and ∠2 ≅ ∠6. Therefore the two triangles are congruent.

We can use the fact that corresponding angles must be equal to construct a line parallel to a given line through a given point:

To construct a line parallel to a given line through a given point:

1. Draw a line segment that intersects the given line and that passes through the given point, P.

2. This line segment forms an angle with the given line. Construct an angle at P congruent to this angle in the corresponding position.

3. Because the corresponding angles are equal, the line constructed is parallel to the given line and it contains the point P. Therefore, it is the desired line.

EXAMPLE 7   Construct a line parallel to m through the point P.

Solution   Step 1:   Draw a line segment that intersects m and contains P.

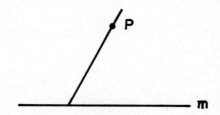

Step 2:   Construct an angle at P congruent to the angle formed by the given line and the line segment.

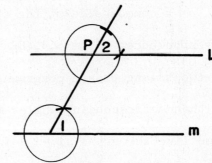

Step 3:   Because ∠1 and ∠2 are congruent and are corresponding angles, lines m and ℓ are parallel. Line ℓ contains the point P. Therefore, line ℓ is the desired line.

EXERCISE 3.1

1. In the drawing at the right give the pairs of angles that are:

   a) alternate interior
   b) alternate exterior
   c) corresponding
   d) vertical

   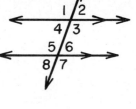

2. In the drawing at the right give the pairs of angles that are:

   a) alternate interior
   b) alternate exterior
   c) corresponding
   d) vertical

   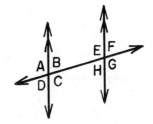

3. List the set(s) of equal angles in Problem 1.

4. List the set(s) of equal angles in Problem 2.

5. List the set(s) of supplementary angles in Problem 1.

6. List the set(s) of supplementary angles in Problem 2.

7. Determine the value of x that will make the lines parallel.

   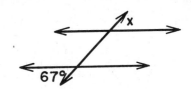

8. Determine the value of x that will make the lines parallel.

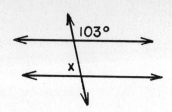

9. Find the number of degrees in each angle.

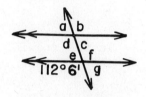

10. Find the number of radians in each angle.

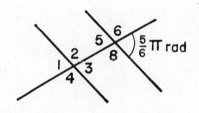

For Problems 11 to 14, refer to Figure 3.2.

11. Find the number of degrees in $\angle c$ and $\angle f$ if $\angle b = 5x + 7°$ and $\angle h = 2x + 79°$.

12. Find the number of degrees in $\angle b$ and $\angle e$ if $\angle d = 9x + 12°$ and $\angle h = 5x + 56°$.

13. Find the number of degrees in $\angle a$ and $\angle f$ if $\angle b = 7x - 37°$ and $\angle g = 2x + 19°$.

14. Find the number of degrees in ∠b and ∠g if ∠d = 11x − 58° and ∠e = 4x + 13°.

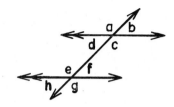

FIGURE 3.2

For Problems 15 and 16, refer to Figure 3.3. $\overline{CB}$ and $\overline{AB}$ are segments.

15. If ∠CAD = 43° and ∠DEB = 56°, find the number of degrees in each angle.

16. If ∠DBE = 69° and ∠ACE = 82°, find the number of degrees in each angle.

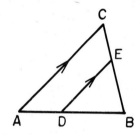

FIGURE 3.3

For Problems 17 and 18, refer to Figure 3.4. $\overline{AC}$ and $\overline{BD}$ are segments.

17. If ∠9 = 49° and ∠2 = 88°, find the number of degrees in the other angles.

18. If ∠4 = 77° and ∠5 = 38°, find the number of degrees in the other angles.

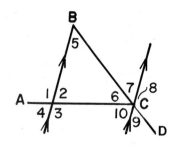

FIGURE 3.4

19. In the drawing to the right, is △ABC ≅ △ADC? Support your answer.

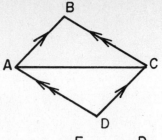

20. In the drawing to the right, if B bisects $\overline{AC}$, is △ABE ≅ △BCD?

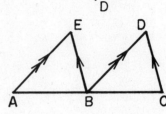

21. Given the drawing to the right where B bisects $\overline{AC}$, is △DCB ≅ △FBA? Support your answer.

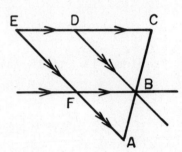

22. Given the drawing to the right, $\overline{AC}$, $\overline{DF}$, and $\overline{DB} \cong \overline{EA}$, is △DCB = △EBA? Support your answer.

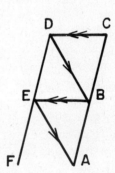

23. Given point P and line segment m, in the drawing below, construct a line parallel to m containing P.

24. Given point O and line segment m in the drawing above, construct a line parallel to m containing O.

3.2 PROOFS

We can use the facts from the preceding section to write proofs. In addition to the statements given on pages 116 and 117, we may now assume the following statements are true.

1. Through a given point not on a given line, one and only one line can be drawn parallel to a given line.

2. Two lines are parallel if a pair of corresponding angles are equal.

3. Two lines are parallel if a pair of alternate interior angles are equal.

4. Two lines are parallel if a pair of alternate exterior angles are equal.

5. If two lines are parallel, then corresponding angles are equal.

6. If two lines are parallel, then alternate interior angles are equal.

7. If two lines are parallel, then a pair of alternate exterior angles are equal.

8. Two lines are parallel if a pair of interior angles on the same side of the transversal are supplementary.

9. If two lines are parallel, then interior angles on the same side of the transversal are supplementary.

Because we have more statements to use, sometimes the proofs may require more steps to reach the desired conclusion.

EXAMPLE 1   Given: Figure to the right.

Prove: △AEB ≅ △BDE

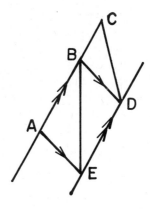

Solution   Because we are given no facts other than what we can see in the drawing, we must study it carefully. We are not given the lengths of any line segments, so we may need to use a common side. With possibly only one pair of sides congruent, we should see if an ASA correspondence is possible. Certainly parallel lines give us much information about angles. Let us organize the facts from the drawing.

		Statement	Reason
	1.	$\overline{AB} \parallel \overline{DE}$	Given
A	2.	$\angle ABE \cong \angle BED$	1 & Alternate interior $\angle$s
S	3.	$\overline{BE} \cong \overline{BE}$	Common side
	4.	$\overline{AE} \parallel \overline{BD}$	Given
A	5.	$\angle AEB \cong \angle DBE$	4 & Alternate interior $\angle$s.
	6.	$\triangle AEB \cong \triangle BDE$	2, 3, 5 & ASA

EXAMPLE 2   Given: $\overline{BD}$ and $\overline{AE}$ bisect each other

Prove: $\overline{AB} \parallel \overline{DE}$

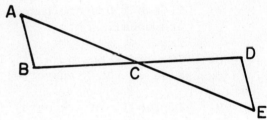

Solution   In order to prove that $\overline{AB} \parallel \overline{DE}$, we need to show either $\angle A \cong \angle E$ or $\angle B \cong \angle D$ so that alternate interior angles will be equal, giving the line segments as parallel. In order to show that the angles are congruent, we must show that $\triangle ACB \cong \triangle ECD$ so we will have corresponding angles congruent. Thus we must start by determining the facts we know about the two triangles.

	Statement	Reason
1.	$\overline{BD}$ and $\overline{AE}$ bisect each other	Given
S 2.	$\overline{AC} \cong \overline{CE}$	1 & Def. of bisect
S 3.	$\overline{BC} \cong \overline{CD}$	1 & Def. of bisect
A 4.	$\angle ACB \cong \angle ECD$	Vertical angles
5.	$\triangle ACB \cong \triangle ECD$	2, 4, 3 & SAS
6.	$\angle B \cong \angle D$	5 & CPCTC
7.	$\overline{AB} \parallel \overline{DE}$	6 & Alt. interior $\angle$s are $\cong$

EXERCISE 3.2

1. Given: $\overline{AD} \parallel \overline{BC}$

    $\overline{AB} \parallel \overline{DC}$

    Prove: $\overline{AB} \cong \overline{CD}$

2. Given: Figure from Problem 1,

    $\overline{AD} \parallel \overline{CB}$

    $\overline{AD} \cong \overline{CB}$

    Prove: $\angle D \cong \angle B$

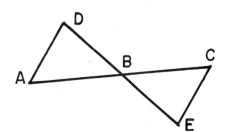

3. Given: $\overline{AC}$ and $\overline{DE}$ intersecting at B,

    $\overline{AB} \cong \overline{CB}$

    $\overline{AD} \parallel \overline{CE}$

    Prove: $\overline{AD} \cong \overline{CE}$

4. Given: C bisects $\overline{AE}$

    Prove: $\triangle ABC \cong \triangle CDE$

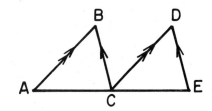

5. Given: $\overline{DC}$, $\overline{AB}$ and $\overline{EF}$ intersect at R

$\overline{DR} \cong \overline{CR}$

$\overline{ER} \cong \overline{FR}$

Prove: $\overline{DA} \parallel \overline{CB}$

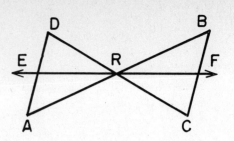

6. Given: Segment $\overline{AF}$

$\overline{AD} \cong \overline{FE}$

$\overline{AB} \cong \overline{FC}$

$\overline{BE} \cong \overline{CD}$

Prove: $\overline{AB} \parallel \overline{CF}$

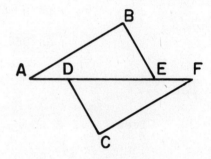

7. Given: Segment $\overline{AC}$

$\overline{AB} \parallel \overline{CD}$

$\overline{AB} \cong \overline{CD}$

$\overline{AF} \cong \overline{EC}$

Prove: $\overline{BE} \cong \overline{FD}$

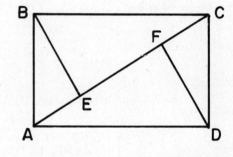

8. Given: $\ell_1 \parallel \ell_2$

O is midpoint of $\overline{AB}$

Prove: O is midpoint of $\overline{CD}$

9. Refer to the drawing as given in Problem 8.

Given: O is the midpoint of $\overline{AB}$

O is the midpoint of $\overline{CD}$

Prove: $\ell_1 \parallel \ell_2$

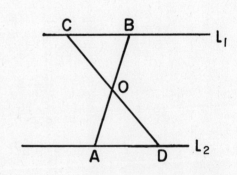

10. Given: Segments $\overline{PS}$ and $\overline{RT}$
    $\overline{PQ} \cong \overline{QS}$
    $\overline{RQ} \cong \overline{QT}$
    Prove: $\overline{PT} \parallel \overline{RS}$
    $\overline{RP} \parallel \overline{ST}$

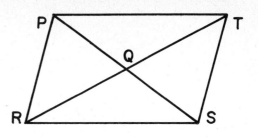

11. A method to construct a line parallel to a given line through a point not on that line was given in Section 3.1. Prove that this procedure will always produce a parallel line.

12. Prove that the sum of the angles in a triangle is 180° or π rad. (Hint: Draw a line through a vertex parallel to the side opposite that vertex.)

3.3 SIMILAR TRIANGLES

We have worked with congruent figures, figures that are the same shape and size. We will now examine figures that are the same shape, but which may not be the same size. Such figures are called SIMILAR FIGURES. Model trains, photographs, and maps are all designed to be smaller versions of what they represent and therefore are similar to the object modeled. On maps we are given the scale to which the map is drawn, for instance 1 in. = 200 mi. This is the ratio between the two.

We can tell when two triangles are similar by comparing the angles.

> Triangles are similar if and only if there is a correspondence between angles for the two triangles and each set of corresponding angles is congruent. This is abbreviated: AAA correspondence is a similarity.

Thus all congruent triangles are similar, but not all similar triangles are congruent.

We use the symbol $\sim$ to indicate that two triangles are similar. Because $\triangle ABC$ is similar to $\triangle RST$, we write: $\triangle ABC \sim \triangle RST$.

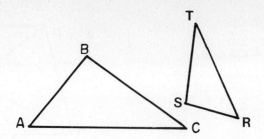

If triangles are similar, then their sides are in proportion just as the measurements on a map are in proportion to the actual distances mapped. We can use this relationship to determine the lengths of unknown sides in two similar triangles. For example, the longest side in the first will be proportional to the longest side in the second; the shortest side in the first will be proportional to the shortest side in the second. The basic pattern to follow when setting up the proportion is:

$$\frac{\text{Side of first triangle}}{\text{Corresponding side of second triangle}} = \frac{\text{Another side of first triangle}}{\text{Corresponding side of second triangle}}$$

For △ABC and △RST, we have:

$$\frac{\overline{AB}}{\overline{RS}} = \frac{\overline{BC}}{\overline{ST}} = \frac{\overline{AC}}{\overline{RT}}$$

Note that the sides for one triangle are always in the numerator while the sides for the other triangle are always in the denominator.

In Section 2.3, when we studied the 30°-60°-90°, the 45°-45°-90°, and the 3-4-5 right triangles, we were working with three specific kinds of similar triangles. All triangles that have angles of 30°, 60°, and 90° are similar. Therefore, the sides will always be in proportion. The same is true for all triangles with angles of 45°, 45°, and 90°, and for all triangles with sides in the ratio 3:4:5.

> If the corresponding sides in two triangles are in proportion, then the two triangles are similar.

We can use this fact to determine if two triangles are similar.

EXAMPLE 1 Determine if the given triangles are similar.

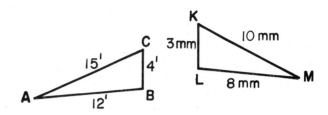

Solution If the triangles are similar, then the corresponding sides must be in proportion. The longest sides of each must correspond: $\overline{AC}$ to $\overline{KM}$; the shortest sides must correspond: $\overline{BC}$ to $\overline{KL}$; therefore, $\overline{AB}$ must correspond to $\overline{LM}$. Let us see if these sides are all proportional. That is, is the following true:

$$\frac{AC}{KM} \stackrel{?}{=} \frac{BC}{KL} \stackrel{?}{=} \frac{AB}{LM}$$

$$\frac{15}{10} \stackrel{?}{=} \frac{4}{3} \stackrel{?}{=} \frac{12}{8}$$

Reducing each fraction to lowest terms we have:

$$\frac{3}{2} \stackrel{?}{=} \frac{4}{3} \stackrel{?}{=} \frac{3}{2}$$

Because $\frac{3}{2} \neq \frac{4}{3}$, the sides are not proportional. Thus, the triangles are not similar.

We can also use the fact that sides of similar triangles are proportional to determine the lengths of unknown sides.

EXAMPLE 2   △ABC is similar to △RST.
Determine the values of x and y.

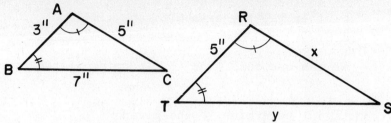

Solution   We are given ∠A ≅ ∠R, ∠B ≅ ∠T, therefore, ∠C ≅ ∠S.

Thus the following sides must be in proportion:

$$\frac{AB}{RT} = \frac{AC}{RS} = \frac{BC}{TS}$$

In order to solve for y, we must choose a proportion for which we know three values with y as the fourth. That gives us the proportion:

$$\frac{AB}{RT} = \frac{BC}{TS}$$

$$\frac{3}{5} = \frac{7}{y}$$

Solving for y we get:

$$y = \frac{35}{3} = 11\frac{2}{3} \text{ in.}$$

We repeat the process to determine the value for x. The necessary proportion is:

$$\frac{AB}{RT} = \frac{AC}{RS}$$

$$\frac{3}{5} = \frac{5}{x}$$

Solving for x, we get:

$$x = \frac{25}{3} = 8\frac{1}{3} \text{ in.}$$

The values are $x = 8\frac{1}{3}$ in. and $y = 11\frac{2}{3}$ in.

EXAMPLE 3   △GHI ~ △KJI
            Find the values of a and b.

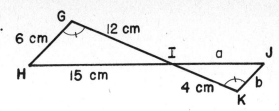

Solution   The proportions for this problem are:

$$\frac{GH}{KJ} = \frac{HI}{JI} = \frac{GI}{KI}$$

Note that we did NOT use $\frac{GI}{IJ}$. We must determine the proper correspondence between angles before we can determine the correspondence between the sides. ∠G corresponds to ∠K, not ∠J.

To solve for a:

$$\frac{HI}{JI} = \frac{GI}{KI}$$

$$\frac{15}{a} = \frac{\cancel{12}^{3}}{\cancel{4}_{1}}$$

$$a = \frac{15}{3} = 5 \text{ cm}$$

To solve for b:

$$\frac{GH}{KJ} = \frac{GI}{KI}$$

$$\frac{6}{b} = \frac{\cancel{12}^{3}}{\cancel{4}_{1}}$$

$$b = \frac{6}{3} = 2 \text{ cm}$$

The solution is a = 5 cm and b = 2 cm.

As you finish a problem, you should always check to be sure that the answer seems reasonable.

A triangle cut by a segment parallel to one side of the triangle is similar to the smaller triangle created.

EXAMPLE 4   Show △ADE ~ △ABC.

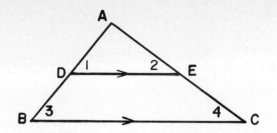

Solution   Because DE∥BC, ∠1 and ∠3, and ∠2 and ∠4 are pairs of corresponding angles. Therefore ∠1 ≅ ∠3 and ∠2 ≅ ∠4. Both triangles share a common angle, A. Thus we have an AAA correspondence between the two triangles, and they are similar triangles.

EXAMPLE 5   For the figure in Example 4, AB = 12 m, AD = 5 m, AE = 7 m, and BC = 15 m. Find the lengths of the segments not given.

Solution   In order to set up the proportions properly, we redraw the similar triangles and label each.

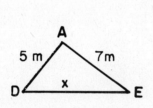 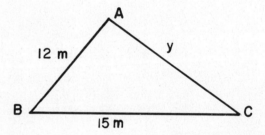

To find the length of $\overline{DE}$ we need a pair of corresponding sides for which we know the length. That means we must use $\overline{AD}$ and $\overline{AB}$. The side corresponding to $\overline{DE}$ is $\overline{BC}$, so we can set up the proportion:

$$\frac{AD}{AB} = \frac{DE}{BC}$$

$$\frac{5}{12} = \frac{x}{15}$$

$$x = \frac{75}{12} = 6\frac{1}{4} = 6.25 \text{ m}$$

We must repeat the process to find the length of $\overline{AC}$. Here the needed proportion is:

$$\frac{AD}{AB} = \frac{AE}{AC}$$

$$\frac{5}{12} = \frac{7}{y}$$

$$y = \frac{84}{5} = 16\frac{4}{5} = 16.8 \text{ m}$$

The solution is: $x = 6.25$ m and $y = 16.8$ m.

Some word problems can be solved using similar triangles.

EXAMPLE 6   A 30 ft tower casts a 48 ft shadow. At the same time a telephone pole casts a 32 ft shadow. What is the height of the telephone pole?

Solution   First we need drawings representing the tower and the telephone pole. We have similar triangles because the position of the sun is the same and both the pole and the tower are upright. We can then set up the proportion:

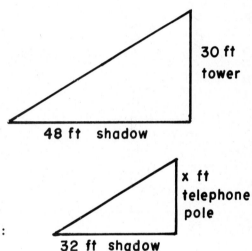

$$\frac{30}{x} = \frac{\cancel{48}^{3}}{\cancel{32}_{2}}$$

$$x = \frac{60}{3} = 20$$

The telephone pole is 20 ft tall.

Sometimes the similar triangles are not oriented the same way. This can make problems confusing. When this happens, be sure to redraw the two triangles so that they have the same orientation (corresponding parts are in the same position in each drawing).

EXAMPLE 7   If ∠ADC ≅ ∠BCA,
            show △ABC ~ △ACD

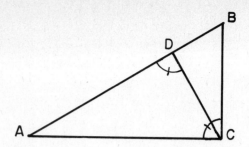

Solution   From the drawing we have that ∠A is a common angle to the two triangles. We also are given ∠ADC = ∠BCA. If two angles in a triangle are congruent to two angles in another triangle, then the third angles are congruent. Thus we have an AAA correspondence and the two triangles are similar.

EXAMPLE 8   For the figure in Example 7, $\overline{AC}$ = 10 in., $\overline{BC}$ = 6 in., and $\overline{CD}$ = 4 in. Find the length of each remaining side in each triangle.

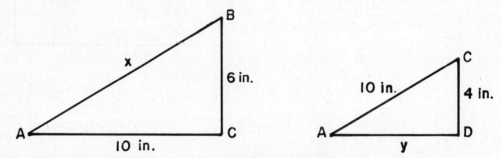

Solution   This figure is particularly tricky, so it is important to redraw the two separate triangles and orient them so corresponding parts are in similar positions. That means we must "turn over" △ADC. Now we can enter the given facts on the drawings and set up the proportions. To find the length of $\overline{AB}$:

$$\frac{AB}{AC} = \frac{BC}{CD}$$

$$\frac{x}{10} = \frac{\cancel{3}^{3}}{\cancel{4}_{2}}$$

$$x = \frac{30}{2} = 15 \text{ in.}$$

To find the length of $\overline{AD}$:

$$\frac{AC}{AD} = \frac{BC}{CD}$$

$$\frac{10}{y} = \frac{\cancel{3}^{3}}{\cancel{4}_{2}}$$

$$y = \frac{20}{3} = 6\frac{2}{3} \text{ in.}$$

The solution is $AB = 15$ in. and $AD = 6\frac{2}{3}$ in.

When we encounter variables in the problem, we proceed to set up a proportion and solve for the value of the unknown.

*EXAMPLE 9   Find the length of $\overline{DE}$ in $\triangle ABC$.

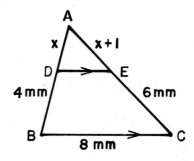

Solution   Because $\overline{DE} \parallel \overline{BC}$, we know that we have similar triangles. Before we can find the length of $\overline{DE}$, however, we must first determine the value of x. To do that we need the proportion:

 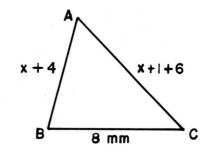

163

$$\frac{x}{x+4} = \frac{x+1}{x+7}$$

$$x(x+7) = (x+1)(x+4)$$

$$x^2 + 7x = x^2 + 5x + 4$$

$$2x = 4$$

$$x = 2$$

Substituting the value of x, we have $AD = 2$ mm, $AE = 3$ mm, $AB = 6$ mm, and $AC = 9$ mm. We enter these values on the

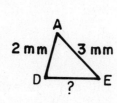
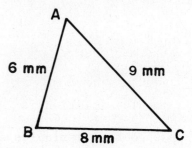

drawings and now can set up the proportion to determine the length of $\overline{DE}$.

$$\frac{AD}{AB} = \frac{DE}{BC}$$

$$\frac{\cancel{2}^1}{\cancel{6}_3} = \frac{DE}{8}$$

$$DE = \frac{8}{3} = 2\frac{2}{3} \text{ mm} = 2.\overline{6} \text{ mm}$$

EXAMPLE 10 Find the value of x in $\triangle ABC$ if $\angle ACB \cong \angle ADC$.

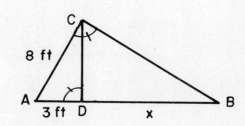

Solution   Once again we have similar triangles so we can set up the proportion:

$$\frac{AC}{AB} = \frac{AD}{AC}$$

$$\frac{8}{3+x} = \frac{3}{8}$$

$$9 + 3x = 64$$

$$3x = 55$$

$$x = 18\frac{1}{3} \text{ ft}$$

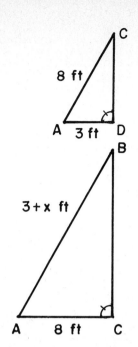

EXAMPLE 11   A girl 3 ft tall walks away from the base of a light pole at the rate of 4 ft per second. If the light is 12 ft above the ground, determine how long her shadow is after 4 seconds; after t seconds.

Solution   First we must make a drawing to represent the problem. After 4 seconds, she will have walked 16 ft. Let x be the length of the shadow. Then we have the proportion:

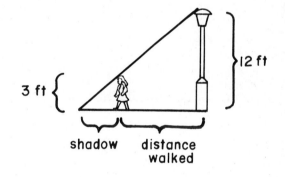

$$\frac{\cancel{12}^{4}}{\cancel{3}_{1}} = \frac{16 + x}{x}$$

$$4x = 16 + x$$

$$3x = 16$$

$$x = 5\frac{1}{3} \text{ ft}$$

165

Her shadow will be $5\frac{1}{3}$ ft long at 4 seconds.

After t seconds, she will have traveled 4t ft. If x is the length of the shadow, we have:

$$\frac{\cancel{12}^{4}}{\cancel{3}_{1}} = \frac{4t + x}{x}$$

$$4x = 4t + x$$

$$3x = 4t$$

$$x = \frac{4}{3}t$$

After t seconds, her shadow will be $\frac{4}{3}t$ ft long.

EXERCISE 3.3

Using the information given, state if the following pairs of triangles can be shown to be congruent, similar, or neither. If they are congruent, state the postulate that will prove them so. If they are similar and not congruent, state the postulate that will prove them similar.

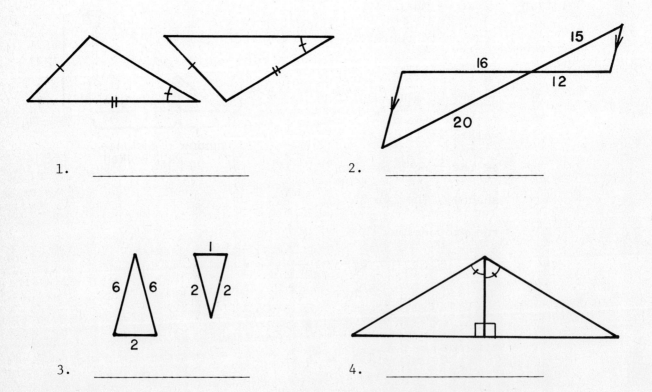

1. _____ 2. _____

3. _____ 4. _____

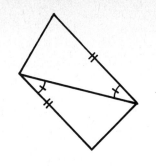

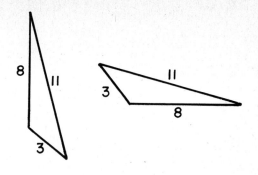

5. _____   6. _____

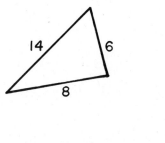 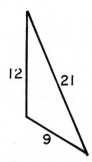   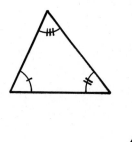 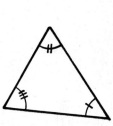

7. _____   8. _____

9. Show △ADE ~ △CBE.
   List the corresponding sides

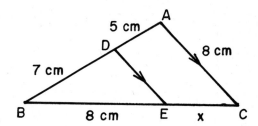

10. Determine the value for x and y in the figure for Problem 9.

11. Show △ABC ~ △DBE.
    List the corresponding sides.

167

12. Determine the value of x and y in the figure for Problem 11.

Use the figure to the right for Problems 13 to 16.

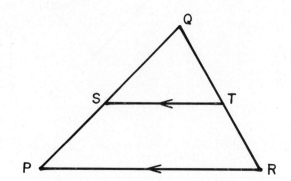

13. If ST = 5 ft, PR = 15 ft, and QT = 4 ft, find the length of QR.

14. If QS = 5 ft, QT = 6 ft, PR = 12 ft, and QP = 15 ft, find the length of QR and ST.

15. If QS = 6 m, QT = x m, TR = 5 m, and SP = x + 7 m, find the length of QT.

16. If QT = 2x m, TR = x m, ST = 3x − 1 m, and PR = 12 m, find the length of QR.

17. A tree casts an 18 ft shadow at the same time a 24 ft flagpole casts a 27 ft shadow. How tall is the tree?

18. A 10 ft tall tetherball pole casts a shadow of 24 ft. How tall is a person who casts a shadow of 13.2 ft?

19. A light is mounted on a pole 15 ft above the level sidewalk. How long is the shadow of a 6 ft tall man who is 10 ft away from the pole?

20. A street light is 12 ft above the level sidewalk. If a woman 5 ft tall casts a 10 ft shadow, how far away is she from the base of the pole?

21. If RQ = 12 in., TQ = 3 in., PQ = 8 in., and ST = 4 in. in △PQR, find the length of the other segments.

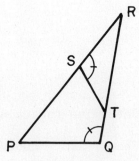

22. If SR = 6 in., PS = 4 in., RT = 5 in., and PQ = 7 in. in △PQR, find the length of the other segments.

23. If AC = 15 cm, AB = 6 cm, and BC = 10 cm in △ABC, find the length of the other segments.

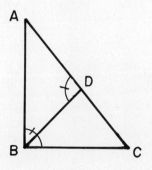

24. If AB = 8 cm, AC = 12 cm, and DB = 6 cm in △ABC, find the length of the other segments.

25. A man 6 ft tall starts walking away from the base of a street light 10 ft above the ground. If he walks at the rate of 5 ft per second, how long is his shadow after 3 seconds?

26. A woman 5 ft tall starts walking away from the base of a street light 15 ft above the ground. If she walks at a constant rate of 4 ft per second, how long is her shadow after 2 seconds?

**27. Determine the equation for the length of the shadow, S, in Problem 25 after t seconds. Solve for S in terms of t.

**28. Determine the equation for the length of the shadow, S, in Problem 26 after t seconds. Solve for S in terms of t.

3.4 PARALLEL LINES AND TRANSVERSALS

We have seen that a line segment parallel to one side of a triangle creates a smaller triangle similar to the original one. We have then used the fact that similar triangles are proportional to solve for unknown sides. A line segment parallel to one side of a triangle also divides the other two sides into proportional parts.

> A line parallel to one side of a triangle divides the other two sides proportionately.

EXAMPLE 1 Given △ABC with $\overline{DE} \parallel \overline{AC}$, give the proportion relating the sides cut by $\overline{DE}$.

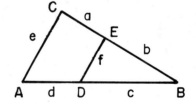

Solution Because $\overline{DE} \parallel \overline{AC}$, then $\overline{DE}$ divides $\overline{CB}$ and $\overline{AB}$ proportionately. That is,

$$\frac{b}{a} = \frac{c}{d}$$

Note that $\overline{DE}$ divides only sides $\overline{CB}$ and $\overline{AB}$ proportionately. It is INCORRECT to use the proportion $\frac{b}{f} = \frac{a}{e}$. The CORRECT proportion would be:

$$\frac{b}{f} = \frac{a+b}{e}.$$

We can extend this result to three or more parallel lines cutting two transversals.

> Three or more parallel lines divide two transversals proportionately.

EXAMPLE 2   Given the drawing to the right with $\overline{AE}$ and $\overline{BF}$, determine the length of $\overline{BF}$.

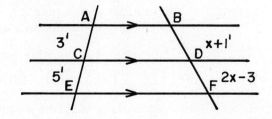

Solution   Because $\overline{AB} \parallel \overline{CD} \parallel \overline{EF}$, we can set up a proportion:

$$\frac{AC}{BD} = \frac{CE}{DF}$$

$$\frac{3}{5} = \frac{x+1}{2x-3}$$

$$6x - 9 = 5x + 5$$

$$x = 14 \text{ ft}$$

$BD = x + 1 = 15 \text{ ft} \qquad DF = 2x - 3 = 25 \text{ ft}$

$BF = BD + DF = 15 + 25 = 40 \text{ ft}$

The answer is $BF = 40$ ft.

Because we know parallel lines will divide a transversal into proportional parts, we can divide a given line segment into a given number of equal segments.

To divide a given line segment into n equal parts, where n is a whole number:

1. Call the endpoints of the given segment A and B. Draw a line segment from A not parallel to $\overline{AB}$.

2. With the compass set a fixed distance, mark off n units on the constructed line segment. Call the last (nth) mark made point C.

3. Draw the line segment from B to C.

4. Construct a line segment parallel to $\overline{BC}$ at each of the other n marks. That is, at each of the marks construct an angle congruent to $\angle BCA$, and extend the side constructed so that it intersects $\overline{AB}$.

5. The intersections of the n parallel line segments divide $\overline{AB}$ into n congruent (equal) parts as desired.

EXAMPLE 3   Using a compass and straightedge, divide $\overline{AB}$ into three equal parts.

Solution   Steps 1 and 2:   Draw a segment from A, not parallel to $\overline{AB}$. With compass set at a fixed distance, mark off three units on the constructed line segment. Call the last mark made, point C.

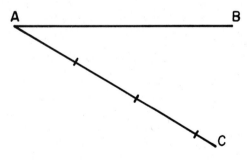

Steps 3
and 4: Draw the line segment from B to C. Construct a line segment parallel to BC at D and E.

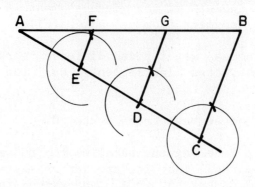

Step 5: The intersections divide $\overline{AB}$ into three equal parts: $\overline{AF}$, $\overline{FG}$, and $\overline{GB}$.

EXAMPLE 4   Using a compass and straightedge, divide $\overline{CD}$ into three parts with the ratio 2:3:4.

Solution   We can use the procedure for dividing a line segment into equal parts. First we determine how many equal parts we need by finding the sum of the ratio: 2 + 3 + 4 = 9. Thus we will need 9 equal parts.

Steps 1
and 2: Draw a segment from C not parallel to CD and mark off nine equal units.

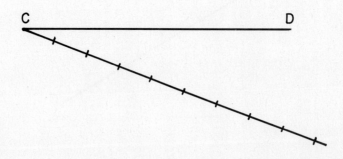

Steps 3
and 4:   Draw the line segment from E to D. Construct a line segment parallel to $\overline{DE}$ at G to form a two-unit length. Go three units to F and construct another line segment parallel to DE to form a three unit length. The remaining length is of four units, as desired.

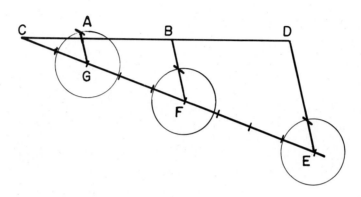

Step 5:   The segment $\overline{CD}$ is divided into proportional parts $\overline{CA}$, $\overline{AB}$, and $\overline{BD}$ with the ratio 2:3:4.

EXERCISE 3.4

Use the figure at the right, △ABC, for Problems 1 through 6.

1. If AD = 7 cm, DB = 8 cm, and CE = 5 cm, find the length of BE.

2. If BE = 10 cm, CE = 7 cm, and BD = 6 cm, find the length of AD.

3. If AB = 14 in., AD = 4 in., and BE = 9 in., find the length of CE.

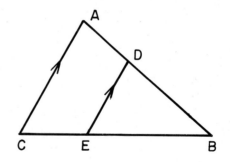

4. If BC = 25 in., BE = 15 in., and AD = 8 in., find the length of BD.

*5. If AD = 2x + 2 ft, CE = 3x, DB = 4x - 1 ft, and EB = 5x - 2 ft, find the length of each segment.

*6. If $BE = 2x + 1$ m, $CE = x + 2$ m, $AD = x - 1$ m, and $DB = x + 3$ m, find the length of each segment.

Use the figure at the right with $\overline{AE}$ and $\overline{BF}$ for Problems 7 to 12.

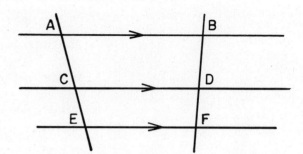

7. If $AC = 9$ in., $CE = 7$ in., and $BD = 11$ in., find the length of DF.

8. If $BD = 4$ in., $DF = 6$ in., and $CE = 5$ in., find the length of AC.

9. If $AE = 15$ mm, $BD = 6$ mm, and $DF = 4$ mm, find the length of AC.

10. If $AC = 8$ cm, $BF = 28$ cm, and $CE = 6$ cm, find the length of DF.

*11. If $AC = x + 3$ in., $CE = 2x$, $BD = 3x + 3$ in., and $DF = 3x$, find the length of each segment.

*12. If $BD = x + 4$ ft, $DF = 4x + 1$ ft, $CE = 2x + 3$ ft, and $AC = x$, find the length of each segment.

**13. A man 6 ft tall walks toward a street light that is 16 ft above the ground. If S is the length of his shadow and D is his distance from the street light, given an expression for S in terms of D.

**14. A 5 ft tall woman is walking away from a street light that is 14 ft above the ground. If S is the length of her shadow and D is her distance from the street light, give an expression for S in terms of D.

For Problems 15 to 20, use a straightedge and a compass.

15. Divide $\overline{AV}$ into three equal parts.

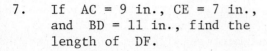

16. Divide $\overline{CG}$ into five equal parts.

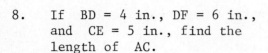

17. Divide $\overline{RS}$ into three parts with the ratio 1:2:4.

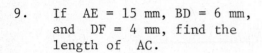

18. Divide $\overline{PQ}$ into three parts with the ratio 1:3:5.

P   Q

19. Divide $\overline{WS}$ into four parts with the ratio 1:1:4:5.

W ──────────────────────────────── S
D ──────────────────────────────── F

20. Divide $\overline{DF}$ into four parts with the ratio 1:2:2:5.

3.5 PROOFS OF SIMILARITY

We have done extensive work with similar triangles, so you should have a good understanding of what facts are needed to guarantee the similarity of two triangles. We can now use that understanding as a guide to write proofs. Let us return to Example 4 of Section 3.3 and give a prooof.

EXAMPLE 1   Prove $\triangle ADE \sim \triangle ABC$.

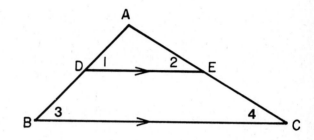

Solution

	Statement	Reason
1.	DE∥BC	Given
A 2.	$\angle 1 \cong \angle 3$	1, & Corresponding angles
A 3.	$\angle 2 \cong \angle 4$	1, & Corresponding angles
A 4.	$\angle A \cong \angle A$	Common angle
5.	$\triangle ADE \sim \triangle ABC$	2, 3, 4 & AAA

All we have done is express the same information in a different, more structured way.

EXAMPLE 2   Given: $\overline{BD}$ intersects $\overline{AE}$ at C

Prove: $\triangle ABC \sim \triangle EDC$.

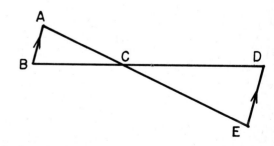

Solution	Statement	Reason
	1. $\overline{AB} \parallel \overline{ED}$	Given
A	2. $\angle B \cong \angle D$	1, & Alternate interior $\angle$s
A	3. $\angle A \cong \angle E$	1, & Alternate interior $\angle$s
A	4. $\angle ACB \cong \angle ECD$	Vertical angles
	5. $\triangle ABC \sim \triangle EDC$	2, 3, 4, & AAA

EXERCISE 3.5

1. Given: $\overline{CE}$ and $\overline{AB}$ intersect at D.

   Prove: $\triangle ADE \sim \triangle CDB$

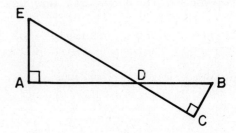

2. Given: $\overline{PS}$ and $\overline{QT}$ intersect at R.

   Prove: $\triangle RPQ \sim \triangle RST$

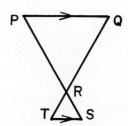

3. Given: $\overline{AC} \parallel \overline{DE}$

   Prove: $\triangle ABC \sim \triangle DBE$

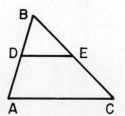

4. Given: $RS \parallel UV$

   $\triangle RST \sim \triangle UVT$

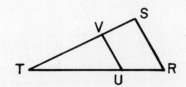

5. Given: Figure to right

   Prove: △ABC ~ △BCD

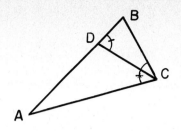

6. Given: ∠RST ≅ ∠RQP

   Prove: △RST ~ △RQP

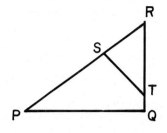

CHAPTER 3 REVIEW

For Problems 1 through 7 refer to Figure 3.5.

1. List pairs of angles that are

    a)   alternate interior

    b)   alternate exterior

    c)   corresponding

    d)   vertical

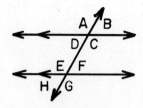

FIGURE 3.5

2. List the angles that are congruent to $\angle A$.

3. List the angles that are supplementary to $\angle F$.

4. If $\angle B = 57.2°$, find the number of degrees in $\angle H$.

5. If $\angle H = \frac{4\pi}{9}$, find the number of radians in $\angle A$.

6. If $\angle C = 7x - 4°$ and $\angle G = 5x + 20°$, find the number of degrees in $\angle F$.

7. If $\angle B = 4x + 0.07$ and $\angle E = 5x + 0.01$, find the number of radians in $\angle A$ (to two decimal places).

8. In $\triangle ABC$ as marked, why is $\overline{PQ} \| \overline{AB}$? $\overline{AC} \| \overline{QR}$? $\overline{PS} \| \overline{BC}$?

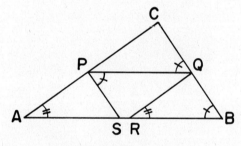

9. If $\ell_1 \| m_1$, $\ell_2 \| m_2$, and $\angle 1 = 125°$, find the number of degrees in $\angle 2$.

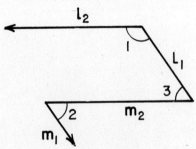

10. If $\overline{AE} \| \overline{CF}$, $\overline{AB} \| \overline{CD}$, and $\angle 1 = 62°$, find the number of degrees in each of the numbered angles.

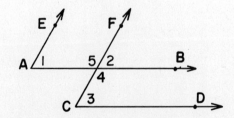

11. Find the number of degrees in ∠1 and ∠2 so that $\ell_1 \parallel \ell_2$.

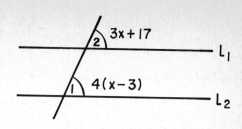

12. Which lines are parallel if:

    a) ∠6 = 90°, ∠1 = 68°, ∠4 = 68°?

    b) ∠3 = 90°, ∠5 = 42°, ∠4 = 48°?

    c) ∠7 = 80°, ∠1 = 36°, ∠2 = 64°, ∠3 = 80°?

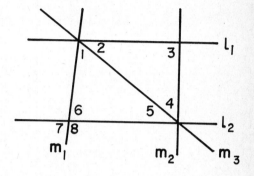

13. If ∠1 = 50° and ∠17 = 110°, find the number of degrees in ∠9 and ∠10.

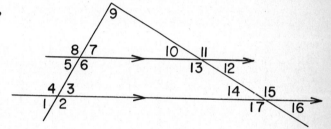

14. Given point Q and line segment m, construct a line parallel to m containing Q.

15. Given: $\overline{FD}$ intersecting $\overline{EC}$ at A

    Prove: △ABC ≅ △CDA

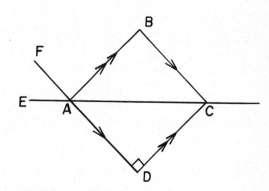

179

16. Given: $\overline{BC} \parallel \overline{AD}$ and $\overline{AB} \parallel \overline{DC}$

    $\overline{BD}$ and $\overline{AC}$ intersecting at O

    Prove: $\overline{AD} \cong \overline{BC}$

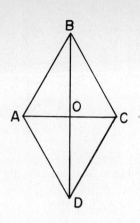

*17. Given: Segment $\overline{AD}$

    $\overline{GC} \parallel \overline{ED}$

    $\overline{AB} \cong \overline{CD}$

    $\overline{GC} \cong \overline{ED}$

    Prove: $\overline{AG} \parallel \overline{BE}$

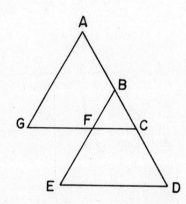

18. Determine the value of x and y.

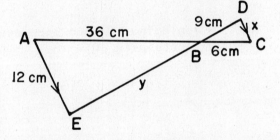

19. Determine the value of x and y.

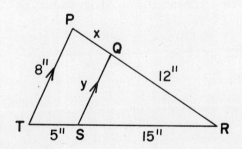

180

20. AB = 9 mm, BC = 3 mm,
    AE = 4x - 1 mm, and
    ED = x + 1 mm.

    a) Find the length of AD.

    b) If BE = 21 mm, find the length of CD.

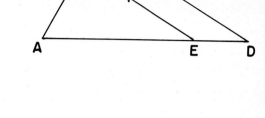

21. If a 33 ft flagpole casts an 18 ft shadow, how tall is a telephone pole that casts a 15 ft shadow?

22. A man 1.8 m tall casts a 3 m shadow when standing 4 m away from the base of a street light. How far above the sidewalk is the street light?

23. ∠ACB ≅ ∠BDC in △ABC. BC = 12 m, BD = 5 m, and AC = 21 m. Determine the length of AD and DC.

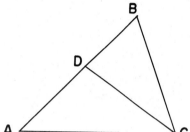

24. A girl 4 feet tall is 20 feet away from the base of a street light 16 feet above the sidewalk. Determine the length of her shadow.

25. The girl in Problem 24 strolls toward the streetlight at a rate of 4 feet per minute. How long is her shadow after 1 minute?

**26. For Problem 25, determine the expression for the length of the girl's shadow after t minutes (t < 5).

27. Divide $\overline{MN}$ into 5 equal parts.

28. Divide $\overline{PQ}$ into parts with the ratio 1:2:2:3.

29. Given: $\overline{AE}$ interesects $\overline{BD}$ at C and $\overline{AB} \parallel \overline{DE}$

    Prove: △ABC ~ △EDC

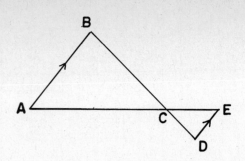

30. Given: △PQR
    $\overline{ST} \parallel \overline{PR}$

    Prove: △PQR ~ △SQT

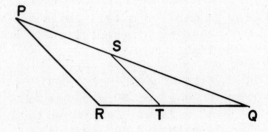

31. Given: △ABC
    ∠ABC ≅ ∠EDC

    Prove: △ABC ~ △EDC

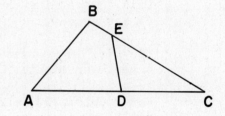

CHAPTER 4
POLYGONS

In Section 2.1 we defined polygons and gave much of the terminology for them. You may wish to review that before proceeding with this chapter.

4.1 DEFINITIONS OF QUADRILATERALS

A QUADRILATERAL is a four-sided polygon. Figure 4.1 illustrates the major types of quadrilaterals. The first category of quadrilaterals, TRAPEZOIDS, requires that at least one pair of sides be parallel. The next category, PARALLELOGRAMS, requires that each pair of sides is parallel. A RECTANGLE is a parallelogram with right angles. A SQUARE is a rectangle with equal sides. A RHOMBUS is a parallelogram with equal sides. Thus a square is a special kind of rectangle, rhombus, parallelogram, trapezoid, and quadrilateral! A rhombus is a special kind of parallelogram, trapezoid, and quadrilateral. This is confusing unless one can visualize the

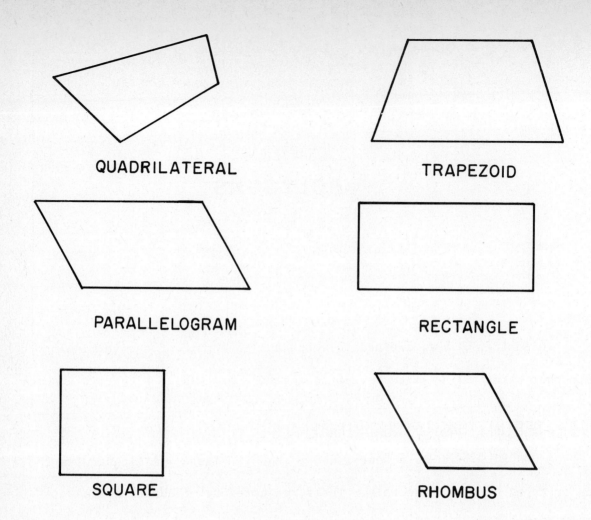

FIGURE 4.1

relationships among the different kinds of quadrilaterals. In Figure 4.2 the rectangle represents all quadrilaterals. Within that group lie all trapezoids. Some trapezoids are parallelograms. Rectangles and rhombuses are special kinds of parallograms. Any figure

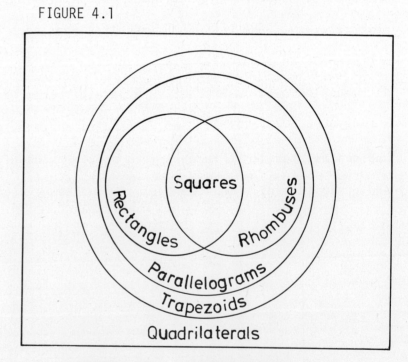

FIGURE 4.2   QUADRILATERALS

184

that is both a rectangle and a rhombus is a square.

All quadrilaterals have four vertices. They also have two DIAGONALS, line segments that connect two nonadjacent (opposite) vertices. Two sides that do not share a common vertex are OPPOSITE SIDES of the quadrilateral.

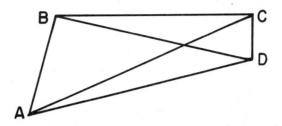

For the quadrilateral ABCD, A and C are opposite vertices, as are B and D. The diagonals are $\overline{AC}$ and $\overline{BD}$. $\overline{BC}$ and $\overline{AD}$ are opposite sides as are $\overline{AB}$ and $\overline{CD}$. Note that the vertices in a polygon are lettered in order clock-wise around the polygon.

Any quadrilateral can be divided into two triangles by drawing in one diagonal.

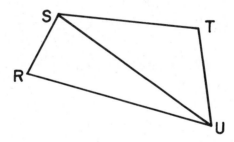

Thus quadrilateral RSTU can be divided into △RSU and △TUS. The sum of the interior angles of each of these triangles is 180° or π rad. From this we see that the sum of the interior angles of a quadrilateral is 360° or 2π rad.

The SUM of the interior ANGLES of a quadrilateral is 360° or 2π rad.

Because the sum of the angles is always 360° or 2π rad for a quadrilateral, knowing three angles gives us the fourth.

EXAMPLE 1   Determine the value of x.

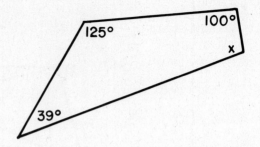

Solution   We know the sum is 360°.

$$39° + 125° + 100° + x = 360°$$
$$264° + x = 360°$$
$$x = 96°$$

EXAMPLE 2   Determine the number of radians in $\angle A$ in a quadrilateral if $\angle B = \frac{5\pi}{7}$, $\angle C = \frac{2\pi}{5}$, and $\angle D = \frac{2\pi}{3}$ rad.

Solution $\angle A + \angle B + \angle C + \angle D = 2\pi$ rad

$$x + \frac{5\pi}{7} + \frac{2\pi}{5} + \frac{2\pi}{3} = 2\pi$$

The lowest common denominator for 3, 5, and 7 is 105 so:

$$x + \frac{5(15)\pi}{105} + \frac{2(21)\pi}{105} + \frac{2(35)\pi}{105} = 2\pi$$

$$x + \frac{75\pi}{105} + \frac{42\pi}{105} + \frac{70\pi}{105} = 2\pi$$

$$x + \frac{187\pi}{105} = 2\pi$$

$$x = \frac{210\pi}{105} - \frac{187\pi}{105}$$

$$x = \frac{23\pi}{105}$$

$\angle A = \frac{23\pi}{105}$ rad.

EXAMPLE 3   Determine the number of degrees in each angle.

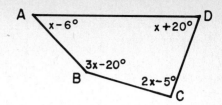

Solution $(x-6) + (3x-20) + (x+20) + (2x-5) = 360°$

$$7x - 11 = 360°$$

$$x = 53°$$

∠A = x − 6 = 53 − 6 = 47°

∠B = 3x − 20 = 159 − 20 = 139°

∠C = 2x − 5 = 106 − 5 = 101°

∠D = x + 20 = 53 + 20 = 73°

The solution is ∠A = 47°, ∠B = 139°, ∠C = 101°, and ∠D = 73°.

EXAMPLE 4   Determine the number of radians in each angle of a quadrilateral if ∠P = 2x − 0.38
∠Q = 3x − 0.08
∠R = x + 0.01
∠S = x + 0.43

Solution $(2x-0.38) + (3x-0.08) + (x+0.01) + x + 0.43 = 6.28$

$$7x - .02 = 6.28$$

$$x = 0.9$$

∠P = 2x − 0.38 = 1.8 − 0.38 = 1.42

∠Q = 3x − 0.08 = 2.7 − 0.08 = 2.62

∠R = x + 0.01 = 0.9 + 0.01 = 0.91

∠S = x + 0.43 = 0.9 + 0.43 = 1.33

The solution is ∠P = 1.42, ∠Q = 2.62, ∠R = 0.91, and ∠S = 1.33.

EXERCISE 4.1

1. Each figure below may be one or more of the following, or none of them: trapezoid, parallelogram, rectangle, square, or rhombus. **List all classifications that apply.**

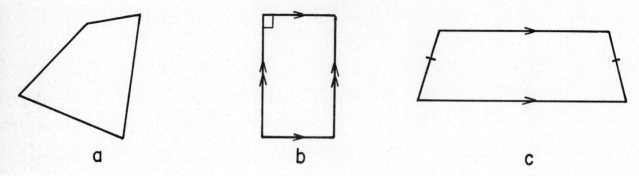

2. Each figure below may be one or more of the following, or none of them: trapezoid, parallelogram, rectangle, square, or rhombus. List all classifications that apply.

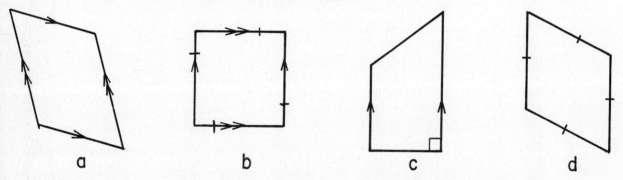

3. For each of the figures below, list the pairs of opposite sides and the opposite vertices.

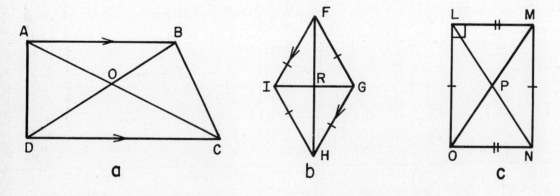

4. List the diagonals for each of the figures in Problem 3.

188

5. Determine the value of x.

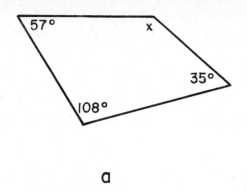

a

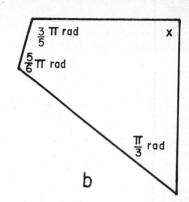

b

6. Determine the value of x.

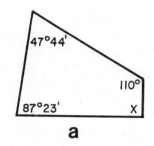

a

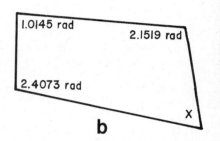

b

7. For quadrilateral ABCD, determine the size of ∠D if:

   a) ∠A = 41.27°, ∠B = 73.59°, ∠C = 112.43°

   b) ∠A = 73°11', ∠B = 131°14', ∠C = 101°54'

8. For quadrilateral ABCD, determine the size of ∠A if:

   a) ∠B = 37.17°, ∠C = 131.47°, ∠D = 64.36°

   b) ∠B = 104°19', ∠C = 67°22', ∠D = 87°39'

9. For quadrilateral PQRS, determine the size of ∠S if:

   a) $\angle P = \frac{5\pi}{7}$, $\angle Q = \frac{2\pi}{7}$, $\angle R = \frac{4\pi}{7}$

   b) ∠P = 0.58, ∠Q = 1.89, ∠R = 2.06

10. For quadrilateral PQRS, determine the size of ∠Q if:

    a) ∠P = 2.40, ∠R = 0.73, ∠S = 1.12

    b) $\angle P = \frac{5\pi}{9}$, $\angle R = \frac{7\pi}{9}$, $\angle S = \frac{2\pi}{9}$

11. Determine the size of each angle in Figure 4.3.

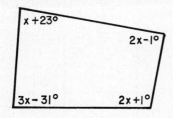

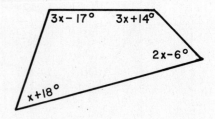

FIGURE 4.3   FIGURE 4.4

12. Determine the size of each angle in Figure 4.4.

13. The largest angle in a quadrilateral is 11° more than twice the smallest angle. The remaining angles are equal and are each 32° more than the smallest angle. Determine the number of degrees in each angle.

14. In a quadrilateral, the first angle is 3° more than twice the third angle; the second angle is 22° more than the third; and the fourth angle is 17° more than the first. Find the number of degrees in each angle.

15. If the angles in a quadrilateral are in the ratio 3:2:4:1, find the number of degrees in each angle.

16. If the angles in a quadrilateral are in the ratio 3:2:2:1, find the number of radians in each angle. (Leave answer exact.)

4.2 TRAPEZOIDS AND PARALLELOGRAMS

A quadrilateral with at least one pair of sides parallel is a TRAPEZOID. The two parallel sides are the BASES of the trapezoid. The ALTITUDE is the perpendicular distance between the bases. The LEGS are the two sides

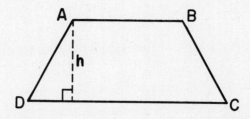

joining the parallel sides; they may be nonparallel. For trapezoid ABCD above, $\overline{AB}$ and $\overline{DC}$ are the bases, $\overline{AD}$ and $\overline{BC}$ are the legs, and h is the altitude.

190

If a trapezoid has two nonparallel sides which are the same length, then it is called an ISOSCELES TRAPEZOID. The diagonals of an isosceles trapezoid are the same length. For isosceles trapezoid PQRS, we have

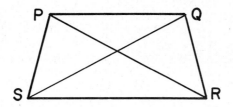

$$\overline{PS} \cong \overline{QR}$$
$$\overline{PR} \cong \overline{QS}$$
$$\overline{PQ} \parallel \overline{SR}$$

EXAMPLE 1 Given the isosceles trapezoid to the right, determine the length of the legs.

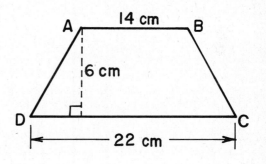

Solution Because the figure is an isosceles trapezoid, we can form the 6 by 14 cm rectangle ABFE. Because $\overline{AB} \parallel \overline{DC}$, the perpendicular distance between them is constant and $\overline{AE} \cong \overline{BF}$. We have constructed $\overline{AE}$ and $\overline{BF}$ so that $\angle AED = 90° = \angle BFC$. Because ABCD is an isosceles trapezoid,

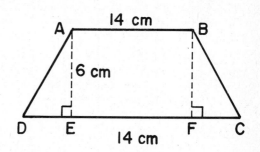

$\overline{AD} \cong \overline{BC}$. We thus have two right triangles $\triangle AED$ and $\triangle BFC$ where the hypotenuse and leg of one are, respectively, congruent to the hypotenuse and one leg of the second. We know that the Pythagorean theorem will give the common length of $\overline{DE}$ and $\overline{FC}$. Therefore $\overline{DE} \cong \overline{FC}$. We also know:

$$DE + FC + 14 \text{ cm} = 22 \text{ cm}$$

$$DE + FC = 8 \text{ cm}$$

$$DE = 4 \text{ cm}$$

$$FC = 4 \text{ cm}$$

We now have a right triangle with legs of 4 cm and 6 cm. We can use the Pythagorean theorem to determine the length of $\overline{AD}$.

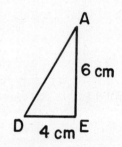

$$c^2 = a^2 + b^2$$
$$x^2 = 6^2 + 4^2$$
$$= 36 + 16$$
$$= 52$$
$$x = \pm \sqrt{13 \cdot 4}$$
$$x = \pm 2\sqrt{13}$$
$$x = 2\sqrt{13} \text{ cm or } 7.2111 \text{ cm (to four places)}$$

The length of the legs is $2\sqrt{13}$ cm or 7.2111 cm.

When a trapezoid has two pairs of parallel sides, it is a PARALLELOGRAM. In a parallelogram, opposite sides are congruent and opposite angles are congruent.

We can see this by first drawing diagonal $\overline{AC}$ in parallelogram ABCD. By

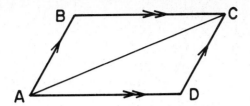

definition of a parallelogram, we have $\overline{AB} \| \overline{DC}$ and $\overline{BC} \| \overline{AD}$. We know then that $\angle CAD \cong \angle ACB$ and that $\angle BAC \cong \angle DCA$ because each pair is a set of alternate interior angles formed by a transversal cutting two parallel lines. We have $\overline{AC}$ as a common side for $\triangle ABC$ and $\triangle CDA$. We thus have an ASA congruence

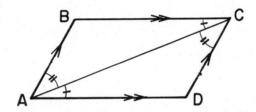

for the triangle, which shows that $\triangle ABC \cong \triangle CDA$. We therefore have corresponding parts congruent or $\overline{AB} \cong \overline{CD}$, $\overline{BC} \cong \overline{DA}$, and $\angle B \cong \angle D$. Restating these results, the opposite sides of a parallelogram are congruent and a pair of opposite angles are congruent. By drawing the other diagonal and using a similar argument, we can show $\angle A \cong \angle C$. We have also seen that either diagonal will form two congruent triangles.

We can also show that the diagonals of a parallelogram bisect each other. We first draw both diagonals and mark the pairs of alternate interior angles as congruent (because the opposite sides of a parallelogram are parallel).

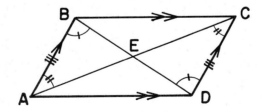

We have seen that opposite sides are congruent so $\overline{AB} \cong \overline{CD}$. We now have an ASA congruence for the triangle $\triangle ABE$ and $\triangle CDE$; therefore, $\triangle ABE \cong \triangle CDE$. We now know the corresponding parts of these two triangles are congruent, or $\overline{BE} \cong \overline{DE}$ and $\overline{AE} \cong \overline{CE}$. Now $\overline{BE} \cong \overline{DE}$ means E is the midpoint of $\overline{BD}$. Similarly we have E as the midpoint of $\overline{AC}$. Because E is the midpoint of both $\overline{AC}$ and $\overline{BD}$, the two diagonals bisect each other (at E).

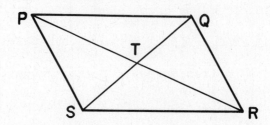

In parallelogram PQRS,

$$\overline{PQ} \cong \overline{SR} \quad \text{and} \quad \overline{PS} \cong \overline{QR}$$

$$\overline{PT} \cong \overline{TR} \quad \text{and} \quad \overline{ST} \cong \overline{TQ}$$

$$\triangle SPQ \cong \triangle QRS \quad \text{and} \quad \triangle PRS \cong \triangle RPQ$$

$$\overline{PQ} \| \overline{SR} \quad \text{and} \quad \overline{PS} \| \overline{QR}$$

EXAMPLE 2 In parallelogram ABCD, determine the lengths of $\overline{AC}$ and $\overline{BD}$.

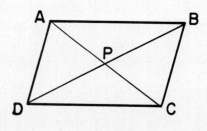

$$AP = x + 1 \text{ ft}$$
$$PC = 2x - 2 \text{ ft}$$
$$DP = 2x + 2 \text{ ft}$$
$$BP = 3x - 1 \text{ ft}$$

Solution Because we know the diagonals bisect each other, we may set either $AP = PC$ or $DP = BP$ and solve for x. Let us set:

$$AP = PC$$

$$x + 1 = 2x - 2$$

$$3 = x$$

$$AP = x + 1 \qquad PC = 2x - 2$$

$$AP = 4 \text{ ft} \qquad PC = 4 \text{ ft}$$

$$AC = 8 \text{ ft}$$

$$DP = 2x + 2 \qquad BP = 3x - 1$$

$$DP = 8 \text{ ft} \qquad BP = 8 \text{ ft}$$

$$BD = 16 \text{ ft}$$

The lengths are $AC = 8$ ft and $BD = 16$ ft.

EXAMPLE 3 In parallelogram WXYZ, $\angle W = 14x + 9°$ and $\angle Z = 2x + 27°$. Find the number of degrees in each angle.

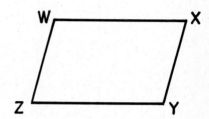

Solution: The sum of the angles in a quadrilateral is 360°, and opposite angles are equal in a parallelogram. Therefore:

$$2\angle W + 2\angle Z = 360°$$

$$\angle W + \angle Z = 180°$$

$$14x + 9 + 2x + 27 = 180$$

$$16x + 36 = 180$$

$$16x = 144$$

$$x = 9°$$

$$\angle W = 14x + 9 \qquad \angle Z = 2x + 27°$$

$$= 14(9) + 9 \qquad = 2(9) + 27$$

$$\angle W = 135° \qquad \angle Z = 45°$$

Therefore $\angle W = \angle Y = 135°$ and $\angle Z = \angle X = 45°$.

We check by seeing if the total is 360°.

$$135 + 135 + 45 + 45 \stackrel{?}{=} 360$$

$$360 \stackrel{\checkmark}{=} 360$$

EXAMPLE 4   Find the number of degrees in each angle
of parallelogram ABCD if:

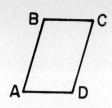

$$\angle A = 2x + y + 5°$$

$$\angle B = 2x + 3y + 7°$$

$$\angle C = 4x - y - 7°$$

$$\angle D = 3x + 2y + 1°$$

Solution   We have two variables, x and y, so we need two equations.

We know that $\angle A + \angle B = 180°$ and that $\angle C + \angle D = 180°$.

This gives:

$\angle A + \angle B = 180°$ $\qquad\qquad$ $\angle C + \angle D = 180°$

$2x + y + 5° + 2x + 3y + 7° = 180°$ $\qquad$ $4x - y - 7° + 3x + 2y + 1° = 180°$

$4x + 4y = 168°$ $\qquad\qquad\qquad\qquad$ $7x + y = 186°$

We now have two equations in two unknowns which we can solve simultaneously:

$$4x + 4y = 168°$$

$$7x + y = 186°$$

Multiplying the second by -4 gives:

$$4x + 4y = 168°$$

$$-28x - 4y = -744°$$

Adding the two equations:

$$-24x = -576°$$

$$x = 24°$$

Substituting $x = 24°$ into the second equation:

$$7x + y = 186°$$

$$168° + y = 186°$$

$$y = 18°$$

Checking in the first equation:

$$4x + 4y = 168°$$

$$4(24) + 4(18) \stackrel{?}{=} 168$$

$$96 + 72 \stackrel{?}{=} 168$$

$$168 \stackrel{\checkmark}{=} 168$$

Therefore $x = 24°$ and $y = 18°$.

$$\angle A = 2x + y + 5° \qquad \angle B = 2x + 3y + 7°$$
$$= 48 + 18 + 5 \qquad\quad = 48 + 54 + 7$$
$$\angle A = 71° \qquad\qquad\qquad \angle B = 109°$$

$$\angle C = 4x - y - 7° \qquad \angle D = 3x + 2y + 1°$$
$$= 96 - 18 - 7 \qquad\quad = 72 + 36 + 1$$
$$\angle C = 71° \qquad\qquad\qquad \angle D = 109°$$

Adding $\angle A + \angle B + \angle C + \angle D$ gives a sum of $360°$, as desired.

Therefore $\angle A = \angle C = 71°$ and $\angle B = \angle D = 109°$.

EXERCISE 4.2

C 1. Given an isosceles trapezoid with bases of 10 cm and 25 cm, and legs of 10 cm, find the altitude.

C 2. Given an isosceles trapezoid with bases of 17 in. and 6 in., and legs of 8 in., find the altitude.

3. An isosceles trapezoid with bases of 50 mm and 70 mm, has legs that form a 60° angle with the base. Determine the length of the legs.

4. An isosceles trapezoid with bases of 17 in. and 23 in. has legs that form a 45° angle with the base. Determine the altitude of the trapezoid.

5. An isosceles trapezoid with a shorter base of 10 ft has legs that form a $\frac{\pi}{6}$ rad angle with the longer base. If the altitude of the trapezoid is 6 ft, determine the length of the longer base. (Give an exact answer.)

6. An isosceles trapezoid with a longer base of 8 m has legs that form a $\frac{\pi}{4}$ rad angle with the longer base. If the legs of the trapezoid are 2 m, find the length of the shorter base. (Give an exact answer.)

7. Given parallelogram ABCD, where $\angle A = 4x + 7°$ and $\angle B = 3x - 2°$, determine the size of each angle.

8. Given parallelogram ABCD where $\angle B = 4x + 15°$ and $\angle D = 6x - 21°$, determine the size of each angle.

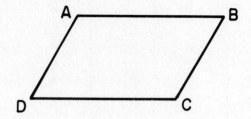

9. Given parallelogram ABCD where
   $\angle A = 5x + 4y$
   $\angle B = 3y + 2x - 2°$
   $\angle C = 3y + 8x - 2°$
   $\angle D = 5y - 3x - 5°$
   Determine the size of each angle.

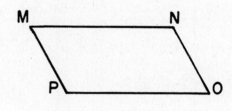

10. Given parallelogram MNOP where
    $\angle M = 3x - y + 0.22$ rad
    $\angle N = 2x + y - 0.58$ rad
    $\angle O = x + 2y - 1.08$ rad
    $\angle P = x + y + 0.12$ rad.
    Determine the size of each angle in rad, correct to two decimal places.

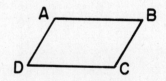

11. Given parallelogram ABCD where AD = 6 cm and $\angle ADC = 60°$, find an altitude. (Give an exact answer.)

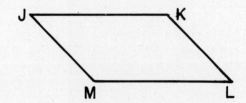

12. Given parallelogram JKLM where JM = 15 in. and $\angle JML = 135°$, find an altitude. (Give an exact answer.)

4.3 RECTANGLES, RHOMBUSES, AND SQUARES

A parallelogram with right angles is a RECTANGLE. Because the sides form a right angle, an altitude of a rectangle is the length of a side adjacent to the base. Note that any side of a rectangle can be considered a base. We can show that the diagonals in a rectangle are equal (congruent).

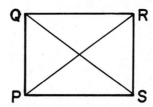

First draw the diagonals for rectangle PQRS. Because a rectangle is a special case of a parallelogram, we know the opposite sides are congruent, or $\overline{QP} \cong \overline{RS}$. We also have $\overline{PS} \cong \overline{SP}$. Because the figure is a rectangle, we know $\angle QPS = 90° = \angle RSP$. Entering this information on the drawing we have an

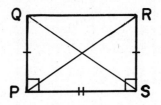

SAS congruence for the triangles $\triangle QPS$ and $\triangle RSP$, so they are congruent. Because $\triangle QPS \cong \triangle RSP$, corresponding parts are congruent, or $\overline{QS} \cong \overline{RP}$. Thus the diagonals are congruent.

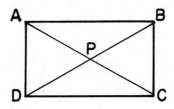

For rectangle ABCD,

$$\overline{AC} \cong \overline{BD}$$

$$\overline{AD} \cong \overline{BC} \quad \text{and} \quad \overline{AB} \cong \overline{CD}$$

$$\overline{AP} \cong \overline{PC} \cong \overline{DP} \cong \overline{PB}$$

$$\angle BAD = \angle ADC = \angle DCB = \angle CBA = 90° = \frac{\pi}{2} \text{ rad}$$

$$\triangle ABC \cong \triangle ADC \cong \triangle ABD \cong \triangle BCD$$

$$\overline{AD} \| \overline{BC} \quad \text{and} \quad \overline{AB} \| \overline{DC}$$

EXAMPLE 1   Given rectangle MNOP where MP = 7 m and MN = 14 m, find the length of the diagonal.

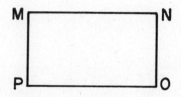

Solution   Because the figure is a rectangle, we know that $\angle PMN = 90°$. We thus have a right triangle, $\triangle PMN$, and can use the Pythagorean theorem.

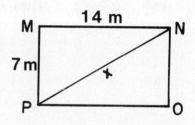

$$c^2 = a^2 + b^2$$

$$x^2 = 7^2 + 14^2$$

$$= 49 + 196$$

$$x^2 = 245$$

$$x = \pm \sqrt{245}$$

$$x = \pm 7\sqrt{5}$$

$$= 7\sqrt{5} \text{ m} \quad \text{or} \quad 15.65 \text{ m}$$

*EXAMPLE 2   Determine the dimensions of the rectangle given if
AC = 3x + 1 in.,
AD = x + 1 in., and
DC = 3x in.

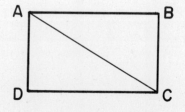

Solution   Using the Pythagorean theorem, we have:

$$c^2 = a^2 + b^2$$

$$(AC)^2 = (AD)^2 + (DC)^2$$

$$(3x+1)^2 = (x+1)^2 + (3x)^2$$

$$9x^2 + 6x + 1 = x^2 + 2x + 1 + 9x^2$$

$$0 = x^2 - 4x$$

$$0 = x(x-4)$$

$x = 0$	or	$x = 4$
$AC = 3x + 1 = 1$ in.		$AC = 3x + 1 = 13$ in.
$AD = x + 1 = 1$ in.		$AD = x + 1 = 5$ in.
$DC = 3x = 0$ in.		$DC = 3x = 12$ in.

Because the value of DC is zero when $x = 0$, we must discard this value since we would not have a rectangle. Therefore the dimensions of the rectangle are 12 in. by 5 in.

EXAMPLE 3   Determine the length of the diagonals if
$AT = 3x + y$ cm,
$TC = 2y + 2$ cm,
$DT = 5x + 1$ cm, and
$TB = 2y + x - 1$ cm.

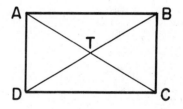

Solution   Because the diagonals in a rectangle are equal and bisect each other, we know that $AT = TC = DT = TB$. Because we have two variables, x and y, we can form two equations:

$AT = TC$ and $DT = TB$

$3x + y = 2y + 2$ $5x + 1 = 2y + x - 1$

$3x - y = 2$ $4x - 2y = -2$

We now have a system of two linear equations in two unknowns for which we want a simultaneous solution.

$$3x - y = 2$$
$$4x - 2y = -2$$

Multiplying the first by -2 gives:

$$-6x + 2y = -4$$
$$4x - 2y = -2$$

Adding the two equations:

$$-2x = -6$$
$$x = 3$$

Substituting $x = 3$ into the first equation,

$$3x - y = 2$$
$$9 - y = 2$$
$$-y = -7$$
$$y = 7$$

Checking in the second equation,

$$4x - 2y = -2$$
$$4(3) - 2(7) \stackrel{?}{=} -2$$
$$12 - 14 \stackrel{?}{=} -2$$
$$-2 \stackrel{\checkmark}{=} -2$$

Therefore $x = 3$ and $y = 7$.

| AT = $3x + y$ | TC = $2y + 2$ | DT = $5x + 1$ | TB = $2y + x - 1$ |
| = 16 cm | = 16 cm | = 16 cm | = 16 cm |

The diagonals therefore are 32 cm.

While it was not necessary to substitute to find the length of each side, we did so to verify the answer. If the values for AT, TC, DT, and TB had not been equal, we would know to go back to check our work.

A parallelogram with all sides of the same length is a RHOMBUS. A more common name for a rhombus is a diamond. The diagonals in a rhombus are NOT necessarily the same length, but they DO bisect each other because a rhombus is a parallelogram. The diagonals are perpendicular and they bisect the angles of the vertices that they join. Let us see why. In rhombus PQRS, we know that

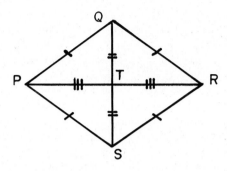

diagonals $\overline{QS}$ and $\overline{PR}$ bisect each other at T because a rhombus is a parallelogram. We can see from the drawing that $\triangle QTP \cong \triangle QTR \cong \triangle STR \cong \triangle STP$ because there is an SSS congruence among all four triangles. In particular, that gives $\angle QPT \cong \angle SPT$ as corresponding parts of congruent triangles $\triangle QTP$ and $\triangle STP$. Thus the diagonal PR bisects $\angle QPS$. Similar arguments show that each of the diagonals bisects the angle whose vertices it joins.

Because $\triangle PTS \cong \triangle PTQ$, we know that corresponding angles $\angle PTS$ and $\angle PTQ$ are congruent. However we also know that $\angle QTS$ is a straight angle and $\angle PTS$ and $\angle PTQ$ are supplementary. If two angles are both supplementary and congruent, each is a right angle. We thus know that the diagonals of a rhombus are perpendicular.

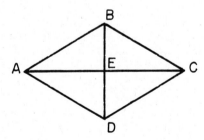

For rhombus ABCD,

$$\overline{AB} \cong \overline{BC} \cong \overline{CD} \cong \overline{DA}$$

$$\overline{AC} \perp \overline{BD}$$

$$\angle BAE \cong \angle DAE \cong \angle BCE \cong \angle DCE$$

$$\angle ABE \cong \angle CBE \cong \angle CDE \cong \angle ADE$$

$$\overline{AE} \cong \overline{EC} \text{ and } \overline{DE} \cong \overline{BE}$$

$$\triangle AED \cong \triangle AEB \cong \triangle BEC \cong \triangle DEC$$

$$\overline{AB} \| \overline{DC} \text{ and } \overline{AD} \| \overline{BC}$$

EXAMPLE 4   Given diagonals of 18 cm and 24 cm for a rhombus, find the length of each side.

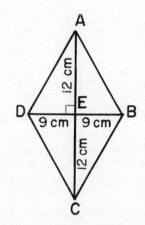

Solution   First make a drawing and enter the given information. AC = 24 cm, so AE = EC = 12 cm. Because DB = 18 cm, DE = EB = 9 cm. We have a right triangle where the legs are in the ratio 3:4 (with a common factor of 3), so the triangle must be a 3:4:5 triangle. The hypotenuse is then 3·5 or 15 cm long. Therefore each side of the rhombus is 15 cm. If we do not recognize that the right triangle is a 3:4:5 right triangle, we can always use the Pythagorean theorem to determine the length of DC. $(DC)^2 = 9^2 + 12^2$; DC = 15.

EXAMPLE 5   Find the length of each side of the given rhombus.

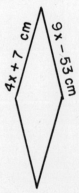

Solution   Because the sides of a rhombus are all the same length, we have the following equation:

$$9x - 53 = 4x + 7$$

$$5x = 60$$

$$x = 12 \text{ cm}$$

$$9x - 53 = 12(9) - 53 \qquad 4x + 7 = 4(12) + 7$$

$$= 55 \text{ cm} \qquad\qquad = 55 \text{ cm}$$

The length of each side is 55 cm.

EXAMPLE 6 Find the length of the diagonals for a rhombus with sides of 8 cm if a vertex angle (the interior angle at a vertex) is $\frac{\pi}{3}$ rad.

Solution First, make a drawing and enter the given information. Because $\angle ABC = \frac{\pi}{3}$ rad, $\angle ABE = \frac{\pi}{6}$ rad. Therefore $\triangle AEB$ is a $\frac{\pi}{6} - \frac{\pi}{3} - \frac{\pi}{2}$ rad triangle with sides in the proportion $1:\sqrt{3}:2$. We draw the two similar triangles and label them as shown in the drawing at the right; we get the proportions

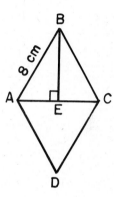

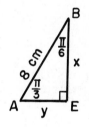

$$\frac{8}{2} = \frac{y}{1}$$

$$y = 4 \text{ cm}$$

and

$$\frac{8}{2} = \frac{x}{\sqrt{3}}$$

$$x = 4\sqrt{3} \text{ cm}$$

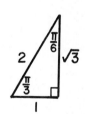

The diagonals of the rhombus are thus 8 cm and $8\sqrt{3}$ cm long.

A SQUARE is a parallelogram that has sides of equal length and right angles. That is, a square is both a rectangle AND a rhombus, and hence possesses the properties of each: the sides are congruent, the diagonals are congruent, perpendicular, and bisect each other and the vertex angles they join. For square PQRS

$\overline{PQ} \cong \overline{QR} \cong \overline{RS} \cong \overline{SP}$

$PR \perp SQ$

$\angle QPT = \angle SPT = \angle TSP = \angle TSR = \angle TRS = \angle TRQ = \angle TQR \cong \angle TQP = 45° = \dfrac{\pi}{4}$ rad

$\angle QPS \cong \angle PSR \cong \angle SRQ \cong \angle RQP \cong 90° \cong \dfrac{\pi}{2}$ rad

$\overline{PR} \cong \overline{SQ}$

$\overline{PT} \cong \overline{QT} \cong \overline{ST} \cong \overline{RT}$

$\triangle STP \cong \triangle STR \cong \triangle RTQ \cong \triangle QTP$

$\overline{PQ} \| \overline{SR}$ and $\overline{PS} \| \overline{QR}$

EXAMPLE 7   If a square has diagonals of 3.4 in., find the length of each side.

Solution   First, make a drawing and enter the known information. Because the diagonal in a square bisects the vertices it joins, $\triangle ADE$ is a $\dfrac{\pi}{4} - \dfrac{\pi}{4} - \dfrac{\pi}{2}$ rad triangle with sides in the ratio $1:1:\sqrt{2}$. We can then let x represent the length of a side of the square and set up the proportion:

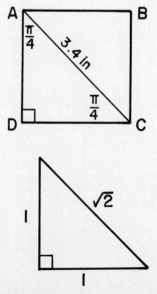

206

$$\frac{3.4}{\sqrt{2}} = \frac{x}{1}$$

$$x = \frac{3.4\sqrt{2}}{2} = 2.4042 \text{ in.}$$

The sides are 2.4042 in. long.

EXAMPLE 8   If two sides of a square are 3x + 19 ft and 7x - 53 ft, find the dimensions of the square.

   Solution   Because all sides in a square are equal, we have the equation:

$$7x - 53 = 3x + 19$$

$$4x = 72$$

$$x = 18 \text{ ft}$$

3x + 19 = 3(18) + 19 = 73 ft and 7x - 53 = 7(18) - 53 = 73 ft

The square is 73 ft on each side.

EXERCISE 4.3

1. For each of the figures below:

    i)   give the appropriate type (or types) of quadrilateral.

    ii)   list the equal sides (if any).

    iii) list the equal angles (if any).

    iv)   state whether or not the diagonals will be equal.

Do not assume any facts that are <u>not</u> given in the drawings. You may use any facts that are known as a result of the facts given.

a.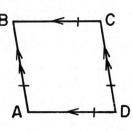

b.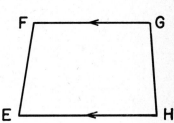

c.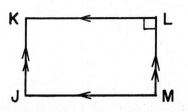

d.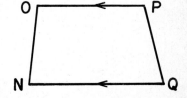

e. 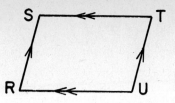 f.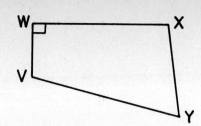

g. 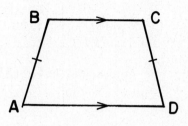 h.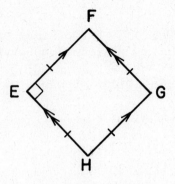

2. Complete the following chart by entering "Yes" or "No" in each rectangle.

TYPE OF QUADRILATERAL / PROPERTY	Parallelogram	Rectangle	Rhombus	Square	Trapezoid	Isosceles Trapezoid
Opposite sides always parallel						
Opposite angles always equal						
Opposite sides always equal						
All sides always equal						
Always has a right angle						
At least one pair of opposite sides always equal						
Diagonals always equal						
Diagonals always bisect each other						
Diagonals always perpendicular						
Diagonal always bisects angles of the vertices it joins						
Diagonal always divides it into congruent triangles						

3. Given a rectangle with sides of 10 in. and 15 in., determine the length of the diagonal. (Give an exact answer.)

4. Given a rectangle with a 16 cm diagonal and one side of 12 cm, determine the length of the other side. (Give an exact answer.)

C 5. Given a rectangle with a 2.8 m diagonal and one side of 1.73 m, determine the length of the other side.

C 6. Given a rectangle with sides of 11.3 in. and 17.6 in., determine the length of the diagonal.

*7. Determine the dimensions of the rectangle given if $AB = x$ ft, $AD = 2x - 1$ ft, and $AC = 2x + 1$ ft.

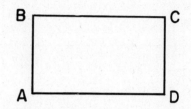

*8. Determine the dimensions of the rectangle given if $AB = x$ meters, $AD = 3x + 3$ meters, and $AC = 4x - 3$ meters.

9. If the diagonals of a rectangle are $2x + 13$ in. and $6x - 19$ in., find the length of the diagonals.

10. If the diagonals of a rectangle are $5x - 7$ cm and $3x + 11$ cm, find the length of the diagonals.

11. If the diagonals of a rhombus are 10 in. and 24 in., determine the length of the sides.

12. If one diagonal of a rhombus is 24 in. and a side is 15 in., determine the length of the other diagonal.

C 13. If one diagonal of a rhombus is 13.5 cm and a side is 8.23 cm, determine the length of the other diagonal.

C 14. If the diagonals of a rhombus are 17.2 cm and 21.3 cm, determine the length of the sides.

15. Determine the length of the diagonals for rectangle ABCD if $AP = 2x + y$ cm, $BP = x + 2y - 1$ cm, $CP = 3x + y$ cm, and $DP = 3y - 2$ cm.

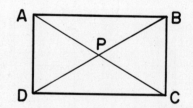

16. Determine the length of the diagonals in rectangle ABCD if $AP = 5x$ ft, $BP = x + 2y - 2$ ft, $CP = 3x + y - 1$ ft, and $DP = 3y - 6$ ft.

17. Find the length of the diagonals for a rhombus with sides of 12 cm if a vertex angle is $\frac{\pi}{3}$ rad. (Give exact answers.)

18. Find the length of the diagonals for a rhombus with sides of 15 cm if a vertex angle is $\frac{2\pi}{3}$ rad. (Give exact answers.)

19. Find the length of the sides of a rhombus ABCD if side AB = 7x + 13 in. and side BC = 9x - 5 in.

20. Find the length of the sides of a rhombus RSTU if side ST = 4x + 11 ft and side RS = 7x - 13 ft.

21. Find the length of the diagonal of a square with sides of 18 cm. (Give an exact answer.)

22. Find the length of the sides of a square with diagonal of 24 cm. (Give an exact answer.)

C 23. Find the length of the sides of a square with diagonal of 129.2 in.

C 24. Find the length of the diagonal of a square with sides of 109.6 in.

*25. A carpenter has been asked to build a rectangular frame with one side 7 in. less than the first side and with the diagonal 2 in. more than the first side. What should the dimensions of the frame be?

*26. An architect is designing a rectangular window for a museum. She is required to have the length 1 ft greater than the width and the diagonal 8 ft more than the length. What dimensions should she use for the window?

4.4 PROOFS

We have many more facts available to use in a proof if we know that the figure is a particular type of quadrilateral.

EXAMPLE 1   Prove that a diagonal of a parallelogram divides it into two congruent triangles. Show that the opposite sides are congruent.

Solution   Given: Parallelogram ABCD

Prove: $\triangle ADC \cong \triangle CBA$

$\overline{AD} \cong \overline{BC}$

$\overline{AB} \cong \overline{DC}$

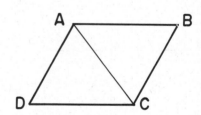

	Statement	Reason
1.	$\overline{AB} \parallel \overline{DC}$	Def. of parallelogram
2.	$\overline{AD} \parallel \overline{BC}$	Def. of parallelogram
A 3.	$\angle BAC \cong \angle ACD$	1, & Alt. interior $\angle$s of $\parallel$ lines are $\cong$
A 4.	$\angle DAC \cong \angle ACB$	2 & Alt. interior $\angle$s of $\parallel$ lines are $\cong$
S 5.	$\overline{AC} \cong \overline{AC}$	Common side
6.	$\triangle ADC \cong \triangle CAB$	3, 5, 4 & ASA
7.	$\overline{AD} \cong \overline{BC}$	6 & CPCTC
8.	$\overline{AB} \cong \overline{DC}$	6 & CPCTC

EXAMPLE 2   Given: Parallelogram ABCD and segment $\overline{DF}$

Prove: $\triangle AED \cong \triangle BFC$
         ABFE is a rectangle

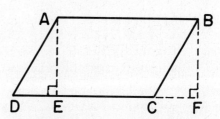

Solution

	Statement	Reason
1.	$\overline{AD} \parallel \overline{BC}$	Opposite sides of a parallelogram are $\parallel$
A 2.	$\angle ADE \cong \angle BCF$	Corresponding $\angle$s of $\parallel$ lines
S 3.	$\overline{AD} \cong \overline{BC}$	Opposite sides of a parallelogram are =
4.	$\angle AED \cong \angle BFC$	Given $\overline{AE} \perp \overline{DE}$ and $\overline{BF} \perp \overline{CF}$, both are right $\angle$s
A 5.	$\angle DAE \cong \angle FBC$	If 2 $\angle$s are = in two $\triangle$s, then the third $\angle$s are $\cong$
6.	$\triangle AED \cong \triangle BFC$	2, 3, 5, & ASA
7.	$\overline{AB} \parallel \overline{EF}$	Opposite sides of a parallelogram are $\parallel$
8.	$\overline{AE} \parallel \overline{BF}$	4 & Corresponding $\angle$s formed by transversal are $\cong$
9.	ABFE is a parallelogram	7, 8 & Def. of parallelogram

10. $\angle BFC = \frac{\pi}{2}$ rad   Def. of right$\angle$

11.   ABFE is a rectangle   9, 10 & Def. of rectangle

EXERCISE 4.4

Use only the definitions of the quadrilaterals given unless other results are suggested.

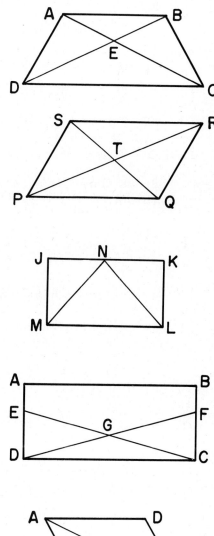

1. Given an isosceles trapezoid ABCD, show $\triangle ABE \sim \triangle DCE$.

2. Given isos. trap. ABCD, $\angle ADC \cong \angle BCD$, show $\angle EDC \cong \angle ECD$.

3. Given parallelogram PQRS, show that $\overline{ST} \cong \overline{QT}$. (Use the results from Example 1.)

4. Given parallelogram PQRS, show that $\overline{PT} \cong \overline{RT}$. (Use the results from Example 1.)

5. Given rectangle JKLM, and N as the midpoint of $\overline{JK}$, show that $\angle JMN \cong \angle KLN$.

6. Given rectangle JKLM with $\angle NML \cong \angle NLM$, show that N is the midpoint of $\overline{JK}$.

*7. Given rectangle ABCD, $\overline{EC}$ and $\overline{DF}$ intersecting at G, and $\overline{ED} \cong \overline{FC}$, prove $\triangle DGC$ is an isosceles triangle.

*8. Given rectangle ABCD, $\overline{EC}$ and $\overline{DF}$ intersecting at G, and $\overline{GD} \cong \overline{GC}$, prove $\overline{AE} \cong \overline{BF}$.

9. Given: $\triangle ABC$
   $\overline{AD}\|\overline{BC}$ and $\overline{DC}\|\overline{AB}$

   Prove: ABCD is a parallelogram
   $\triangle ABC \cong \triangle CDA$

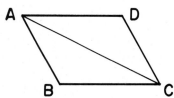

10. Prove that the sum of the angles in any quadrilateral is 2π rad. (Hint: Draw a diagonal.)

11. Prove that the opposite angles in a parallelogram are congruent.

12. Prove that the diagonals of a parallelogram bisect each other. (Use the results from Example 1.)

13. Prove that the diagonals in a rectangle are equal.

*14. Prove that the diagonals of a rhombus bisect the angles at the vertices they join.

*15. Prove that the diagonals of a rhombus bisect each other.

*16. Prove that the diagonals of a rhombus are perpendicular to each other.

*17. Prove that a diagonal of a rhombus divides the rhombus into two congruent triangles. Also prove that the altitude of each triangle is one half the other diagonal.

*18. Prove that the diagonals of a square divide the square into four isosceles triangles.

*19. Prove that two parallel lines are always the same distance apart (the perpendicular distance is the same). (Hint: Use a rectangle and a diagonal.)

*20. Prove that the base angles for an isosceles trapezoid are congruent. (Hint: Use the results from Section 2.7, Problem 15.)

*21. Prove that the diagonals of an isosceles trapezoid are congruent. (Hint: Use the result from Problem 20.)

22. Given the quadrilateral ABCD with right angles ∠A and ∠B and $\overline{AD} \cong \overline{BC}$, prove that $\overline{DC} \| \overline{AB}$. Is it possible to prove that ∠D and ∠C are right angles?

4.5 OTHER POLYGONS

We have studied extensively three-sided and four-sided polygons. We now turn to polygons with more sides. The different polygons and their numbers of sides are listed in Table 4.1. All of the names of the polygons that are not in parentheses should be memorized.

TABLE 4.1   Polygons.

Name of Polygon	Number of Sides
Triangle	3
Quadrilateral	4
Pentagon	5
Hexagon	6
(Heptagon) 7-gon	7
Octagon	8
(Nonagon) 9-gon	9
(Decagon) 10-gon	10
(Undecagon) 11-gon	11
(Dodecagon) 12-gon	12
⋮	⋮
n-gon	n

Most people are familiar with the pentagon because of the Pentagon Building in Washington, D.C. As we walk or drive, we frequently encounter octagons-- stop signs. Many nuts and boltheads are hexagonal. The other types of polygons we encounter rarely.

In order to determine the number of degrees or radians contained in the interior angles of a polygon we can draw all the diagonals from one vertex to form triangles. We know that each triangle contains π rad or 180°. By counting the number of triangles formed and multiplying by π rad or 180°, we can determine the sum of the angles for the polygon.

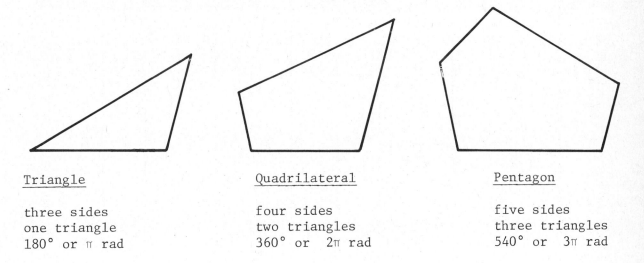

Triangle

three sides
one triangle
180° or π rad

Quadrilateral

four sides
two triangles
360° or 2π rad

Pentagon

five sides
three triangles
540° or 3π rad

As we can see above, in any polygon there are two fewer triangles than the number of sides. So we have a general rule to use.

> The sum of the interior angles in a polygon equals
>
> $$(n - 2)(180° \underline{\text{ or }} \pi \text{ rad}),$$
>
> where n is the number of sides.

EXAMPLE 1   Determine the number of degrees in the sum of the interior angles of a 15-gon.

Solution   A 15-gon has 15 sides, so we can use the formula with $n = 15$.

$$\text{Sum of interior angles} = (n - 2)(180°)$$
$$= (13)(180°)$$
$$= 2340°$$

A 15-gon contains 2340°.

Working backwards we can determine the number of sides of a polygon if we know the sum of the interior angles.

EXAMPLE 2   Determine the number of sides for a polygon with 27π radians as the sum of the interior angles.

Solution
$$\text{Sum of interior angles} = (n - 2)(\pi)$$
$$27\pi = (n - 2)(\pi)$$
$$27 = n - 2$$
$$29 = n$$

The polygon has 29 sides.

A REGULAR POLYGON has equal sides and equal angles. That is, a regular polygon is equilateral and equiangular. If we know that a polygon is regular and know how many sides it has, then we can calculate the size of each angle.

> Each interior angle in a regular polygon equals
>
> $$\frac{(n-2)(180° \text{ or } \pi \text{ rad})}{n},$$
>
> where n is the number of sides.

EXAMPLE 3   Determine the number of radians in each interior angle of a regular 13-gon.

   Solution   For a 13-gon, $n = 13$; we use the formula:

$$\text{Each interior angle} = \frac{(n-2)(\pi)}{n}$$

$$= \frac{(13-2)(\pi)}{13}$$

$$= \frac{11\pi}{13} \text{ rad} \quad \text{or} \quad 2.6583 \text{ rad}$$

As before, we can determine the number of sides in a regular polygon if we know the number of degrees in each interior angle.

EXAMPLE 4   If each interior angle contains 160°, determine how many sides the regular polygon has.

   Solution

$$\text{Each interior angle} = \frac{(n-2)(180°)}{n}$$

$$160° = \frac{(n-2)(180°)}{n}$$

$$160n = 180n - 360$$

$$360 = 20n$$

$$18 = n$$

The polygon has 18 sides.

An EXTERIOR ANGLE of a polygon is the angle formed by one side and the adjacent side extended. In the polygon ABCD, $\angle$BDE is the exterior angle at D.

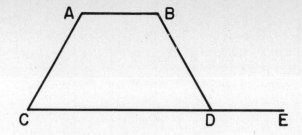

For a regular polygon, all exterior angles are equal.

For a regular polygon, one interior angle can be found by using the formula stated above. An exterior angle will be the supplement of an interior angle or:

$$\text{Exterior angle} = 180° - \text{interior angle}$$

$$= 180 - \frac{(n-2)(180)}{n}$$

$$= \frac{180n - 180n + 360}{n}$$

$$= \frac{360}{n}$$

Each exterior angle of a regular polygon equals

$$\frac{360°}{n} = \frac{2\pi \text{ rad}}{n}$$

where n is the number of sides.

EXAMPLE 5 Determine the number of degrees in each exterior angle of a regular 20-gon.

Solution
$$\text{Exterior angle} = \frac{360°}{n}$$
$$= \frac{360}{20}$$
$$= 18°$$

Each exterior angle of a regular 20-gon contains 18°.

EXERCISE 4.5

1. Complete the following table for regular polygons.

Number of sides	Sum of interior angles (in degrees)	Number of degrees in each interior angle
3		
5		
8		
9		
12		

2. Complete the following table for regular polygons.

Number of sides	Sum of interior angles (in radians)	Number of radians in each interior angle
4		
6		
7		
10		
11		

3. Find the total number of degrees in the interior angles for a 13-gon.

4. Find the total number of degrees in the interior angles for a 50-gon.

5. Find the total number of radians in the interior angles for a 20-gon.

6. Find the total number of radians in the interior angles for a 15-gon.

C 7. Find the number of degrees in one interior angle of a regular 13-gon.

C 8. Find the number of degrees in one interior angle of a regular 50-gon.

C 9. Find the number of radians in each interior angle for a regular 20-gon.

C 10. Find the number of radians in each interior angle for a regular 15-gon.

11. How many sides does a polygon have if the sum of its interior angles is 900°?

12. How many sides does a polygon have if the sum of its interior angles is 4140°?

13. How many sides does a polygon have if the sum of its interior angles is 7π rad?

14. How many sides does a polygon have if the sum of its interior angles is 17π rad?

15. How many sides does a regular polygon have if the measure of one of its interior angles is 144°?

16. How many sides does a regular polygon have if the measure of one of its interior angles is 168°?

17. How many sides does a regular polygon have if the measure of one of its interior angles is $7\pi/9$ rad?

18. How many sides does a regular polygon have if the measure of one of its interior angles is $6\pi/7$ rad?

19. Determine the number of degrees in one exterior angle of a regular pentagon.

20. Determine the number of degrees in one exterior angle of a regular dodecagon.

21. Determine the number of radians in one exterior angle of a regular hexagon.

22. Determine the number of radians in one interior angle of a regular octagon.

23. How many sides does a regular polygon have if one exterior angle contains 45°?

24. How many sides does a regular polygon have if one exterior angle contains 24°?

25. How many sides does a regular polygon have if one exterior angle contains 0.4π rad?

26. How many sides does a regular polygon have if one exterior angle contains $\pi/5$ rad?

27. Determine the number of degrees in each angle.

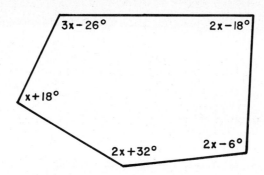

C 28. Determine the number of radians in each angle.

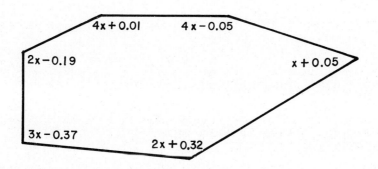

CHAPTER 4 REVIEW

1. Match each item in Column I with the appropriate item(s) from Column II. An item in Column II can be used once, more than once, or not at all.

   Column I

   a) Diagonals are same length.

   b) All sides are congruent.

   c) Diagonals are perpendicular.

   d) Opposite sides are congruent.

   e) Diagonals bisect each other

   f) All angles are right angles.

   g) At least two sies are parallel.

   Column II

   A. isosceles trapezoid

   B. parallelogram

   C. rectangle

   D. rhombus

   E. square

   F. trapezoid

   G. none of the above

2. Given a quadrilateral where $\angle A = 103°14'$, $\angle B = 71°35'$, $\angle C = 57°19'$, find the number of degrees in $\angle D$.

3. Given a quadrilateral where $\angle A = 1.34$ rad, $\angle B = 2.13$ rad, $\angle C = 0.75$ rad, find the number of radians in $\angle D$.

4. Find the number of radians in each angle of a quadrilateral if the angles are in the ratio 2:2:3:5. (Leave answer exact.)

5. Find the number of degrees in each angle of a quadrilateral if the second angle is 25° more than the first, the third is 10° less than the first, and the fourth is 20° less than twice the first.

C 6. Find the length of the legs of an isosceles trapezoid with bases of 16 m and 12 m and altitude of 4.8 m.

7. Determine the altitude of an isosceles trapezoid with $2\sqrt{5}$ in. legs that form a $\pi/3$ rad angle with the longer base. (Give an exact answer.)

8. Given a parallelogram ABCD where

   $\angle A = 2x + y - 9°$

   $\angle B = 3x + y + 4°$

   $\angle C = x + y + 16°$

   $\angle D = 2x + 2y - 1°$

   Determine the size of each angle.

9. Find the length of the shorter side of a parallelogram that forms a 45° angle with the base if the altitude is 12 cm. (Give an exact answer.)

C 10. Determine the length of the diagonal of a rectangle with 17 cm length and 13 cm height.

*11. Determine the dimensions of a rectangle if one side is $2x + 2$ ft, the other is $3x - 4$ ft, and the diagonal is $3x + 10$ ft.

12. If the diagonals of a rhombus are 8.4 cm and 11.2 cm, find the length of a side.

13. Determine the length of the diagonals in a rectangle if one diagonal is $3x + 17$ in. and the other is $4x + 5$ in.

14. If the 15 in. sides of a rhombus form a 60° angle at one vertex, find the length of the diagonals. (Give exact answers.)

15. Find the length of the sides of a square with 14 in. diagonals. (Give an exact answer.)

C 16. Find the length of the diagonals of a square if the sides are 4.3 mm.

C 17. A landscape architect is designing a rectangular pool that must have a diagonal of 12 ft. If the architect wishes the sides of the pool to be in a ratio of 5:8, what should the length of the sides be (to the nearest hundredth of a foot)?

18. Given: △ABC

$\overline{DE} \parallel \overline{AC}$

∠DAF ≅ ∠EFC

Prove: $\overline{DA} \cong \overline{EF}$

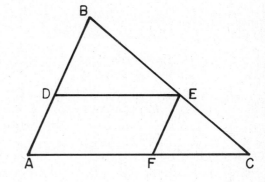

*19. Given: Parallelogram ABCD where the diagonals are perpendicular.

Prove: ABCD is a rhombus

20. Determine the number of degrees in each interior angle of a regular nonagon (9 sides).

21. Determine the number of sides of a regular polygon when each interior angle contains $\frac{5\pi}{6}$ rad.

C 22. Determine the number of radians in each exterior angle of a regular heptagon (7 sides). (Give answer correct to two decimal places.)

23. How many sides does a regular polygon have if each exterior angle contains 22°30'?

24. If the interior angles of a polygon are given by $\angle 1 = x + 22°$, $\angle 2 = x + 10°$, $\angle 3 = 2x - 35°$, $\angle 4 = 2x - 46°$ and $\angle 5 = x + 36°$, find the number of degrees in each angle.

CHAPTER 5
PERIMETER AND AREA OF POLYGONS

Now that we have defined the basic polygons, we can calculate the perimeter and area for each.

5.1 SQUARES AND RECTANGLES

The PERIMETER of a two-dimensional figure is the distance "around" the figure. The PERIMETER of a POLYGON is the sum of the lengths of the sides of the polygon.

$$\text{Perimeter}_{polygon} = \text{Sum of length of sides}$$

Because perimeter is a measure of length, it is <u>always</u> measured in <u>linear</u> units (e.g., centimeters, meters, feet, inches, yards, and miles).

The AREA of a two-dimensional figure is the number of square units needed to cover the figure exactly. Alternatively, the area of a figure is the number of square units contained in its surface. Because area is a measure of surface, <u>not</u> length, it is always measured in <u>square</u> units, e.g., square centimeters (cm^2), square meters (m^2), square feet (ft^2), and square inches (in^2). Because linear units are <u>not</u> the same as square units we must <u>always</u> indicate the unit of measure in our calculations. If the unit is not known, we can use "unit" for linear measure and "square unit" for area.

We can count the number of square units in a figure to determine its area.

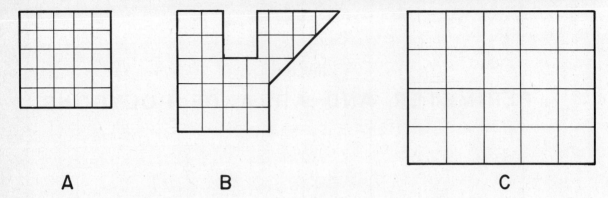

A   B   C

Figure A, a square, contains 5 × 5 = 25 square units. Figure B, an irregular shape, contains 19 whole-square units and 5 half-square units, for a total of 21.5 square units. Figure C, a rectangle, contains 4 × 5 = 20 square units. For a regular figure, it is much easier to calculate the area rather than to draw in and count up the number of square units contained in it.

$$Area_{square} = s^2,$$
where s is the length of a side.

$$Area_{rectangle} = \ell \cdot w,$$
where ℓ is the length and w is the width.

The perimeter of Figure A is $5 + 5 + 5 + 5 = 4(5) = 20$ units. The perimeter of Figure B is $3.5 + 2 + 1.5 + 2 + 2 + 5 + 4 + 2 + 3\sqrt{2} = 22 + 3\sqrt{2}$ units. The perimeter of Figure C is $4 + 5 + 4 + 5 = 2(4 + 5) = 18$ units. For **regular** figures it is easier to calculate the perimeter rather than count up the number of units around the figure.

$$Perimeter_{square} = 4s,$$
where s is the length of a side.

$$Perimeter_{rectangle} = 2(\ell + w),$$
where ℓ is the length and w is the width.

EXAMPLE 1   Find the area and the perimeter for a 9 cm by 19 cm rectangle.

Solution
$$A_{rectangle} = \ell \cdot w$$
$$= 9 \text{ cm} \cdot 19 \text{ cm}$$
$$= 171 \text{ cm}^2$$
$$P_{rectangle} = 2(\ell + w)$$
$$= 2(9 \text{ cm} + 19 \text{ cm})$$
$$= 56 \text{ cm}$$

The area is 171 cm^2 and the perimeter is 56 cm.

Note that in the example above the units were written in the problem to be sure that the answer has the proper units. In the area problem, cm·cm = cm^2, which is an appropriate unit for measuring area. In calculating perimeter, we had cm + cm = cm, which is an appropriate measure of length. By writing in the units we also check to be sure that we are using the <u>same</u> unit, and not multiplying or adding feet and inches, for instance.

EXAMPLE 2   Assuming that opposite sides are parallel, calculate the perimeter of the given figure.

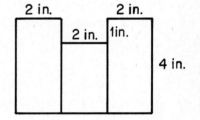

Solution   The perimeter is the distance around the outside of the figure, so we can ignore the inner lines. In the second drawing, the inner lines are dashed to show that they can be ignored. The lengths for the other sides have been entered. We can now calculate the perimeter.

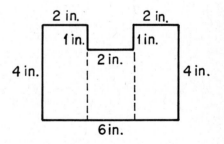

$$P_{polygon} = \text{Sum of length of sides}$$

$$= 4 \text{ in.} + 6 \text{ in.} + 4 \text{ in.} + 2 \text{ in.} + 1 \text{ in.} + 2 \text{ in.} + 1 \text{ in.}$$

$$+ 2 \text{ in.}$$

$$= 22 \text{ in.}$$

The perimeter for the given figure is 22 in.

EXAMPLE 3   Find the dimensions of a square with an area of 14 ft^2.

Solution

$$A_{square} = s^2$$

$$14 \text{ ft}^2 = s^2$$

$$\pm\sqrt{14} \text{ ft} = s$$

$$3.74 \text{ ft} = \sqrt{14} \text{ ft} = s$$

Because we introduced the square root to the problem (by taking the square root of both sides of the equation), we must use the $\pm$ sign. However, since we are measuring length, which is always positive, we can discard the minus sign.

*EXAMPLE 4   A rectangle has a perimeter of 66 in. and an area of 260 in^2. Find the dimensions of the rectangle.

Solution   We know $A_{rectangle} = \ell \cdot w$ and $P_{rectangle} = 2(\ell + w)$. Using these two equations, we get

$$260 \text{ in}^2 = \ell \cdot w \quad \text{and} \quad 66 \text{ in.} = 2(\ell + w)$$

Solving the second for ℓ, we get:

$$66 \text{ in.} = 2\ell + 2w$$

$$33 \text{ in.} - w = \ell$$

Substituting this into the first equation gives:

$$260 \text{ in}^2 = (33 \text{ in.} - w)w$$

$$260 \text{ in}^2 = 33w \text{ in.} - w^2$$

$$w^2 - 33w \text{ in.} + 260 \text{ in}^2 = 0$$

$$(w - 13 \text{ in.})(w - 20 \text{ in.}) = 0$$

$$w - 13 \text{ in.} = 0 \quad \text{or} \quad w - 20 \text{ in.} = 0$$

$$w = 13 \text{ in.} \quad\quad\quad\quad w = 20 \text{ in.}$$

If $w = 13$ in.,   If $w = 20$ in.,

$$260 \text{ in}^2 = \ell \cdot w \quad\quad\quad 260 \text{ in}^2 = \ell \cdot w$$

$$260 \text{ in}^2 = \ell \cdot (13 \text{ in.}) \quad\quad 260 \text{ in}^2 = \ell \cdot (20 \text{ in.})$$

$$20 \text{ in.} = \ell \quad\quad\quad\quad 13 \text{ in.} = \ell$$

The rectangle is 20 in. by 13 in. We obtained two sets of answers because either 20 or 13 can be considered the width and the other number will be the length.

EXAMPLE 5   A rectangle is three times as long as it is wide. If the area of the rectangle is 147 m^2, find its perimeter.

Solution   Let x = width

$$3x = \text{length}$$

$$A_{\text{rectangle}} = \ell \cdot w$$

$$147 \text{ m}^2 = 3x(x)$$

$$147 \text{ m}^2 = 3x^2$$

$$49 \text{ m}^2 = x^2$$

$$\pm 7 \text{ m} = x$$

width = x = 7 m   length = $3x$ = 21 m

Because we are measuring length, we discard the -7 as a value for the width, and obtain the dimensions of 21 m by 7 m for the rectangle.

The perimeter is:

$$P_{\text{rectangle}} = 2(\ell + w)$$

$$= 56 \text{ m}$$

The perimeter for the rectangle is 56 m.

We can use our knowledge about squares and rectangles to find the area of irregular figures. In this chapter <u>we will assume that figures that appear to</u>

have square corners (right angles) do, and that sides that appear parallel, are. Thus if a figure looks like a rectangle, we will assume that it is one.

EXAMPLE 6   Find the area of the given figure.

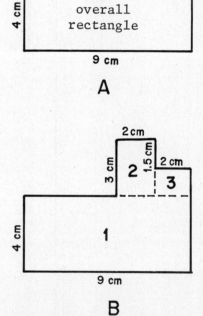

Solution   There are several ways to approach this problem. We can think of it as one large rectangle with two smaller rectangles removed as illustrated in A. Or we can think of it as the sum of several smaller rectangles. One way of dividing the figure into smaller rectangles is shown in B. There are many different ways it could have been divided. Let us look at the two different approaches. With the first, the area is given by:

$A = A_I - A_{II} - A_{III}$

We need to determine the dimensions for regions I and II before we can calculate the

230

areas. Region I is 4 cm + 3 cm = 7 cm wide. Region I, therefore is 9 cm by 7 cm. The length of Region II is 9 cm - 2 cm - 2 cm = 5 cm. Thus Region II is 5 cm by 3 cm. Region III is 2 cm by 1.5 cm. We can now calculate the desired area.

$$A = A_I - A_{II} - A_{III}$$
$$= (9 \text{ cm})(7 \text{ cm}) - (5 \text{ cm})(3 \text{ cm}) - (2 \text{ cm})(1.5 \text{ cm})$$
$$= 63 \text{ cm}^2 - 15 \text{ cm}^2 - 3 \text{ cm}^2$$
$$= 45 \text{ cm}^2$$

The area is 45 cm^2.

Using the second approach we get:

$$A = A_1 + A_2 + A_3$$

Once again we must determine the dimensions for each region before we can calculate the areas. Region 1 is 9 cm by 4 cm. Region 2 is 2 cm by 3 cm. The width of Region 3 is 3 cm - 1.5 cm = 1.5 cm. Thus Region 3 is 2 cm by 1.5 cm.

The desired area is:

$$A = A_1 + A_2 + A_3$$
$$= (9 \text{ cm})(4 \text{ cm}) + (2 \text{ cm})(3 \text{ cm}) + (2 \text{ cm})(1.5 \text{ cm})$$
$$= 36 \text{ cm}^2 + 6 \text{ cm}^2 + 3 \text{ cm}^2$$
$$= 45 \text{ cm}^2$$

We once again find the desired area is 45 cm^2.

Because both methods obtain the correct answer, we may use whichever method we prefer. In general, we would probably use the method that requires the fewest calculations. In Example 6, we had to calculate the areas of three regions no matter which method we chose. Other problems may be simpler using one method or the other.

EXAMPLE 7 A store manager wishes to fence in a rectangular parking lot adjacent to her store. The building will serve as one side of the lot so she needs to erect fence for only three sides. She has 400 m of fencing to use. Give the expression for the area of the lot in terms of x, where x is the length of the side perpendicular to the building.

Solution First, we must make a drawing.

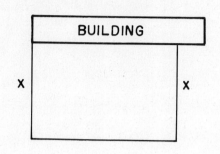

Because x is the length of the side perpendicular to the building, the remaining side to be fenced is 400 - 2x.

The area then is:

$$A = \ell \cdot w$$
$$= (400-2x)x$$
$$= 400x - 2x^2$$

The expression for the area is $400x - 2x^2$.

EXERCISE 5.1

1. Calculate the perimeter of each figure.

   a)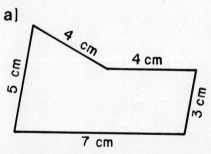

   b)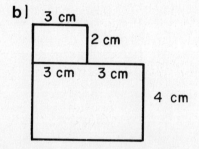

2. Calculate the perimeter of each figure.

   a)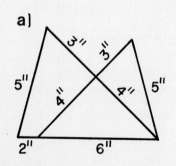

   b)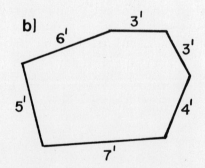

3. Find the area and perimeter of a 15 ft square.

4. Find the perimeter and area of a 18 cm square.

5. Find the area and perimeter of a 1.9 m by 2.2 m rectangle.

6. Find the perimeter and area of a 1.1 in. by 2.9 in. rectangle.

7. How many square inches are in a square foot?

8. Which is larger, 1/2 a square meter or a 1/2 meter square?

C 9. Find the perimeter of a square with an area of 219.2 cm^2.

C 10. Find the perimeter of a square with area of 91.06 in^2.

C 11. Find the area of a rectangle with one side of 37 cm and a diagonal of 47 cm.

C 12. Find the area of a rectangle with a diagonal of 52 in. and one side of 27 in.

*13. A rectangle has a perimeter of 44 in. and an area of 105 in^2. Find the dimensions of the rectangle.

*14. A rectangle has a perimeter of 46 cm and an area of 130 cm^2. Find the dimensions of the rectangle.

*15. A rectangle has a length of 3 ft more than twice its width. If the perimeter is 120 ft, find the area.

*16. A rectangle has a length of 7 m less than three times its width. If the perimeter is 106 m, find the area.

*17. A rectangle has a length of twice the sum of its width and 5 ft. If the area is 168 ft^2, find its dimensions.

*18. A rectangle has a length of one-half the sum of its width and 13 mm. If the area is 99 mm^2, find its dimensions.

19. A farmer is making an additional rectangular holding pen for her horses. By using an existing fence for one side, she can minimize her expense. The side of the pen parallel to the existing fence is 5 ft more than twice the other side. If she uses 65 ft of new fencing, what are the dimensions of the pen?

20. A dog trainer is adding another rectangular run to his kennels. He is able to use the kennel building for one side of the new run. The side adjacent to the kennel building is three times the sum of the other side and 2 ft. If he uses 40 ft of fencing to build the new run, what are its dimensions?

21. Find the unshaded area of each figure. Assume that sides are parallel or perpendicular if they appear parallel or perpendicular in the drawing.

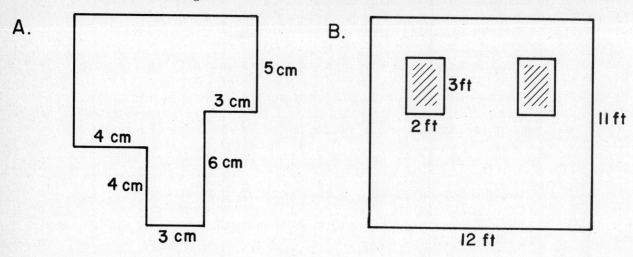

22. Find the area of each figure.

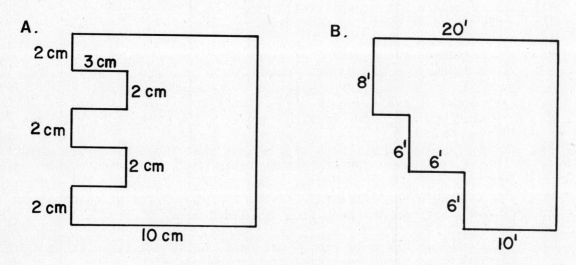

**23. A man has 300 ft of fencing, which he wishes to use to fence in a rectangular area. He can use the 20 ft long wall of his garage as part of one side of the fence. If the side perpendicular to the garage is x, give the expression for the area of the yard, A, in terms of x.

**24. A woman wishes to make a garden by fencing in a rectangular area. In addition, she wishes to divide the garden in half by erecting another fence parallel to one of the sides. She has 400 ft of fencing to use. Give the expression for the area of the garden, if x is the length of the fence dividing the garden in half.

***25. If a rectangle has length $3x^2 + 2$ and width Δx, give the expression for the area of the rectangle. (Δx is a symbol from calculus, which is read "delta x.")

**26. If a rectangle has width Δx and length $5x^2 - 2x + 7$, give the expression for the area of the rectangle. (Δx is a symbol from calculus which is read "delta x.")

**27. A printed page has 1 in. margins at the top and bottom and 3/4 in. margins at the sides. The area of the printed portion is to be 44 in². Give the expression for the area of the sheet of paper in terms of x, where x is the width of the sheet.

**28. A printed page has 1 1/2 in. margins at the top and bottom and 1 in. margins at the sides. The area of the printed portion is to be 20 in². Give the expression for the area of the sheet of paper in terms of x, where x is the length of the sheet.

5.2 PARALLELOGRAMS AND TRIANGLES

In Section 4.4, we proved in Example 2 that for any parallelogram ABCD, △AED = △BFC and that ABFE is a rectangle. Therefore, as we can see from the drawing below, the area of

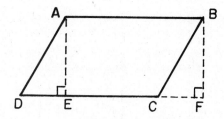

rectangle ABFE is equal to the area of parallelogram ABCD. The area of the rectangle is given by length × width or (AE)(AB). The area of the parallelogram then is given by (AE)(AB), or altitude × base, where the ALTITUDE is the perpendicular distance from a base to the opposite side.

$$\text{Area}_{parallelogram} = h \cdot b,$$
where b is a base and h is an altitude to that base.

EXAMPLE 1   Find the area and perimeter of the given parallelogram.

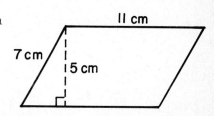

Solution   The altitude is 5 cm. (Note that we do NOT use the 7 cm. We need the perpendicular distance from the base to the opposite side.) The base is 11 cm.

$$A_{parallelogram} = h \cdot b$$
$$= (5 \text{ cm})(11 \text{ cm})$$
$$= 55 \text{ cm}^2$$

The parallelogram has an area of 55 cm^2.

$$P_{polygon} = \text{Sum of length of sides}$$
$$= 2(7 \text{ cm}) + 2(11 \text{ cm})$$
$$= 36 \text{ cm}$$

The parallelogram has a 36 cm perimeter.

EXAMPLE 2   Find the area of the given parallelogram.

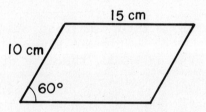

Solution   We are not given the altitude, but we do know the measure of the angle formed by the leg and the base, so we can calculate the altitude.

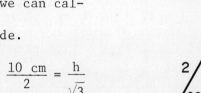

$$\frac{10 \text{ cm}}{2} = \frac{h}{\sqrt{3}}$$

$$5\sqrt{3} \text{ cm} = h$$

Now that we know the lengths of the altitude and the base we can calculate the area.

$$A_{parallelogram} = h \cdot b$$
$$= (5\sqrt{3} \text{ cm})(15 \text{ cm})$$
$$= 75\sqrt{3} \text{ cm}^2 = 129.90 \text{ cm}^2$$

The area of the given parallelogram is 129.90 cm^2.

*EXAMPLE 3 In a particular parallelogram, the altitude is one half the sum of the base and 1 cm. The area is 13 cm^2 more than six times the base. Find the area of the parallelogram.

Solution

$$x = \text{base}$$

$$\tfrac{1}{2}(x + 1) = \text{altitude}$$

$$13 + 6x = \text{area}$$

$$A_{\text{parallelogram}} = h \cdot b$$

$$13 + 6x = \tfrac{1}{2}(x + 1)(x)$$

$$26 + 12x = (x + 1)(x)$$

$$26 + 12x = x^2 + x$$

$$0 = x^2 - 11x - 26$$

$$0 = (x - 13)(x + 2)$$

Either $x - 13 = 0$ or $x + 2 = 0$

$\qquad\quad x = 13 \qquad\qquad\quad x = -2$

We discard -2 because we cannot have a negative length for the base.

$$\text{Base} = x = 13 \text{ cm}$$

$$\text{Altitude} = \tfrac{1}{2}(x + 1) = 7 \text{ cm}$$

$$\text{Area} = 13 + 6x = 91 \text{ cm}^2$$

Because $(13)(7) = 91 \text{ cm}^2$, which is the value we obtained for the area when 13 cm was substituted for x, we can be confident that the area of the parallelogram is 91 cm^2.

To find the area of a triangle, we can construct a parallelogram with the given triangle as one of the triangles formed by a diagonal of the parallelogram. Because $\triangle ABC \cong \triangle ADC$, we can find the area of $\triangle ABC$ by taking one half

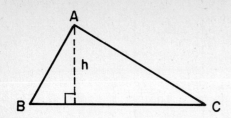 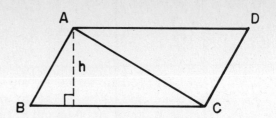

of the area of parallelogram ABCD. (See Problem 9 in Section 4.4 for the proof that ABCD is a parallelogram and that △ABC ≅ △ADC.) The area of the parallelogram is altitude × base or (h)(BC). The area for the triangle then is given by $\frac{1}{2}$(h)(BC), or $\frac{1}{2}$ altitude × base where the ALTITUDE is the perpendicular distance from a base to the opposite vertex.

$$\text{Area}_{triangle} = \frac{1}{2} h \cdot b,$$

where b is a base and h is the altitude to the base.

EXAMPLE 4   Find the area and perimeter of the given triangle.

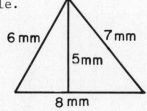

Solution
$$A_{triangle} = \frac{1}{2} h \cdot b$$
$$= \frac{1}{2} (5 \text{ mm})(8 \text{ mm})$$
$$= 20 \text{ mm}^2$$

$$P_{polygon} = \text{sum of the length of sides}$$
$$= 6 \text{ mm} + 7 \text{ mm} + 8 \text{ mm}$$
$$= 21 \text{ mm}$$

EXAMPLE 5   Find the area of a right triangle with legs of 6 cm and 10 cm.

Solution   Because the legs of a right triangle form a right angle, we can use one leg for the base and the other for the altitude.

$$A_{triangle} = \frac{1}{2} h \cdot b$$

$$= \frac{1}{2} (6 \text{ cm})(10 \text{ cm})$$

$$= 30 \text{ cm}^2$$

The area of the triangle is 30 cm^2.

EXAMPLE 6   Find the area of the given triangle.

Solution   The altitude is 5.7' and the base is 4'. Even though we must extend the base to obtain the altitude, the measure we use for the base is the actual base of the triangle.

$$A_{triangle} = \frac{1}{2} h \cdot b$$

$$= \frac{1}{2} (5.7')(4')$$

$$= 11.4 \text{ ft}^2$$

The area of the triangle is 11.4 ft^2.

We can demonstrate that 11.4 ft^2 is indeed the correct area of △ABC by the following subtraction:

$$\text{Area}_{\triangle ABC} = \text{Area}_{\triangle ABD} - \text{Area}_{\triangle CBD}$$

$$= \frac{1}{2}(5.7)(4+3) - \frac{1}{2}(5.7)(3)$$

$$= \frac{1}{2}(5.7)(7) - \frac{1}{2}(5.7)(3)$$

$$= 19.95 - 8.55$$

$$\text{Area}_{\triangle ABC} = 11.4 \text{ ft}^2$$

Thus either procedure results in the same area for △ABC.

We find the area of an equilateral triangle with side of length s.

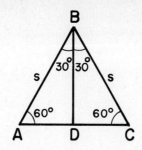

We know $\triangle ABD \cong \triangle CBD$ (how?), therefore $\overline{AD} \cong \overline{DC}$, so each must be of length $\frac{s}{2}$. We can use the Pythagorean theorem to determine the length of BD, or the altitude, h, of the triangle.

$$a^2 + b^2 = c^2$$

$$\left(\frac{s}{2}\right)^2 + h^2 = s^2$$

$$h^2 = s^2 - \frac{s^2}{4}$$

$$h^2 = \frac{3s^2}{4}$$

$$h = \pm \frac{s}{2}\sqrt{3}$$

Because we are measuring length, we discard the negative root, giving us an altitude of $\left(\frac{s}{2}\right)\sqrt{3}$. We can calculate the area.

$$A_{triangle} = \frac{1}{2} h \cdot b$$

$$= \frac{1}{2} \left(\frac{s}{2}\sqrt{3}\right)(s)$$

$$= \frac{s^2}{4} \cdot \sqrt{3}$$

$$\boxed{\text{Area}_{equilateral\ triangle} = \frac{s^2}{4} \cdot \sqrt{3}}$$

where s is the side.

EXAMPLE 7   Find the area of an equilateral triangle with sides of 18 cm.

Solution

$$A_{\text{equilateral triangle}} = \frac{s^2}{4} \cdot \sqrt{3}$$

$$= \frac{(18 \text{ cm})^2}{4} \sqrt{3}$$

$$= \frac{2^{\cancel{2}} \cdot 9^2 \text{ cm}^2}{\cancel{4}_1} \sqrt{3}$$

$$= 81\sqrt{3} \text{ cm}^2$$

The area of the triangle is $81\sqrt{3}$ cm^2 or 140.30 cm^2.

EXERCISE 5.2

For Problems 1 to 6, find the area and perimeter of each parallelogram. (Give exact answers.)

1.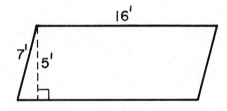

2.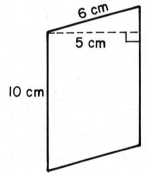

3.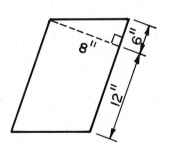

4.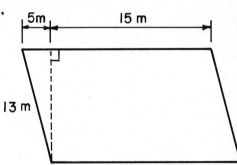

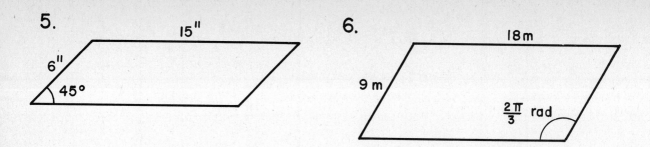

For Problems 7 to 16, find the area and perimeter of each triangle. Give exact answers, unless a C indicates otherwise.

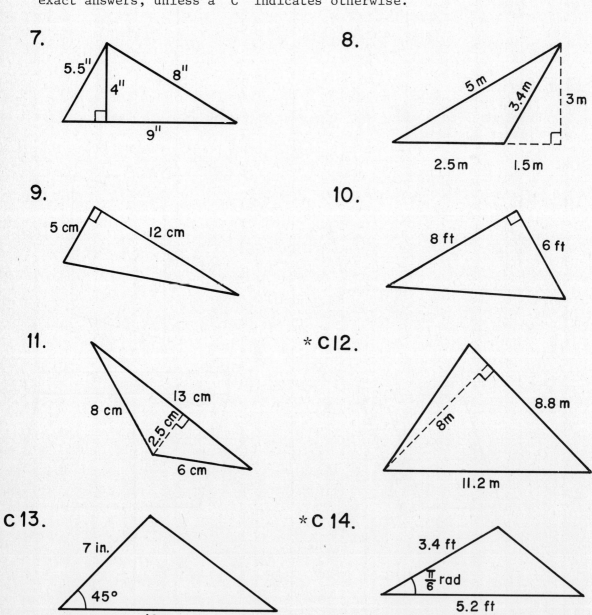

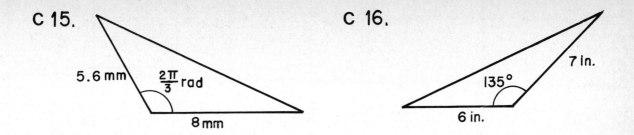

17. Find the area of an equilateral triangle with 14 cm sides.

18. Find the area of an equilateral triangle with 22 in. sides.

19. Find the length of the sides of an equilateral triangle with an area of $64\sqrt{3}$ ft^2.

20. Find the length of the sides of an equilateral triangle with an area of $49\sqrt{3}$ m^2.

Find the area for each figure given in Problems 21 through 24. Assume that sides are parallel or perpendicular if they appear parallel or perpendicular in the drawing.

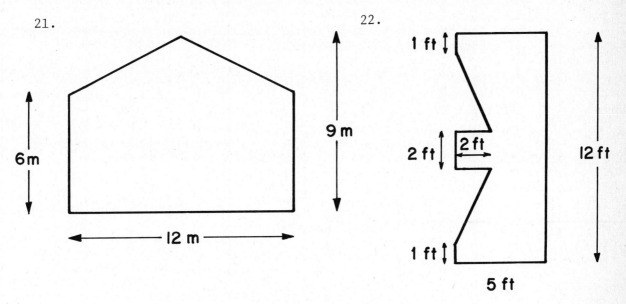

23. 24.

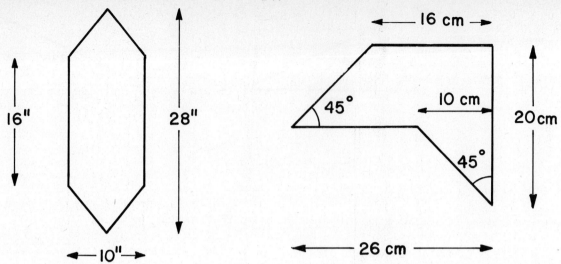

25. Find the base of a parallelogram with altitude of 8 cm and area of 152 cm^2.

26. Find the altitude of a parallelogram with base of 16 in. and area of 176 in^2.

27. Find the area of a square that has the same perimeter as an equilateral triangle with area of $16\sqrt{3}$ in^2.

28. Find the area of an equilateral triangle that has the same perimeter as a square with area of 81 yd^2.

29. Find the area of an isosceles triangle with 14 in. equal legs and another leg of $14\sqrt{3}$ in., if the angle included between the legs of different length is 30°.

30. Determine the formula for the area of an equilateral triangle of height h.

*31. The altitude of a parallelogram is one less than half the sum of the base and −1 in. The area is 21 in^2 less than ten times the base. Find the area of the parallelogram.

*32. The base of a parallelogram is 3 cm more than twice the altitude. The area is 18 cm^2 less than three times the square of the altitude. Find the area of the parallelogram.

*33. The base of a triangle is 4 m more than twice the altitude. The area of the triangle is 14 m^2 more than the square of the altitude. Find the area of the triangle.

*34. The altitude of a triangle is one half the sum of the base and 4 ft. The area is 7 ft^2 more than four times the base. Find the area of the triangle.

**35. A 32 cm piece of wire is to be cut into two pieces. One will form a square and the other an equilateral triangle. If x is the length of the side of the square, give the expression for the sum of the areas of the two figures.

**36. A 44 in. wire is to be cut into two pieces. One will form a square and the other an equilateral triangle. If x is the length of the side of the triangle, give the expression for the sum of the areas of the two figures.

5.3 RHOMBUSES AND TRAPEZOIDS

A diagonal of a rhombus divides it into two congruent triangles, each with an altitude equal to one half of the other diagonal. (See Problem 17 of Section 4.4 for the proof.) Therefore in order to find the area of a rhombus, we can consider it to be equal to the area of the two triangles composing the rhombus. The area of rhombus ABCD, is equal to the area of $\triangle ABC$ plus the

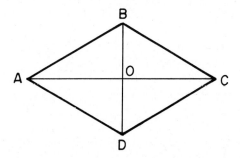

area of $\triangle ACD$. Because the two triangles are congruent, however, the area of the rhombus is twice the area of $\triangle ABC$.

$$\text{Area}_{\text{rhombus ABCD}} = 2 \cdot \text{Area}_{\triangle ABC}$$

$$= 2(\tfrac{1}{2} h \cdot b)$$

$$= 2(\tfrac{1}{2})(BO)(AC)$$

$$= (BO)(AC)$$

$$= \tfrac{1}{2}(AC)(BD)$$

when $\overline{AC}$ and $\overline{BD}$ are the two diagonals of the rhombus.

$$\text{Area}_{\text{rhombus}} = \frac{1}{2} \cdot d_1 \cdot d_2,$$

where d_1 and d_2 are the diagonals.

Many people prefer to think of the area of a rhombus as the sum of the areas of the two triangles rather than remember the special formula for the area of a rhombus. You should determine for yourself which approach you prefer.

EXAMPLE 1   Find the area of a rhombus with diagonals of 6.2 cm and 9.3 cm.

Solution
$$\text{Area}_{\text{rhombus}} = \frac{1}{2} \cdot d_1 \cdot d_2$$

$$= \frac{1}{2}(\overset{3.1}{\cancel{6.2}} \text{ cm})(9.3 \text{ cm})$$

$$= 28.83 \text{ cm}^2$$

EXAMPLE 2   Find the area of a rhombus with sides of 20 in. and one 32 in. diagonal.

Solution   First we make a drawing.

Because AC = 32 in., AO = 16 in.

Because the diagonals in a rhombus are perpendicular and AB = 20 in. and AO = 16 in., we know that BO = 12 in. (Why?) Thus BD = 24 in., and we can now compute the area.

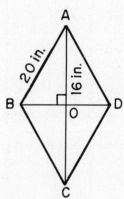

$$A_{\text{rhombus}} = \frac{1}{2} \cdot d_1 \cdot d_2$$

$$= \frac{1}{2}(32 \text{ in.})(24 \text{ in.})$$

$$= 384 \text{ in}^2$$

The area of the rhombus is 384 in^2.

EXAMPLE 3 Find the perimeter of a rhombus if it has an area of 126 cm^2 and one diagonal of 14 cm.

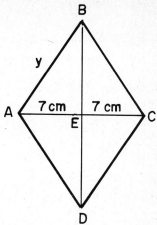

Solution First we enter the known information on a drawing. We can then solve for the other diagonal.

$$A_{rhombus} = \frac{1}{2} d_1 \cdot d_2$$

$$126 \text{ cm}^2 = \frac{1}{2}(14 \text{ cm})(x)$$

$$18 \text{ cm} = x$$

The other diagonal $(\overline{BD})$ is 18 cm long, which gives 9 in. for the length of $\overline{BE}$.

We can use the Pythagorean theorem to determine the length of the side.

$$c^2 = a^2 + b^2$$

$$y^2 = (9)^2 + (7)^2$$

$$y^2 = 130$$

$$y = \pm \sqrt{130}$$

$$y = \sqrt{130} \text{ cm}$$

The perimeter of the rhombus is given by:

$$P_{rhombus} = 4 \cdot y$$

$$= 4\sqrt{130} \text{ cm or } 45.61 \text{ cm}$$

The perimeter of the rhombus is $4\sqrt{130}$ cm or about 45.61 cm.

EXAMPLE 4 A rhombus has an area of 240 in^2. One diagonal is given by $2x$, and the other by $4x - 2$ in. Find the length of the sides of the rhombus.

Solution We first must determine the length of each diagonal. We can then use the Pythagorean theorem to find the length of the sides.

$$A_{rhombus} = \frac{1}{2} d_1 \cdot d_2$$

$$240 \text{ in}^2 = \frac{1}{2}(2x)(4x - 2)$$

$$240 = 4x^2 - 2x$$

$$0 = 4x^2 - 2x - 240$$

$$= 2(2x^2 - x - 120)$$

$$0 = 2(2x + 15)(x - 8)$$

$$2x + 15 = 0 \quad \text{or} \quad x - 8 = 0$$

$$x = -\frac{15}{2} \text{ in.} \quad x = 8 \text{ in.}$$

Because $x = -\frac{15}{2}$ in. will result in negative values for the the diagonals, we discard that root and use only $x = 8$ in.

When $x = 8$ in.:

One diagonal $= 2x$ Other diagonal $= 4x - 2$

$= 16$ in. $= 30$ in.

Using the Pythagorean theorem to find the length of the side, we get:

$$c^2 = a^2 + b^2$$

$$= \left(\frac{16}{2}\right)^2 + \left(\frac{30}{2}\right)^2$$

$$= 64 + 225$$

$$c^2 = 289$$

$$c = \pm \sqrt{289}$$

$$c = \pm 17 \text{ in.}$$

The length of the side is 17 in.

A diagonal of a trapezoid divides it into two triangles each with altitude, h, the altitude of the trapezoid. The area of a trapezoid ABCD

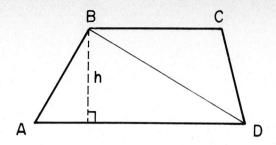

is then equal to the area of △ABD plus the area of △BCD, or:

$$\text{Area}_{\text{trapezoid ABCD}} = \text{Area}_{\triangle ABD} + \text{Area}_{\triangle BCD}$$

$$= \frac{1}{2} h(AD) + \frac{1}{2} h(BC)$$

$$= \frac{1}{2} h (AD + BC)$$

where $\overline{AD}$ and $\overline{BC}$ are the bases of the trapezoid.

$$\text{Area}_{\text{trapezoid}} = \frac{1}{2} h(b_1 + b_2),$$

where b_1 and b_2 are the bases of the trapezoid and h is the altitude between the bases.

EXAMPLE 5 Determine the area of the given trapezoid.

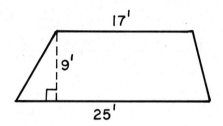

Solution
$$A_{\text{trapezoid}} = \frac{1}{2} h(b_1 + b_2)$$

$$= \frac{1}{2}(9 \text{ ft})(17 \text{ ft} + 25 \text{ ft})$$

$$= \frac{1}{2}(9 \text{ ft})(42 \text{ ft})$$

$$= 9 \text{ ft}(21 \text{ ft})$$

$$= 189 \text{ ft}^2$$

The area of the given trapezoid is 189 ft^2.

EXAMPLE 6   Find the altitude of a trapezoid with a 252 in^2 area and bases of 18 in. and 24 in.

Solution
$$A_{trapezoid} = \frac{1}{2} h(b_1 + b_2)$$

$$252 \text{ in}^2 = \frac{1}{2} h(18 \text{ in.} + 24 \text{ in.})$$

$$252 \text{ in}^2 = \frac{1}{2} h(42 \text{ in.})$$

$$252 \text{ in}^2 = (21 \text{ in.})h$$

$$12 \text{ in.} = h$$

The altitude is 12 in.

EXAMPLE 7   Find the area of an isosceles trapezoid with bases of 17 in. and 27 in. and legs of 13 in.

Solution   Enter the given information on a drawing.

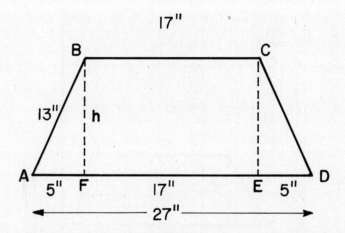

Because we have an isosceles trapezoid we know that

AD − BC = AF + ED.   Because AF = ED we have AF = ED = 5 in.

We now know we have a right triangle, △ABF = h = 12 in. (Why?) and we can compute the area of the trapezoid.

$$A_{trapezoid} = \tfrac{1}{2} h(b_1 + b_2)$$

$$= \tfrac{1}{2}(12 \text{ in.})(17 \text{ in.} + 27 \text{ in.})$$

$$= (6 \text{ in.})(44 \text{ in.})$$

$$= 264 \text{ in}^2$$

The area of the trapezoid is 264 in^2.

EXAMPLE 8 Find the length of the bases of a trapezoid with area of 120 cm^2 if the altitude is given by x, one base by $3x - 1$ cm and the other by $4x - 1$ cm.

Solution

$$A_{trapezoid} = \tfrac{1}{2} h(b_1 + b_2)$$

$$120 \text{ cm}^2 = \tfrac{1}{2}(x)[(3x - 1) + (4x - 1)]$$

$$120 = \tfrac{1}{2} x[7x - 2]$$

$$240 = 7x^2 - 2x$$

$$0 = 7x^2 - 2x - 240$$

$$= (7x + 40)(x - 6)$$

$7x + 40 = 0$ or $x - 6 = 0$

$x = \dfrac{-40}{7}$ cm $\qquad x = 6$ cm

We must discard the root $x = \dfrac{-40}{7}$ cm because it gives a negative value for the altitude.

When $x = 6$ cm,

One base $= 3x - 1$ cm $\qquad$ Other base $= 4x - 1$ cm

$\qquad\qquad = 17$ cm $\qquad\qquad\qquad\quad = 23$ cm

The dimensions of the trapezoid are altitude $= 6$ cm, one base $= 17$ cm, and other base $= 23$ cm.

EXAMPLE 9 Find the perimeter of the given trapezoid. (Give an exact answer.)

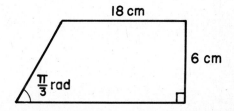

251

Solution We have a $\frac{\pi}{6} - \frac{\pi}{3} - \frac{\pi}{2}$ rad triangle, so we may set up proportions to determine the lengths of x and y.

$$\frac{2}{x} = \frac{\sqrt{3}}{6 \text{ cm}}$$

$$\frac{12 \text{ cm}}{\sqrt{3}} = x$$

$$4\sqrt{3} \text{ cm} = x$$

$$\frac{1}{y} = \frac{\sqrt{3}}{6 \text{ cm}}$$

$$\frac{6 \text{ cm}}{\sqrt{3}} = y$$

$$2\sqrt{3} \text{ cm} = y$$

The perimeter of the trapezoid is given by:

$$P_{\text{trapezoid}} = 18 \text{ cm} + 6 \text{ cm} + 18 \text{ cm} + y + x$$

$$= 42 \text{ cm} + 2\sqrt{3} \text{ cm} + 4\sqrt{3} \text{ cm}$$

$$= 42 + 6\sqrt{3} \text{ cm}$$

The perimeter of the trapezoid is $42 + 6\sqrt{3}$ cm.

EXERCISE 5.3

1. Determine the area of a rhombus with diagonals of 18 cm and 13 cm.

2. Determine the area of a rhombus with diagonals of 9 cm and 10 cm.

C 3. Determine the perimeter of the rhombus in Problem 1.

C 4. Determine the perimeter of the rhombus in Problem 2.

5. Find the area of a rhombus with sides of 25 in. and one 48 in. diagonal.

6. Find the area of a rhombus with sides of 41 cm and one 80 cm diagonal.

C 7. Find the area of a rhombus with sides of 13 cm and one 19 cm diagonal.

C 8. Find the area of a rhombus with sides of 23 in. and one 37 in. diagonal.

9. Find the other diagonal of a rhombus with an area of 203 ft^2 and one diagonal of 29 ft. (Given exact answer.)

10. Find the other diagonal of a rhombus with an area of 279 m^2 and one diagonal of 31 m. (Given exact answer.)

11. Find the perimeter of a rhombus with an area of 96 in^2 and one diagonal of 16 in. (Give exact answer.)

12. Find the perimeter of a rhombus with an area of 336 in^2 and one diagonal of 48 in. (Give exact answer.)

*13. Find the length of the diagonals of a rhombus with an area of 104 mm^2 and one diagonal given by 2x + 3 mm and the other by 3x + 1 mm.

*14. Find the length of the diagonals of a rhombus with 115 mm^2 area and one diagonal given by x mm and the other by 2x + 3 mm.

C 15. Find the perimeter for the rhombus in Problem 13.

C 16. Find the perimeter for the rhombus in Problem 14.

17. Find the area for a rhombus with sides of 14 cm if one vertex angle is $\frac{\pi}{3}$ rad. (Give exact answer.)

18. Find the area for a rhombus with sides of 10 in. if one vertex angle is 60°. (Give exact answer.)

*19. A rhombus has area of 864 in^2. One diagonal is given by x + 16 in., and the other by 2x + 8 in. Find the length of the sides of the rhombus.

*20. A rhombus has area of 1536 in^2. One diagonal is given by x + 18 in. and the other by 2x + 4 in. Find the length of the sides of the rhombus.

21. Find the area of a trapezoid with an altitude of 5 in. and bases of 15 in. and 5 in.

22. Find the area of a trapezoid with an altitude of 7 cm and bases of 11 cm and 13 cm.

23. Find the area of a trapezoid with an altitude of $4\frac{1}{2}$ in. and bases of $6\frac{1}{2}$ in. and $8\frac{3}{4}$ in.

24. Find the area of a trapezoid with an altitude of 9.6 m and bases of 3.4 m and 12.9 m.

For Problems 25 to 30, find the missing values.

	h	b_1	b_2	$A_{trapezoid}$
25.		10 ft	12 ft	66 ft^2
26.	5 cm		15 cm	40 cm^2
27.		$4\frac{1}{2}$ ft	$7\frac{1}{2}$ ft	96 ft^2
28.	2 ft	$3\frac{1}{2}$ ft		8 ft^2
29.	4.6 yd		19.9 yd	85.1 yd^2
30.		1.8 in.	2.4 in.	2.52 in^2

Problems 31 to 36 refer to Figure 5.1, an isosceles trapezoid.

31. If AB = 15 cm, DC = 21 cm, and AD = 5 cm, determine the area of the trapezoid.

32. If AB = 22 in., AD = 10 in., and DR = 8 in., determine the area of the trapezoid.

33. If AB = 30 in., BC = 13 in., and AR = 12 in., determine the area of the trapezoid.

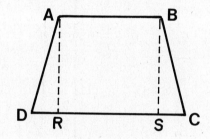

FIGURE 5.1

34. If DC = 34 cm, CB = 13 cm, and CS = 12 cm, determine the area of the trapezoid.

35. If AD = 10 cm, AB = 8 cm, and $\angle D = \frac{\pi}{4}$ rad, determine the area of the trapezoid. (Give an exact answer.)

36. If AD = 6 in., AB = 20 in., and $\angle D = 60°$, determine the area of the trapezoid. (Give an exact answer.)

37. Find each base of an isosceles trapezoid if the larger base is 14 in. more than the smaller and the legs are twice the smaller base. The perimeter is 68 in.

38. In a given trapezoid, the larger base is 2.5 in. more than three times the shorter leg, and the shorter base is 2 in. more than twice the shorter leg. The longer leg is 1 in. more than the shorter leg, and the perimeter is 51 in. Find the length of each base.

*39. A trapezoid has area of 66 in^2. The longer base is given by 2x - 1 in., the shorter base by x + 4 in., and the altitude by x - 6 in. Find the dimensions of the trapezoid.

*40. A trapezoid has area of 66 in^2. The longer base is given by x + 4 in., the shorter base by x - 2 in., and the altitude by x - 4 in. Find the dimensions of the trapezoid.

**41. Give the expression for the area of a trapezoid where one base is given by y_i, the other base by y_{i+1}, and the altitude by Δx.

**42. Give the expression for the area of a trapezoid where one base is given by $f(x_i)$, the other by $f(x_{i+1})$, and the altitude by Δx. (The expression $f(x_1)$ is read "f of x sub i.")

5.4 REGULAR POLYGONS AND COMPOSITE FIGURES

Recall that a regular polygon has equal sides and equal interior angles. The CENTER of a regular polygon is the point that is the center of both the circumscribed circle and the inscribed circle of the polygon (fully discussed in Section 6.2). The APOTHEM of any regular polygon is the perpendicular distance from the center to a side. For the regular pentagon ABCDE, $\overline{OF}$ is the apothem.

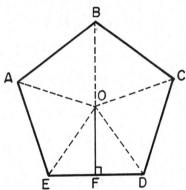

The pentagon may be divided into five congruent triangles, each with an area $\frac{1}{2}$ (ED)(OF).

The area of the pentagon will be:

$$\text{Area}_{\text{pentagon}} = 5 \cdot \frac{1}{2}(ED)(OF)$$

$$= 5 \cdot \frac{1}{2}(s)(a)$$

$$= \frac{1}{2}(5s)(a)$$

where s is the length of one side and a is the apothem. The perimeter of the pentagon equals 5s, so we have:

$$\text{Area}_{\text{pentagon}} = \frac{1}{2} p \cdot a,$$

where p is the perimeter and a is the apothem. This rule will hold for any regular polygon.

$$\text{Area}_{\text{polygon}} = \frac{1}{2} p \cdot a,$$

where p is the perimeter and a is the apothem.

EXAMPLE 1   Find the area of a regular octagon with 2 in. sides and apothem of 2.414 in.

Solution   An octagon has eight sides, so the perimeter = 8(2 in.) = 16 in.

$$A_{\text{polygon}} = \frac{1}{2} p \cdot a$$

$$= \frac{1}{2} (16 \text{ in.})(2.414 \text{ in.})$$

$$= 19.312 \text{ in}^2$$

EXAMPLE 2   Find the area of a regular hexagon with 8 cm sides.

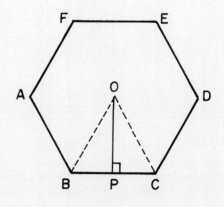

Solution   A regular hexagon has interior angles of $\frac{4(180)}{6} = 120°$. That is, $\angle ABC = 120°$, and $\angle OBC = 60°$. We can construct $\triangle OPB$, a

30°-60°-90° triangle, to calculate the apothem. We can draw the similar triangles and set up our proportion.

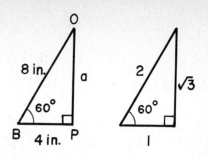

$$\frac{8 \text{ in.}}{2} = \frac{a}{\sqrt{3}}$$

$$a = 4\sqrt{3} \text{ in.}$$

The apothem is $4\sqrt{3}$ in. and the perimeter is $6 \cdot (8 \text{ in.}) = 48$ in. We can now calculate the area:

$$A_{polygon} = \frac{1}{2} p \cdot a$$

$$= \frac{1}{2} (48 \text{ in.})(4\sqrt{3} \text{ in.})$$

$$= 96\sqrt{3} \text{ in}^2 \text{ or } 166.28 \text{ in}^2$$

We can use the knowledge we have gained to determine the areas of irregular figures.

EXAMPLE 3   Determine the area of the given figure.

Solution   One way to calculate the area is to find the area of the rectangle and subtract the area of the triangle. One half of the triangle is a $\frac{\pi}{6} - \frac{\pi}{3} - \frac{\pi}{2}$ rad triangle, so we may set up a proportion to determine x, the altitude of the triangle.

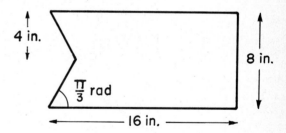

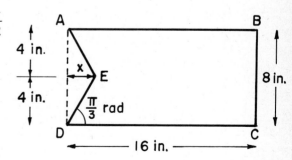

257

$$\frac{x}{1} = \frac{4}{\sqrt{3}}$$

$$x = \frac{4}{3}\sqrt{3} \text{ in.}$$

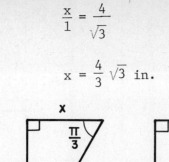

The area of the larger triangle $\triangle AED$ is given by:

$$A_{triangle} = \frac{1}{2} h \cdot b$$

$$= \frac{1}{2} \left(\frac{4\sqrt{3} \text{ in.}}{3}\right)(8 \text{ in.})$$

$$= \frac{16}{3} \cdot \sqrt{3} \text{ in}^2 \text{ or } 9.238 \text{ in}^2$$

The area of the rectangle is:

$$A_{rectangle} = \ell \cdot w$$

$$= (16 \text{ in.})(8 \text{ in.})$$

$$= 128 \text{ in}^2$$

$$A_{figure} = A_{rectangle} - A_{triangle}$$

$$= 128 \text{ in}^2 - 9.238 \text{ in}^2$$

$$= 118.762 \text{ in}^2$$

To the nearest square inch, the desired area is 119 in^2.

EXERCISE 5.4

1. Find the area of a regular polygon with a perimeter of 80 in. and an apothem of 12 in.

2. Find the area of a regular polygon with a perimeter of 17 in. and an apothem of 1.4 in.

3. The area of a regular polygon is 144 ft^2 and its perimeter is 48 ft. Find the apothem.

4. The area of a regular polygon is 216 cm^2 and its apothem is 3 cm. Find its perimeter.

5. Find the area of a regular octagon with 7 cm sides and an apothem of 8.45 cm.

6. Find the area of a regular decagon with 9 in. sides and an apothem of 13.85 in.

C 7. If the area of a regular pentagon is 170.33 cm^2 and it has a 6.19 cm apothem, find the length of each side.

C 8. If the area of a regular heptagon is 294.32 mm^2 and it has a 9.34 mm apothem, find the length of each side.

9. Find the area of a regular hexagon with 4 in. sides.

10. Find the area of a regular hexagon with $6\sqrt{3}$ ft sides.

11. Find the area of a regular hexagon with an apothem of 10 m.

12. Find the area of a regular hexagon with an apothem $8\sqrt{2}$ in.

For Problems 13 to 18 determine the area of the given figure. Assume that sides are parallel or perpendicular if they appear parallel or perpendicular in the drawing. If a drawing appears symmetric, you may assume that it is. (Give answer to nearest hundredth.)

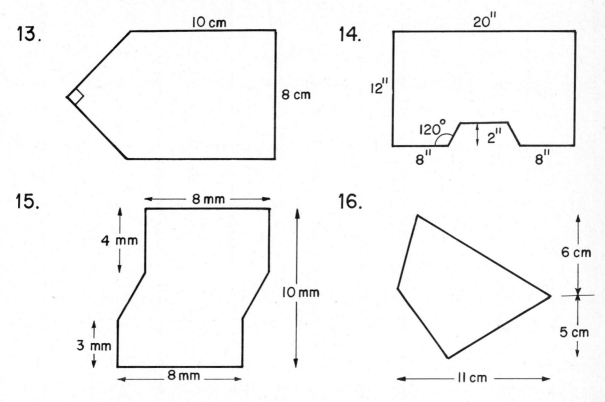

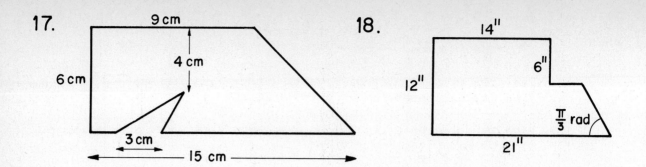

CHAPTER 5 REVIEW

1. Calculate the perimeter and the area of the figure on the right.

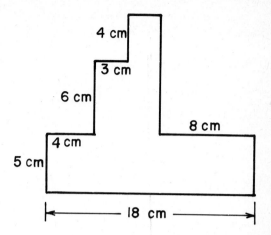

2. Find the perimeter of a square with a $7\sqrt{2}$ ft diagonal.

3. Find the area of a square with 36 ft perimeter.

C 4. Find the perimeter of a square with area of 32.94 cm^2.

*5. A rectangle has perimeter of 66 in. and area of 260 in^2. Find the dimensions of the rectangle.

*6. A rectangle is 3 ft longer than three times its width. If its area is 90 ft^2, find its dimensions.

**7. A man wishes to build some dog runs using 600 ft of fencing. He will fence in a rectangular area and then erect two fences parallel to one side. If x is the length of the dividing fences, give the expression for the area of the dog runs.

**8. A printed page has 3 cm margins at the sides and 4 cm margins at the top and bottom. The area of the printed portion is to be 450 cm^2. Give the expression for the area of the sheet of paper in terms of x, where x is the width of the sheet.

9. Find the area and perimeter of each figure.

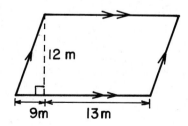 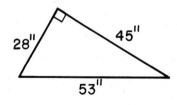

C 10. Find the area and perimeter of each figure.

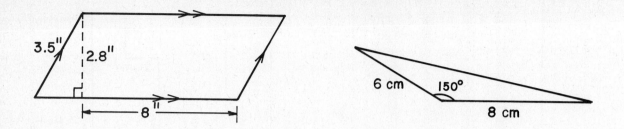

11. Find the area of an equilateral triangle with 36 mm perimeter.

12. Find the area of a square that has the same perimeter as an equilateral triangle with area of $64\sqrt{3}$ cm^2.

*13. The altitude of a parallelogram is 8 in. less than the base. The area is 18 in^2 less than eleven times the base. Find its dimensions.

*14. The base of a triangle is 1 ft longer than the altitude. The area is 4 ft^2 more than three times the base. Find the area of the triangle.

**15. A 60 cm piece of wire is to be cut into three pieces. One will form a square and the two remaining, of equal length, will form equilateral triangles. If x is the length of the side of the square, give the expression for the sum of the areas of the three figures.

16. Determine the area of a rhombus with diagonals of 14 cm and 21 cm.

C 17. Determine the perimeter of the rhombus in Problem 16.

18. Find the area of a rhombus with 17 in. sides and one diagonal of 30 in.

19. Find the other diagonal of a rhombus with area of 153 cm^2 and one diagonal of 17 cm.

*20. Find the length of the diagonals of a rhombus with area of 230 m^2 if one diagonal is 3 meters longer than the other.

C 21. Find the perimeter of the rhombus in Problem 20.

22. Find the area of a rhombus with 9 in. sides if one vertex angle is 120°. (Give an exact answer.)

23. Find the area of a trapezoid with altitude of 8 in. and bases of 12 in. and 15 in.

24. Find the perimeter and area of an isosceles trapezoid with bases of 15 in. and 23 in. and legs of 5 in.

25. Find the altitude of a trapezoid with area of 161 mm^2 if the bases are 17 mm and 29 mm.

26. Find the other base of a trapezoid with area of 207 ft^2 if the altitude is 9 ft and one base is 15 ft.

27. Find the area of an isosceles trapezoid with legs of 6.1 mm and bases of 5 mm and 7.2 mm.

28. Determine the area and perimeter of an isosceles trapezoid if the bases are 11 in. and 23 in. and each leg forms a $\pi/4$ rad angle with the longer base. (Give exact answers.)

*29. A trapezoid has area of 44 yd^2. The longer base is 1 yd longer than three times the altitude and the shorter base is 1 yd longer than twice the altitude. Find the dimensions of the trapezoid.

**30. Give the expression for the area of a trapezoid where one base is given by $g(x_i)$, the other by $g(x_{i+1})$, and the altitude by Δx.

31. Find the area of a regular polygon with perimeter of 27.9 cm and apothem of 4.2 cm.

32. Find the area of a regular pentagon with 7 in. sides and apothem of 4.82 in.

33. Find the area of a regular hexagon with 4 cm sides.

34. Find the area of a regular hexagon with apothem of 7 in.

*35. Find the length of each side of a regular hexagon if the area of the hexagon is $150\sqrt{3}$ in^2.

*36. Find the apothem of a regular hexagon if the area of the hexagon is $\frac{243\sqrt{3}}{2}$ in^2.

For Problems 37-40 determine the area of the given figure. You may assume that sides are parallel or perpendicular if they appear parallel or perpendicular in the drawing. If a drawing appears symmetric, you may assume it is. (Give answers correct to hundredths.)

37.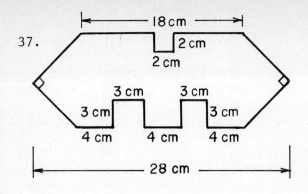

38.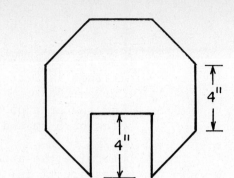

39.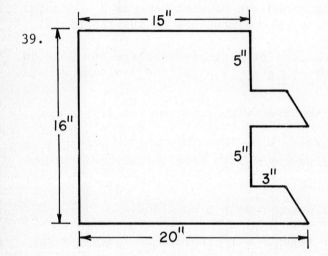

40.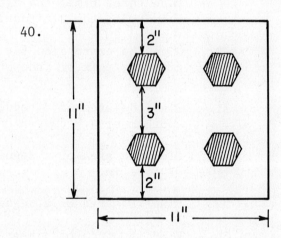

CHAPTER 6
CIRCLES

6.1 BASIC DEFINITIONS

In Chapter 1 we defined a CIRCLE to be the set of points that are the same distance from a fixed point. The fixed point is the CENTER of the circle, and the fixed distance is the RADIUS. A line segment from the center of the circle to a point on the circle is also called a RADIUS (plural, RADII). Thus the word "radius" is used to represent a distance and to represent a particular line segment. All radii of a given circle are the same length. If two circles have the same radius they are EQUAL or CONGRUENT circles. Two circles having the same center and different radii are called CONCENTRIC circles. In Figure A, below,

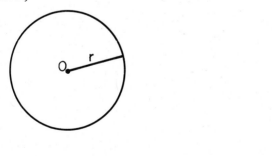
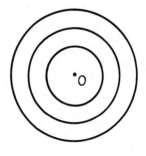

A   B

O is the center of the circle and r is the radius. B illustrates three concentric circles. Note that we are assuming that the given point O in the center of the circle is indeed the center of the circle.

265

Two distinct radii form an angle with the center of the circle as the vertex. Any angle formed by two radii of a circle is called a CENTRAL ANGLE of the circle. The smallest value for a central angle is 0° or 0 rad; the largest angle is 360° or 2π rad. Thus a circle contains 360° or 2π rad. If the two radii form a straight angle (180° or π rad), they form a DIAMETER of the circle. Thus a DIAMETER is a line segment that passes through the center of the circle and has endpoints on the circle. Just as the word radius is used in two ways, so, too, the word diameter has a second meaning; it also means the length of the line segment passing through the center with endpoints on the circle. This gives the following relationship.

The diameter, d, of a circle is given by

$$d = 2r,$$

where r is the radius of the circle.

All diameters in a given circle are the same length.

EXAMPLE 1   If AB = 12 in., find the diameter of the circle.

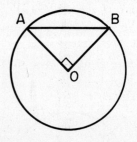

Solution   We are given that ∠AOB is a right angle and we know the radii of a circle are all the same length.

Therefore we have an isosceles right triangle, or a 45°-45°-90° triangle. We can use this fact to determine the radius.

$$\frac{12 \text{ in.}}{\sqrt{2}} = \frac{x}{1}$$

$$x = \frac{12}{\sqrt{2}} = 6\sqrt{2} \text{ in.}$$

The diameter is given by:

$$d = 2r$$
$$= 2(6\sqrt{2})$$
$$= 12\sqrt{2} \text{ in. or } 16.97 \text{ in.}$$

The diameter of the circle is 17 in. (to the nearest inch).

Note that we could have used the Pythagorean theorem to solve this example.

The CIRCUMFERENCE of a circle is the distance around the circle, or, its perimeter. A segment of the circle itself (or a part of its circumference) is called an ARC. A diameter cuts a circle into two equal arcs called SEMICIRCLES. An arc measuring less than a semicircle is a MINOR ARC; one measuring greater than a semicircle is a MAJOR ARC. Two letters are sufficient to indicate a minor arc; three letters are needed to indicate a major arc. In figure A, below, the diameter AB, cuts the circle into two semicircles. In B, the

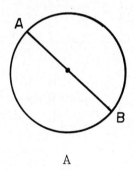

A

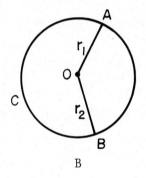

B

radii, r_1 and r_2 form a central angle $\angle AOB$. They also cut the circle into two arcs: minor arc AB, written $\widehat{AB}$, and major arc ACB, written $\widehat{ACB}$. Because $\angle AOB$ cuts the circle to form $\widehat{AB}$, we say that $\angle AOB$ INTERCEPTS $\widehat{AB}$. We also say that $\widehat{AB}$ SUBTENDS central angle $\angle AOB$.

Arcs are measured in degrees or radians. The number of degrees or radians in an arc is equal to the measure of the central angle that intercepts the arc.

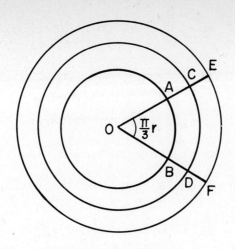

FIGURE 6.1

In Figure 6.1 $\angle EOF = \pi/3$ rad, so $\widehat{AB} = \widehat{CD} = \widehat{EF} = \pi/3$ rad. Note that two arcs that subtend the same central angle will have an equal number of arc degrees (or radians) even though they will NOT be the same length. However, they each __will__ represent the same portion of a circle.

The measure of arc degrees or radians	=	The measure of the central angle subtended by the arc.

EXAMPLE 2   Find $\widehat{AB}$.

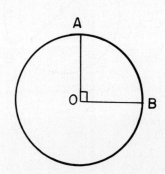

Solution $\widehat{AB}$ subtends $\angle AOB = 90°$. Therefore $\widehat{AB} = 90°$.

A line that intersects a circle at one and only one point is a TANGENT to the circle. A radius drawn to the point of tangency is perpendicular to the tangent. A line that intersects a circle at two points is a SECANT to the circle. The portion of the secant that lies within the circle is called a

CHORD. Two chords that have one common endpoint form an INSCRIBED ANGLE. In Figure 6.2, $\overleftrightarrow{AB}$ is a

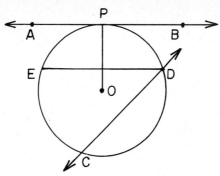

FIGURE 6.2

tangent to the circle, and P is the point of tangency. Radius $\overline{OP}$ is perpendicular to tangent $\overleftrightarrow{AB}$. $\overleftrightarrow{CD}$ is a secant to the circle, and $\overline{CD}$ is a chord. Chords $\overline{CD}$ and $\overline{ED}$ form an inscribed angle, ∠EDC. ∠EDC intercepts $\widehat{EC}$, and $\widehat{EC}$ subtends ∠EDC. Note that $\overline{CD}$ represents a chord (line segment), while $\widehat{CD}$ represents an arc (segment of the circle). The length of $\overline{CD}$ is not the length of $\widehat{CD}$.

When an equilateral triangle is inscribed in a circle, the vertices of the triangle divide the circle into three equal arcs. Because a circle contains 360°, $\widehat{AC} = \widehat{CB} = \widehat{BA} = 120°$ and ∠COB = 120°. We know, however, that ∠CAB = 60°. We therefore conclude that an inscribed angle intercepts an arc that is double the size of the inscribed angle.

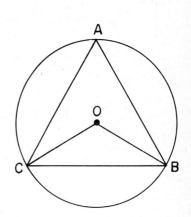

> Inscribed angle is equal to half of the intercepted arc.

Thus any inscribed angle intercepting a semicircle is a right angle.

If two inscribed angles intercept the same arc, the two angles are equal.

EXAMPLE 3 If ∠ABC = 1.02 rad, determine $\widehat{AC}$.

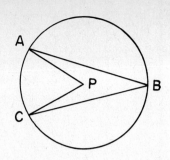

Solution ∠ABC is an inscribed angle so we use the relationship:

$$\text{Inscribed angle} = \frac{1}{2} \text{ intercepted arc}$$

$$1.02 = \frac{1}{2} \widehat{AC}$$

$$2(1.02) = \widehat{AC}$$

$$2.04 \text{ rad} = \widehat{AC}.$$

$\widehat{AC}$ is 2.04 rad.

EXAMPLE 4 Determine the measure of ∠ACD if ∠ABD = π/7 rad.

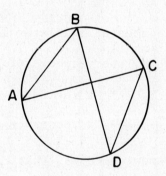

Solution ∠ACD and ∠ABD intercept the same arc, so they are equal.

Hence, ∠ACD = π/7 rad.

EXAMPLE 5 If ∠A = 4x + 4°
∠B = 3x + 1°
∠C = 8x − 5°,

determine $\widehat{AC}$.

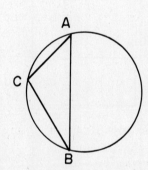

Solution Because the sum of the angles in a triangle is 180°, we have:

$$\angle A + \angle B + \angle C = 180°$$

$$4x + 4° + 3x + 1° + 8x - 5° = 180°$$

$$15x = 180°$$

$$x = 12°$$

$$\angle A = 4x + 4° \qquad \angle B = 3x + 1° \qquad \angle C = 8x - 5°$$

$$\angle A = 52° \qquad \angle B = 37° \qquad \angle C = 91°$$

$\stackrel{\frown}{AC}$ subtends $\angle B$ and $\angle B = 37°$, so we have:

$$\text{Inscribed } \angle = \frac{1}{2} \text{ intercepted arc}$$

$$37° = \frac{1}{2} \stackrel{\frown}{AC}$$

$$74° = \stackrel{\frown}{AC}$$

We have $\stackrel{\frown}{AC} = 74°$.

EXAMPLE 6   Given:
$$\stackrel{\frown}{AD} = 2x + 0.0349$$
$$\stackrel{\frown}{DC} = 2x + 0.3142$$
$$\stackrel{\frown}{CB} = x + 0.3491$$
$$\stackrel{\frown}{BA} = 2x - 0.0349,$$

determine $\angle B$.

Solution   We know that the sum of the arcs is 6.2832 rad, so we have:

$$\stackrel{\frown}{AD} + \stackrel{\frown}{DC} + \stackrel{\frown}{CB} + \stackrel{\frown}{BA} = 6.2832$$

$$2x + 0.0349 + 2x + 0.3142 + x + 0.3491 + 2x - 0.0349 = 6.2832$$

$$7x + 0.6633 = 6.2832$$

$$x = 0.8028$$

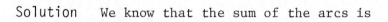

$\stackrel{\frown}{AD} = 1.6406 \qquad \stackrel{\frown}{DC} = 1.9199 \qquad \stackrel{\frown}{CB} = 1.1519 \qquad \stackrel{\frown}{BA} = 1.5708$

As a check, we see if $\stackrel{\frown}{AC} + \stackrel{\frown}{DC} + \stackrel{\frown}{CB} + \stackrel{\frown}{BA} = 6.2832$. It does, so our next step is to determine the size of $\angle B$. We know $\angle B$ intercepts $\stackrel{\frown}{ADC}$, so we have:

$$\angle B = \frac{1}{2}\widehat{ADC}$$

$$= \frac{1}{2}(\widehat{AD} + \widehat{DC})$$

$$= \frac{1}{2}(1.6406 + 1.9199)$$

$$\angle B = 1.7802$$

Thus $\angle B$ contains 1.7802 rad.

Earlier we defined the CIRCUMFERENCE as the distance around a circle, or its perimeter. For every circle, the ratio of the circumference to the diameter is the constant π. We can use this ratio to calculate the circumference.

Circumference of a circle is given by:

$$C = 2\pi r = \pi d,$$

where r is the radius and d is the diameter of the circle.

Since circumference is a measure of length, it is measured in linear units such as millimeters, centimeters, meters, inches, feet, and yards.

EXAMPLE 7   Determine the circumference of a circle with a 3 in. radius.

Solution

$$C = 2\pi r$$

$$= 2\pi \cdot 3 \text{ in.}$$

$$= 6\pi \text{ in.} = 18.85 \text{ in.}$$

The circumference is 18.85 in. or 19 in. (to the nearest inch).

We have worked with arc degrees and arc radians. If $\theta = 90°$, then $\widehat{CD} = 90°$ and $\widehat{AB} = 90°$ in Figure 6.3. However, we realize that $\widehat{AB}$ is longer than $\widehat{CD}$, even though they each represent one quarter of the circle. We can calculate the arc length of an arc if we know the circumference of the circle and the arc degrees or radians of the arc.

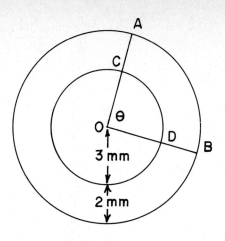

FIGURE 6.3

$$\frac{\text{Arc length}}{\text{Circumference}} = \frac{\text{Arc degrees}}{360°} = \frac{\text{Central angle}}{360°}$$

$$\frac{\text{Arc length}}{\text{Circumference}} = \frac{\text{Arc radians}}{2\pi \text{ rad}} = \frac{\text{Central angle}}{2\pi \text{ rad}}$$

Let us examine the relationship between arc length and the central angle measured in radians for a 1 cm radius circle. We have a circumference of 2π cm, which gives:

$$\frac{\text{Arc length}}{\text{Circumference}} = \frac{\text{Arc radians}}{2\pi \text{ rad}} = \frac{\text{Central angle}}{2\pi \text{ rad}}$$

$$\frac{\text{Arc length}}{2\pi \text{ cm}} = \frac{\text{Arc radians}}{2\pi \text{ rad}} = \frac{\text{Central angle}}{2\pi \text{ rad}}$$

Multiplying each expression by 2π gives:

Arc length = Arc radians = Central angle

Thus, if we use radians instead of degrees to measure angles in our unit circle, we have a simple relationship among the three measures.

For a circle with radius r, we have:

$$\frac{\text{Arc length}}{2\pi r} = \frac{\text{Arc radians}}{2\pi \text{ rad}} = \frac{\text{Central angle}}{2\pi \text{ rad}}$$

Multiplying each expression by 2π, we obtain:

$$\frac{\text{Arc length}}{r} = \text{Arc radians} = \text{Central angle}$$

If the arc length equals the radius, we have:

$$\frac{r}{r} = \text{Arc radians} = \text{Central angle}$$

or

$$1 \text{ rad} = \text{Central angle of arc of length } r.$$

Thus, in any circle an arc with length r subtends a central angle of 1 rad. This is the definition of 1 rad.

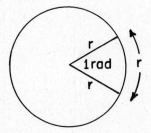

One radian is defined to be the measure of the angle subtended by an arc equal in length to the radius of the circle in which the angle is a central angle.

EXAMPLE 8 Determine the arc length of $\widehat{AB}$ and $\widehat{CD}$ in Figure 6.3.

Solution $\widehat{AB}$ is a 90° arc in a circle with a 5 mm radius.

$$\frac{\text{Length } \widehat{AB}}{\text{Circumference}} = \frac{\text{Arc degrees}}{360°}$$

$$\frac{\text{Length } \widehat{AB}}{2\pi \cdot 5 \text{ mm}} = \frac{\cancel{90°}^{1}}{\cancel{360°}_{4}}$$

$$\text{Length } \widehat{AB} = \frac{10\pi}{4} \text{ mm}$$

$$\text{Length } \widehat{AB} = \frac{5\pi}{2} \text{ mm}$$

$\widehat{CD}$ is a 90° arc in a circle with a 2 mm radius.

$$\frac{\text{Length } \widehat{CD}}{\text{Circumference}} = \frac{\text{Arc degrees}}{360°}$$

$$\frac{\text{Length } \widehat{CD}}{2\pi \cdot 3 \text{ mm}} = \frac{\overset{1}{\cancel{90°}}}{\underset{4}{\cancel{360°}}}$$

$$\text{Length } \widehat{CD} = \frac{6\pi}{4} \text{ mm}$$

$$\text{Length } \widehat{CD} = \frac{3\pi}{2} \text{ mm}$$

Thus the length of $\widehat{AB} = \frac{5\pi}{2}$ mm and the length of $\widehat{CD} = \frac{3\pi}{2}$ mm even though they both are 90° arcs.

EXAMPLE 9   If $\angle CAB = \frac{5\pi}{12}$ rad and the diameter of circle O is 14 in., find the length of $\widehat{BC}$.

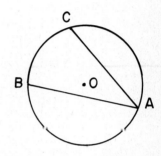

Solution   Because $\angle CAB$ is an inscribed angle, intercepting $\widehat{BC}$,

$$\angle CAB = \frac{1}{2} \widehat{CB}$$

$$\frac{5\pi}{12} \text{ rad} = \frac{1}{2} \widehat{CB}$$

$$\frac{5\pi}{6} \text{ rad} = \widehat{CB}$$

$$\frac{\text{Length } \widehat{CB}}{\text{Circumference}} = \frac{\text{Arc radians}}{2\pi \text{ rad}}$$

$$\frac{\text{Length } \widehat{CB}}{\pi d} = \frac{\frac{5\pi}{6}}{2\pi}$$

$$\frac{\text{Length } \widehat{CB}}{14\pi \text{ in.}} = \frac{5\cancel{\pi}}{6} \cdot \frac{1}{2\cancel{\pi}}$$

$$\text{Length } \widehat{CB} = \frac{5}{\underset{6}{\cancel{12}}} \cdot \frac{\overset{7}{\cancel{14}}\pi}{1} \text{ in.}$$

$$\text{Length } \widehat{CB} = \frac{35\pi}{6} \text{ in.} = 5.8\overline{3} \text{ in.}$$

The length of $\widehat{CB}$ is $\frac{35\pi}{6}$ in. or 5.83 in. (to the nearest hundredth).

EXAMPLE 10   The arc of a 16 ft pendulum is 4 ft. Find the number of radians in the arc.

Solution   First we represent the problem in a drawing. We need the circumference of the circle with radius of 16 ft.

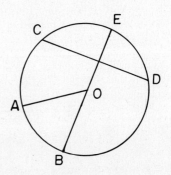

$$C = 2\pi r$$

$$C = 32\pi \text{ ft}$$

The number of radians is given by:

$$\frac{\text{Length of arc}}{\text{Circumference}} = \frac{x}{2\pi \text{ rad}}$$

$$\frac{\overset{1}{\cancel{4}}}{\underset{8}{\cancel{32\pi}}} = \frac{x}{2\pi}$$

$$8\pi x = 2\pi$$

$$x = \frac{1}{4} \text{ rad} = 0.25 \text{ rad}$$

Thus the arc contains 0.25 rad.

EXERCISE 6.1

1. For the figure to the right, give

   a) the radii

   b) a diameter

   c) two minor arcs

   d) two major arcs

   e) a chord

   f) a central angle

   g) a semicircle

2. For the figure to the right, give

   a) a secant

   b) a tangent

   c) the chord or chords

   d) a point of tangency

   e) two major arcs

   f) two minor arcs

   g) an inscribed angle

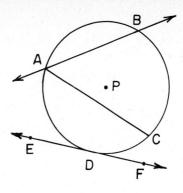

3. For each angle, give the arc that is twice the measure of the angle.

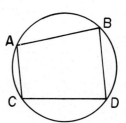

4. For each angle, give the arc that is twice the measure of the angle.

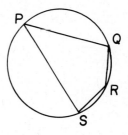

5. If $\widehat{AB} = 29°$, $\widehat{BC} = 73°$, and $\widehat{CD} = 103°$, find $\widehat{CA}$.

6. If $\widehat{AB} = 2.92$ rad, $\widehat{CB} = 1.6$ rad, and $\widehat{AD} = 1.07$ rad, find $\widehat{CD}$.

7. If $\widehat{ABC} = \dfrac{7\pi}{5}$ rad, find $\widehat{AC}$.

8. If $\widehat{BAC} = 203°$, find $\widehat{BC}$.

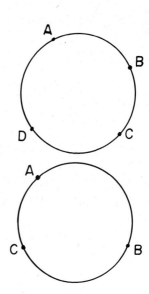

9. If $\overparen{EFG} = 178°$ and $\overparen{EHF} = 235°$, find $\overparen{GF}$.

10. If $\overparen{HEF} = 1.96$ rad and $\overparen{FHE} = 5.21$ rad, find $\overparen{HE}$.

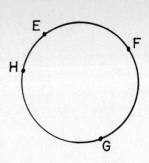

11. If $\overparen{RTS} = 240°$, find $\angle ORS$.

12. If $\angle OSR = 29°$, find $\overparen{RTS}$.

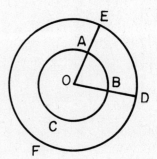

13. If $\overparen{ACB} = \dfrac{19\pi}{12}$ rad, find $\overparen{ED}$.

14. If $\overparen{AB} = \dfrac{\pi}{3}$ rad, find $\overparen{EFD}$.

15. If $\overparen{AC} = 102°$, find $\angle ABC$.

16. If $\angle ABC = 41°$, find $\overparen{AC}$.

17. If $\angle ABC = 72°$, find $\overparen{ABC}$.

18. If $\overparen{ABC} = 288°$, find $\angle ABC$.

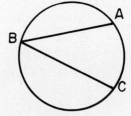

19. If $\overparen{BC} = 122°$, find $\overparen{CA}$.

20. If $\angle CBA = 24°$, find $\overparen{BC}$.

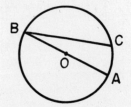

21. If ∠AOB = 100°, find ∠ACB.

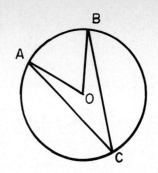

22. If ∠ACB = 44°, find ∠AOB.

23. If ∠ABD = 93°, find ∠ACD.

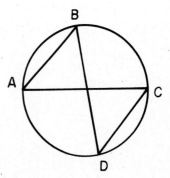

24. If ∠ACD = 78°, find ∠ABD.

25. If ∠C = 3x − 5°, ∠B = x + 2°, and ∠A = x + 8°, find $\widehat{AC}$.

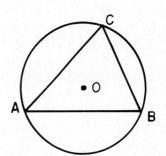

26. If ∠A = x + 6°, ∠B = x + 10°, and ∠C = 3x − 11°, find $\widehat{CB}$.

27. If $\widehat{NK}$ = 1.16 rad, $\widehat{KL}$ = 2.14 rad, and $\widehat{LM}$ = 1.74 rad, find ∠KLM.

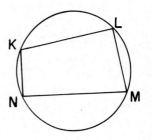

28. If $\widehat{MN}$ = 2.08 rad, $\widehat{LM}$ = 0.94 rad, and $\widehat{KN}$ = 1.86 rad, find ∠KNM.

29. If $\stackrel{\frown}{PQ} = 3x - 12°$, $\stackrel{\frown}{QR} = 2x + 2°$, $\stackrel{\frown}{SR} = 3x + 2°$, and $\stackrel{\frown}{PS} = x + 8°$, find $\angle PQR$.

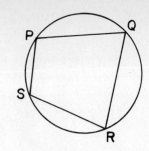

30. If $\stackrel{\frown}{PQ} = 2x + 4°$, $\stackrel{\frown}{QR} = 3x + 5°$, $\stackrel{\frown}{RS} = 3x - 9°$, and $\stackrel{\frown}{PS} = 2x + 10°$, find $\angle SRQ$.

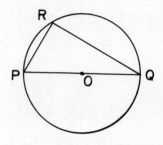

31. Given that $\overline{PQ}$ is a line segment, $RQ = 24$ cm, and $RP = 10$ cm, find the radius of the circle.

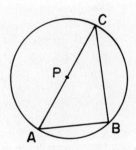

32. Given that $\overline{AC}$ is a line segment, $BC = 8$ in., and the radius is 5 in., determine the length of $\overline{AB}$.

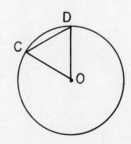

33. Given that $CD = 14$ in. and $\angle COD = \pi/3$ rad, find the radius of the circle.

34. Given that FG = 22 cm and ∠FPG = 60°, find the radius of the circle.

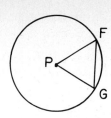

Problem 35 through 38 refer to the drawing to the right. ($\overline{OQ}$ is a segment containing R.)

35. Given that $\overline{PQ}$ is tangent to circle O at point P, PQ = 12 in., and QO = 13 in., find the radius of the circle.

36. Given that $\overline{PQ}$ is tangent to circle O, PQ = 16 cm, and OQ = 20 cm, find the radius of the circle.

37. Given that $\overline{PQ}$ is tangent to circle O, PQ = 24 cm, and RQ = 18 cm, find the radius of the circle.

38. Given that $\overline{PQ}$ is tangent to circle O, PQ = 15 cm, and RQ = 9 cm, find the radius of the circle.

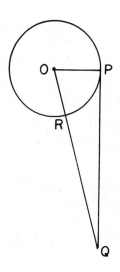

39. Find the circumference of a circle with a 7 in. radius. (Give exact answer.)

40. Find the circumference of a circle with a 16 in. diameter. (Give exact answer.)

41. Find the diameter of a circle if the circumference is 11π mm.

42. Find the radius of a circle if the circumference is 13π cm.

43. Find the length of $\overset{\frown}{AC}$ if the diameter is 12 mm and ∠AOC = 72°.

44. Find the length of $\overset{\frown}{AC}$ if the radius is 4 ft and ∠AOC = 0.34 rad.

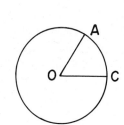

For Problems 45 to 50, give exact answers.

45. Find the length of $\overparen{LM}$ if the radius is 3 cm.

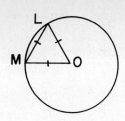

46. Find the length of $\overparen{ACB}$ if the diameter is 7 cm.

47. Find the circumference of the circle if AC = 7 cm and BC = 11 cm.

48. Find the circumference of the circle if AC = 5 cm and BC = 9 cm.

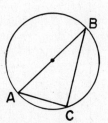

49. Find the length of $\overparen{AC}$ if the radius is 7 cm.

50. Find the length $\overparen{BAC}$ if the radius is 9 cm.

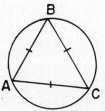

51. A 12 ft pendulum swings through an arc of 28°. Find the length of the arc.

52. A 9 ft pendulum swings through an arc of 0.6 rad. Find the length of the arc.

53. A pendulum swings through an arc of 0.8 rad and the length of the arc is 2.4 in. Find the length of the pendulum.

54. A pendulum swings through an arc of 18° and the length of the arc is 6.3 in. Find the length of the pendulum.

C 55. If a bicycle has 26" diameter wheels, how fast is the bicycle traveling (in miles per hour) when the wheel is making 12 revolutions per minute? (1 mi = 5280 ft)

C 56. If a bicycle has 24" diameter wheels, how fast is the bicycle traveling (in miles per hour) when the wheel is making 12 revolutions per minute? (1 mi = 5280 ft)

C 57. How many revolutions per minute must a 20" diameter wheel on a motorcycle make in order for the motorcycle to go 50 miles per hour? (1 mi = 5280 ft)

C 58. How many revolutions per minute must a 26" diameter wheel on a bicycle make in order for the bike to go 30 miles per hour? (1 mi = 5280 ft)

6.2 CHORDS AND TANGENTS

Recall that a chord is a line segment with both endpoints on the circle. A chord that passes through the center of the circle is a diameter. Because a chord and radii to its endpoints form an isosceles triangle, we also know:

> A radius (or diameter) perpendicular to a chord bisects the chord.
>
> A radius (or diameter) that bisects a chord is perpendicular to the chord.
>
> A perpendicular bisector of a chord passes through the center of the circle.

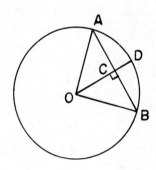

EXAMPLE 1 Given a circle O with a radius of 4 cm where CD = 1 cm, find the length of AB.

Solution We have a right triangle, $\triangle OBC$, where

OB = 4 cm, so

OC = 4 cm − CD

= 4 cm − 1 cm = 3 cm

We can now use the Pythagorean theorem to find the length of BC.

$$a^2 + b^2 = c^2$$

$$x^2 + 3^2 = 4^2$$

$$x^2 + 9 = 16$$

$$x^2 = 7$$

$$x = \pm\sqrt{7}$$

$$x = \sqrt{7} \text{ cm or } 2.6458 \text{ cm}$$

BC = 2.6458 cm

Because C is the midpoint of AB, the length of AB is given by

$$AB = 2 \cdot BC$$
$$= 2(2.6458)$$
$$= 5.2916 \text{ cm}$$

AB = 5.3 cm (to the nearest tenth).

Two chords the same distance from the center of a given circle are congruent.

Two chords that are congruent in a given circle are the same distance from the center of the circle.

EXAMPLE 2 If FO = OC and AC = 6 cm, find the length of DE.

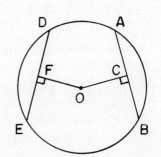

Solution Because FO = OC, we know that

$$AB = DE$$
$$AB = 2 \cdot AC$$
$$= 2 \cdot 6$$
$$AB = 12 \text{ cm}$$
$$DE = AB = 12 \text{ cm}$$

Thus DE = 12 cm.

A polygon for which each side is a chord in a given circle is an INSCRIBED POLYGON for that circle. That is, the polygon is said to be INSCRIBED in the circle. In figure A below, △ABC is inscribed in circle O. In B, a

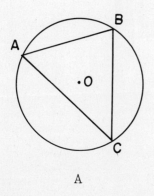

A

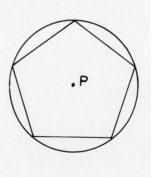

B

pentagon is inscribed in circle P. Because an inscribed angle is one half the intercepted arc, we know all inscribed right triangles must have a diameter for their hypotenuse.

EXAMPLE 3   A $\frac{\pi}{4} - \frac{\pi}{4} - \frac{\pi}{2}$ rad triangle is inscribed in a circle. If one leg of the triangle is 6 in., find the diameter of the circle.

Solution   First we must use the given information to make a drawing. We know that BC = AB = 6 in. We can use the reference $\frac{\pi}{4} - \frac{\pi}{4} - \frac{\pi}{2}$ rad triangle to set up a proportion to determine the length of AC, the hypotenuse.

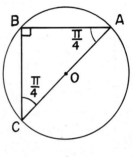

$$\frac{AC}{\sqrt{2}} = \frac{6 \text{ in.}}{1}$$

AC = $6\sqrt{2}$ in. or 8.4853 in.

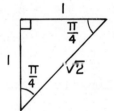

The diameter of the circle, AC, is $6\sqrt{2}$ in. or 8.5 in. (to the nearest tenth).

EXAMPLE 4   Determine the length of AB if the diameter of the circle is 10 cm.

Solution   First we draw in the radius to each vertex of the triangle. Because the sides are all equal, we know we have an equilateral triangle. We know that OB = OC = OA = 5 cm because they are radii. We know that the three triangles △BOC, △BOA, and △COA are congruent by SSS. Thus the

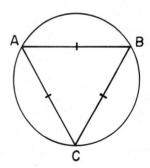

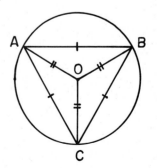

base angles for each of the smaller triangles is one half the angle in △ABC, which means that the base angles of the smaller triangles are each 30°. Likewise the central angles ∠BOC, ∠COA, and ∠AOB are equal and each must contain 120°.

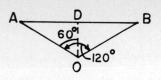

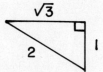

We can construct the altitude for △OBA, which creates two smaller congruent triangles, △OBD and △OAD. The smaller triangles, each are 30°-60°-90° triangles, so we may use our reference 30°-60°-90° triangle to set up the proportion to determine the length of AD:

$$\frac{AD}{\sqrt{3}} = \frac{5 \text{ cm}}{2}$$

$$AD = \frac{5\sqrt{3}}{2} \text{ cm}$$

$$AB = 2 \cdot AD$$

$$= 2(\frac{5\sqrt{3}}{2} \text{ cm})$$

$$AB = 5\sqrt{3} \text{ cm or } 8.6603 \text{ cm}$$

AB is $5\sqrt{3}$ cm or 8.7 cm (to the nearest tenth).

In any given circle, equal chords have equal arcs and equal central angles.

In any given circle, equal arcs and equal central angles have equal chords.

EXAMPLE 5 Determine the diameter of the circle if $AD = 8$ in.

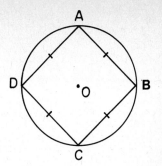

Solution We know that

$$AB = BC = CD = DA$$

so $\quad \widehat{AB} = \widehat{BC} = \widehat{CD} = \widehat{DA}$

We also know that

$$\widehat{AB} + \widehat{BC} + \widehat{CD} + \widehat{DA} = 2\pi \text{ rad}$$

By substitution

$$4 \cdot \widehat{AB} = 2\pi \text{ rad}$$

$$\widehat{AB} = \frac{\pi}{2} \text{ rad}$$

so $\quad \widehat{AB} = \widehat{BC} = \widehat{CD} = \widehat{DA} = \frac{\pi}{2} \text{ rad}$

Therefore,

$$\angle DAB = \frac{1}{2} \widehat{DCB}$$

$$= \frac{1}{2} \cdot \pi \text{ rad}$$

$$\angle DAB = \frac{\pi}{2} \text{ rad}$$

We now know that $\overline{DOB}$ is a diameter. (Because the sides are all congruent, we know we have either a rhombus or a square. Because $\angle DAB = \frac{\pi}{2}$ rad, we now know quadrilateral ABCD is a square.)

We can use our reference $\frac{\pi}{4}, \frac{\pi}{4}, \frac{\pi}{2}$ rad triangle to set up the proportion to determine the length of DOB:

$$\frac{8 \text{ in.}}{1} = \frac{DOB}{\sqrt{2}}$$

$DOB = 8\sqrt{2}$ in. or 11.3137 in.

The diameter of the circle, DOB, is $8\sqrt{2}$ in. or 11.3 in. (to the nearest tenth).

EXAMPLE 6   Determine the number of degrees in $\stackrel{\frown}{DC}$.

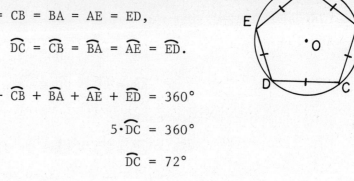

Solution   Because $DC = CB = BA = AE = ED$,

we know that $\stackrel{\frown}{DC} = \stackrel{\frown}{CB} = \stackrel{\frown}{BA} = \stackrel{\frown}{AE} = \stackrel{\frown}{ED}$.

$$\stackrel{\frown}{DC} + \stackrel{\frown}{CB} + \stackrel{\frown}{BA} + \stackrel{\frown}{AE} + \stackrel{\frown}{ED} = 360°$$

$$5 \cdot \stackrel{\frown}{DC} = 360°$$

$$\stackrel{\frown}{DC} = 72°$$

Thus $\stackrel{\frown}{DC} = 72°$.

Recall that a tangent is a line that intersects a circle at one and only one point. The point of intersection is the point of tangency. In

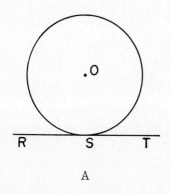

A

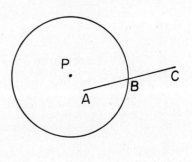

B

figure A above, segment RT is tangent to circle O at S because $\overleftrightarrow{RT}$ intersects O at only one point. In figure B, however, AC is not a tangent to circle P at B because $\overleftrightarrow{AC}$ intersects the circle in a second point.

A radius (or diameter) of a circle is perpendicular to a tangent at the point of tangency.

A line (or line segment) perpendicular to a radius (or diameter) at an outer endpoint is a tangent to the circle at that endpoint.

EXAMPLE 7 $\angle AOC = \frac{\pi}{3}$ rad, $\overline{AB}$ is tangent at A, and the radius of the circle is 5 in. Find the length of $\overline{BC}$.

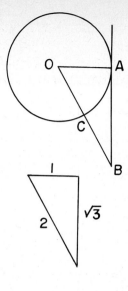

Solution   Because $\overline{AB}$ is tangent at A, we know that $\angle OAB = \frac{\pi}{2}$ rad. Thus we have a $\frac{\pi}{3} - \frac{\pi}{6} - \frac{\pi}{2}$ rad triangle with the shorter leg as a radius of the circle. We can use the ratio $1:\sqrt{3}:2$ to set up a proportion:

$$\frac{OA}{1} = \frac{OB}{2}$$

$$\frac{5 \text{ in.}}{1} = \frac{OB}{2}$$

$$10 \text{ in.} = OB$$

Now we have:

$$CB = OB - OC$$
$$= 10 \text{ in.} - 5 \text{ in.}$$
$$CB = 5 \text{ in.}$$

Thus $CB = 5$ in.

Two tangents to a circle from the same point outside the circle are equal in length (congruent).

EXAMPLE 8   Given tangents $\overline{AC}$ and $\overline{BC}$, $\overparen{ADB} = 240°$, and the radius of 5 in., find the perimeter of APBC.

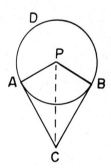

Solution   If $\overparen{ADB} = 240°$, then $\overparen{AB} = 360° - \overparen{ADB} = 120°$. We know that $\angle PAC = 90° = \angle PBC$. We also know that $\triangle PAC = \triangle PBC$ by SAS (or SSS), so $\angle APC \cong \angle BPC$. But $\angle APC + \angle BPC = \angle APB = \overparen{AB} = 120°$.

289

Because the two angles, ∠APC and ∠BPC, are congruent, we know that each must be 60° or

∠APC = 60°

∠BPC = 60°

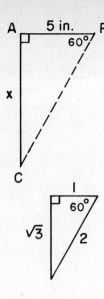

We therefore have 30°-60°-90° triangles, △APC and △BPC. We can then set up the proportion:

$$\frac{AC}{\sqrt{3}} = \frac{5 \text{ in.}}{1}$$

$$AC = 5\sqrt{3} \text{ in.}$$

Because two tangents from the same point are congruent, we know

$$BC = AC = 5\sqrt{3} \text{ in.}$$

The perimeter, P, of the figure is

P = AC + AP + PB + BC

P = 5√3 in. + 5 in. + 5 in. + 5√3 in.

= 10 + 10√3 in. or 27.32 in.

The perimeter of APBC is 10 + 10√3 in. or 27.32 in. (to the nearest hundredth).

A polygon in which each side is a tangent to a given circle is a CIRCUMSCRIBED POLYGON for that circle. That is, the polygon is said to be CIRCUMSCRIBED about the circle. In figure A, below, △ABC is circumscribed about circle O. In figure B, an octagon is circumscribed about circle P.

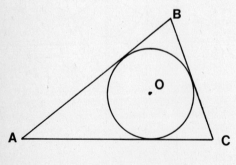

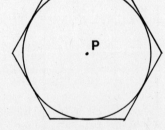

A

B

290

We can also say circle O is inscribed in △ABC and circle P is inscribed in the octagon. That is, we can say either that the inner figure is inscribed in the outer figure, or that the outer figure is circumscribed about the inner figure.

EXAMPLE 9 A quadrilateral is circumscribed about circle P. If AB = 15 m and DC = 21 m, determine the perimeter of the quadrilateral.

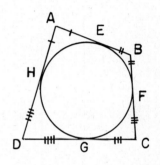

Solution We know that

$$AE = AH$$

and $$EB = BF$$

so $$AE + EB = 15 \text{ m}$$

and $$AH + BF = 15 \text{ m}$$

$$DG = HD$$

and $$GC = CF$$

so $$DG + GC = 21 \text{ m}$$

and $$HD + CF = 21 \text{ m}$$

The perimeter, P, is:

P = (AE + EB) + (AH + BF) + (DG + GC) + (HD + CF)

  = 15 m + 15 m + 21 m + 21 m

  = 72 m

The perimeter of the quadrilateral is 72 m.

EXAMPLE 10 △ABC is circumscribed about a circle with P, Q, and R as the points of tangency. If AB = 14 in., BC = 12 in., and AC = 18 in., determine the lengths of CP, BQ, and AP.

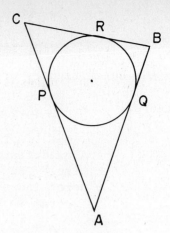

Solution We know AP + PC = 18 in., CR + RB = 12 in., and AQ + QB = 14 in. If we let x = CP, then we have

AP = 18 - x

RB = 12 - x

AQ = AP

and QB = RB

so AB = AQ + QB

AB = AP + RB

14 = (18 - x) + (12 - x)

2x = 16

x = 8 in.

Thus, CP = 8 in.

BQ = RB = 12 - x = 4 in.

AP = 18 - x = 10 in.

Therefore CP = 8 in., BQ = 4 in., and AP = 10 in.

EXAMPLE 11 A regular hexagon is circumscribed about a 7 in. radius circle. Find the area of the hexagon.

Solution   First we need a drawing. The radius is the apothem of the hexagon. We must calculate the length of one side. Using the reference 30°-60°-90° triangle, we set up the proportion:

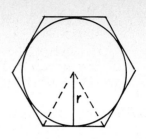
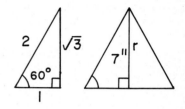

$$\frac{x}{1} = \frac{7 \text{ in.}}{\sqrt{3}}$$

$$x = \frac{7\sqrt{3}}{3} \text{ in.}$$

One side of the hexagon is $2 \cdot \left(\frac{7\sqrt{3}}{3}\right)$ in. or $\frac{14\sqrt{3}}{3}$ in.

The area of the hexagon is given by

$$A = \frac{1}{2} p \cdot a \text{ where } p \text{ is the perimeter and } a \text{ is the apothem.}$$

$$= \frac{1}{2} \cdot 6 \left(\frac{14\sqrt{3} \text{ in.}}{3}\right)(7 \text{ in.})$$

$$= 98\sqrt{3} \text{ in}^2 \text{ or } 169.7410 \text{ in}^2$$

The area of the hexagon is $98\sqrt{3}$ in^2 or 169.7 in^2 (to the nearest tenth).

EXAMPLE 12   A $\frac{\pi}{6} - \frac{\pi}{3} - \frac{\pi}{2}$ rad triangle is circumscribed about a circle. If the shorter leg is 18 cm, find the radius of the circle.

Solution First label a drawing. Because we are given a $\frac{\pi}{6} - \frac{\pi}{3} - \frac{\pi}{2}$ rad triangle, we can determine the length of the other sides as entered on the drawing. We know that $\triangle UPS \cong \triangle VPS$ by SSS. Therefore $\angle PSV \cong \angle PSU = \frac{\pi}{6}$ rad because $\angle VSU = \frac{\pi}{3}$ rad. We thus know that $\triangle UPS$ is a $\frac{\pi}{6} - \frac{\pi}{3} - \frac{\pi}{2}$ rad triangle and can determine the length of PU, a radius, if we can find the length of US. We can follow the procedure of Example 10:

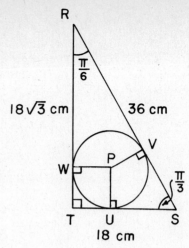

$$x = US$$
$$TU = 18 \text{ cm} - x$$
$$RV = 36 \text{ cm} - x$$
$$RT = WT + RW$$
$$RT = TU + RV$$
$$18\sqrt{3} = (18 - x) + (36 - x)$$
$$2x = 54 - 18\sqrt{3}$$
$$x = 27 - 9\sqrt{3}$$
$$US = 27 - 9\sqrt{3} \text{ cm}$$

We can now use our reference triangle to set up the proportion to determine the radius of the circle, UP.

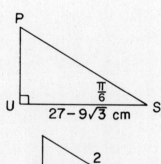

$$\frac{UP}{1} = \frac{27 - 9\sqrt{3}}{\sqrt{3}}$$

$$UP = \frac{(27 - 9\sqrt{3}) \cdot \sqrt{3}}{3}$$

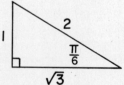

$$UP = 9\sqrt{3} - 9 \text{ cm or } 6.5885 \text{ cm}$$

The radius of the circle is $9\sqrt{3} - 9$ cm or 6.6 cm (to nearest tenth).

EXERCISE 6.2

Give exact answers unless a C is by the problem.

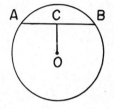

C 1. If $AB \perp CO$, $AB = 13$ in., and $OC = 2.3$ in., find the radius of the circle.

C 2. If $AB \perp CO$, $OC = 3.1$ in., and the radius of the circle is 5.2 in., find the length of AB.

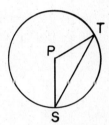

3. If $ST = 16$ in. and the radius of the circle is 10 in., find the area of $\triangle PST$.

4. If $ST = 24$ in. and the radius of the circle is 13 in., find the area of $\triangle PST$.

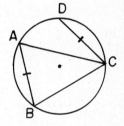

5. If $\angle ACB = 42°$, find $\overarc{CD}$.

6. If $\overarc{CD} = 86°$, find $\angle ACB$.

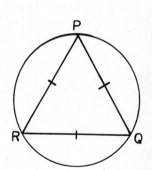

7. Determine the length of $\overline{PQ}$ if the diameter of the circle is 14 in.

8. Determine the length of $\overline{QR}$ if the diameter of the circle is 12 in.

9. Determine the length of $\overarc{PQ}$ if the diameter is 14 in.

10. Determine the length of $\overarc{QR}$ if the diameter is 12 in.

11. If TU is 22 cm, determine the radius of the circle. (Drawing on next page.)

12. If VW is 18 cm, determine the radius of the circle.

13. If the area of TUVW is 144 in^2, determine the diameter of the circle.

14. If the area of TUVW is 256 in^2, determine the diameter of the circle.

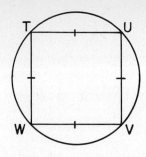

15. A regular octagon is inscribed in a circle. Determine the number of radians in the arc for one side (one chord).

16. A regular hexagon is inscribed in a circle. Determine the number of radians in the arc for one side (one chord).

17. Determine the area of a regular hexagon inscribed in a circle with a 6 in. radius.

18. Determine the area of a regular hexagon inscribed in a circle with an 8 in. radius.

C 19. If PQ = 19 cm and OQ = 23 cm, find the radius of the circle.

C 20. If the radius of the circle is 9 cm and OQ = 14 cm, find PQ.

21. If PQ = 15 in. and RQ = 9 in., find the radius of the circle.

22. If PQ = 20 in. and RQ = 10 in., find the radius of the circle.

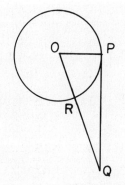

23. Given quadrilateral ABCD circumscribed about circle O, AB = 17 in. and DC = 19 in., find the perimeter of ABCD.

24. Given quadrilateral ABCD circumscribed about circle O, AD = 19 in., and BC = 23 in., find the perimeter of ABCD.

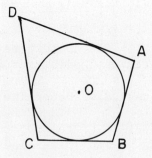

25. Given a square with an area of 289 m^2 circumscribed about a circle, find the radius of the circle.

26. Given a square with an area of 225 yd² circumscribed about a circle, find the radius of the circle.

27. Given △ABC circumscribed about circle P, AB = 24 in., BC = 19 in., CA = 17 in., and D, E, and F the points of tangency, find the length of $\overline{DC}$, $\overline{AE}$, and $\overline{BF}$.

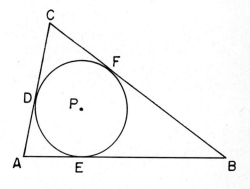

28. Given △ABC circumscribed about circle P, AB = 24 in., BC = 32 in., CA = 17 in., and D, E, and F the points of tangency, find the length of $\overline{CF}$, $\overline{BE}$, and $\overline{AD}$.

29. A regular hexagon is circumscribed about an 11 in. radius circle. Find the area of the hexagon.

30. A regular hexagon is circumscribed about a 9 in. radius circle. Find the area of the hexagon.

31. A $\frac{\pi}{6} - \frac{\pi}{3} - \frac{\pi}{2}$ rad triangle is circumscribed about a circle. If the hypotenuse is 24 cm, find the radius of the circle.

32. A $\frac{\pi}{6} - \frac{\pi}{3} - \frac{\pi}{2}$ rad triangle is circumscribed about a circle. If the longer leg is $11\sqrt{3}$ cm, find the radius of the circle.

33. An isosceles right triangle is circumscribed about a circle. If one leg is 14 cm, find the radius of the circle.

34. An isosceles right triangle is circumscribed about a circle. If one leg is 10 cm, find the radius of the circle.

6.3   AREA, SECTORS, AND COMPOSITE FIGURES

The area of the region enclosed by a circle is commonly referred to as the area of the circle or the area of a circular region. The context of the discussion will indicate whether we are talking about the circle itself or some portion of the region enclosed by the circle.

If we take a circle of radius r, cut it into two semicircles, and

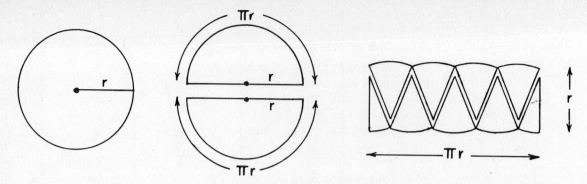

cut into each semicircle along radii so we can "unroll" the semicircle, we can take these two sections and fit them together. As we make more and more cuts into each semicircle, the region formed will resemble more and more a rectangle with length πr and width r. The area would then by $\pi r \cdot r$ or πr^2.

$$\text{Area}_{circle} = \pi r^2, \text{ where } r \text{ is the radius.}$$

The area of a circle, being a measure of surface, is always measured in <u>square</u> units, i.e., m^2, in^2, cm^2, ft^2.

EXAMPLE 1   Find the area of a circle with a 72 in. diameter.

Solution   We must first determine the radius:

$$D = 2r$$
$$72 \text{ in.} = 2r$$
$$36 \text{ in.} = r$$

Now we can calculate the area.

$$A_{circle} = \pi r^2$$
$$= \pi (36 \text{ in.})^2$$
$$= 1296\pi \text{ in}^2 \text{ or } 4071.5041 \text{ in}^2$$

The area of the circle is 1296π in^2 or 4072 in^2 (to the nearest unit).

A SECTOR of a circle is the region formed by two radii and the intercepted arc. A SEGMENT of a circle is the region between a chord and its arc. In figure A below, the shaded portion represents the sector formed by radii $\overline{OA}$

A   B

and $\overline{OB}$ and $\widehat{AB}$. The sector can also be identified by its central angle, $\angle AOB$. In figure B, the shaded portion represents the segment formed by chord $\overline{ST}$ and its arc, $\widehat{ST}$.

The area of a sector is proportional to its central angle.

$$\frac{\text{Area}_{sector}}{\text{Area}_{circle}} = \frac{\text{Central angle in degrees}}{360°}$$

or

$$\frac{\text{Area}_{sector}}{\text{Area}_{circle}} = \frac{\text{Central angle in radians}}{2\pi \text{ rad}}$$

EXAMPLE 2   Find the area of a sector in a 4 cm radius circle if its central angle is $\pi/6$ rad.

Solution

$$\frac{\text{Area}_{sector}}{\text{Area}_{circle}} = \frac{\text{Central angle}}{2\pi \text{ rad}}$$

$$\frac{\text{Area}_{sector}}{\pi(4)^2} = \frac{\frac{\pi}{6}}{2\pi}$$

$$\text{Area}_{sector} = \frac{1}{12} \cdot 16\pi$$

$$= \frac{4\pi}{3} \text{ cm}^2 \text{ or } 4.1888 \text{ cm}^2$$

The area is $\frac{4\pi}{3}$ cm^2 or 4.2 cm^2 (to the nearest tenth).

The area of a segment is the difference between the area of the sector and the area of the triangle formed by the radii and the chord.

$$\text{Area}_{segment} = \text{Area}_{sector} - \text{Area}_{triangle}$$

where the triangle is formed by the two radii and the chord.

EXAMPLE 3   Find the area of segment AB if $\angle AOB = \dfrac{\pi}{3}$ rad and the radius is 20 in.

Solution   First let us find the area of the sector.

$$\dfrac{\text{Area}_{sector}}{\text{Area}_{circle}} = \dfrac{\text{Central angle in radians}}{2\pi \text{ rad}}$$

$$\dfrac{\text{Area}_{sector}}{\pi r^2} = \dfrac{\angle AOB}{2\pi}$$

$$\dfrac{\text{Area}_{sector}}{\pi (20)^2} = \dfrac{\frac{\pi}{3}}{2\pi}$$

$$\text{Area}_{sector} = \left(\dfrac{\pi}{3}\right)\left(\dfrac{1}{2\pi}\right)(400\pi)^{200}$$

$$= \dfrac{200\pi}{3}$$

$$\text{Area}_{sector} = \dfrac{200\pi}{3} \text{ in}^2$$

In order to find the area of the triangle, we need to know both the altitude and the base. We can use a reference $\dfrac{\pi}{6} - \dfrac{\pi}{3} - \dfrac{\pi}{2}$ rad triangle to find these values.

To find the altitude, y:

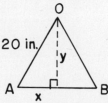

$$\dfrac{y}{\sqrt{3}} = \dfrac{20 \text{ in.}}{2}$$

$$y = 10\sqrt{3} \text{ in.}$$

The altitude is $10\sqrt{3}$ in.

To find x:

$$\frac{x}{1} = \frac{20}{2}$$

$$x = 10 \text{ in.}$$

The base, which is twice x, is 20 in.

$$A_{triangle} = \frac{1}{2} \cdot h \cdot b$$

$$= \frac{1}{2} (10\sqrt{3} \text{ in.})(20 \text{ in.})$$

$$= 100\sqrt{3} \text{ in}^2$$

$$A_{segment} = A_{sector} - A_{triangle}$$

$$A_{segment} = \frac{200\pi}{3} \text{ in}^2 - 100\sqrt{3} \text{ in}^2 \text{ or } 36.2344 \text{ in}^2$$

The area of the segment is $\frac{200\pi}{3} - 100\sqrt{3}$ in² or 36.2 in² (to the nearest tenth).

EXAMPLE 4 Find the area of the shaded portion of the figure at the right if AB = 6 cm and BC = 8 cm.

Solution Because the triangle is inscribed in a semicircle, it is a right triangle. Using the Pythagorean theorem, we find that the hypotenuse is 10 cm. The radius is 5 cm because the hypotenuse is a diameter. To find the desired area, we can take one half the area of the circle and subtract the area of the triangle.

$$A_{circle} = \pi r^2$$

$$= \pi(5)^2$$

$$= 25\pi \text{ cm}^2$$

$$A_{triangle} = \frac{1}{2} h \cdot b$$

$$= \frac{1}{2} (8 \text{ cm})(6 \text{ cm})$$

$$= 24 \text{ cm}^2$$

$$A_{shaded} = \frac{1}{2} A_{circle} - A_{triangle}$$

$$= \frac{1}{2} (25\pi) - 24 \text{ cm}^2$$

$$= \frac{25\pi}{2} - 24 \text{ cm}^2 \text{ or } 15.2699 \text{ cm}^2$$

The shaded area is $\frac{25\pi}{2} - 24 \text{ cm}^2$ or 15.3 cm^2 (to the nearest tenth).

An ANNULUS is the area between two concentric circles. The shaded portion of the figure at the right represents the annulus, or ring. To find the area of an annulus, subtract the area of the smaller circle from that of the larger circle.

EXAMPLE 5 Find the area of an annulus formed by two circles 7 in. and 11 in. in radius.

Solution

$$A_{larger\ circle} = \pi r^2$$

$$= \pi (11)^2$$

$$= 121\pi \text{ in}^2$$

$$A_{smaller\ circle} = \pi r^2$$

$$= \pi (7)^2$$

$$= 49\pi \text{ in}^2$$

$$A_{annulus} = A_{larger\ circle} - A_{smaller\ circle}$$

$$= 121\pi - 49\pi \text{ in}^2$$

$$= 72\pi \text{ in}^2 \text{ or } 226.1947 \text{ in}^2$$

The desired area is $72\pi \text{ in}^2$ or 226.2 in^2 (to the nearest tenth).

We can use our information about determining area to find the area of composite figures.

EXAMPLE 6   Find the area of the shaded portion of the figure at the right if the radius of the larger circle is 4 mm. Assume that all dots represent the centers of arcs or circles.

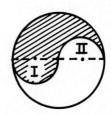

Solution   Because Area I = Area II, the desired area is the area of the larger semicircle.

$$A_{circle} = \pi r^2$$
$$= \pi(4)^2$$
$$= 16\pi \text{ mm}^2$$

$$A_{semicircle} = 8\pi \text{ mm}^2$$

The desired area is 8π mm^2.

EXAMPLE 7   Find the area of the given figure.

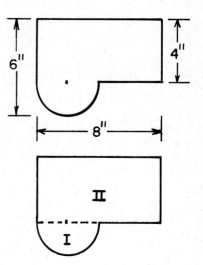

Solution   The figure can be divided into a semicircle, Area I, and a rectangle, Area II.

$$\text{Area}_I = \frac{1}{2} \pi r^2$$
$$= \frac{1}{2} \pi (2)^2$$
$$= 2\pi \text{ in}^2$$

$$\text{Area}_{II} = \ell \cdot w$$
$$= 4 \cdot 8$$
$$= 32 \text{ in}^2$$
$$A_{total} = A_I + A_{II}$$
$$A_{total} = 2\pi + 32 \text{ in}^2 = 38.2832 \text{ in}^2$$

The total area is $2\pi + 32$ in^2 or 38.3 in^2 (to the nearest tenth).

EXERCISE 6.3

Give exact answers unless there is a C by the problem.

1. Determine the area of a 7 cm radius circle.

2. Determine the area of a 3 cm radius circle.

C 3. Find the area of a circle with a diameter of 2.16 m.

C 4. Find the area of a circle with a diameter of 1.93 m.

5. Determine the area of a circle with a circumference of 14π in.

6. Determine the area of a circle with a circumference of 22π in.

C 7. Determine the area of a circle with a circumference of 14.5 mm.

C 8. Determine the area of a circle with a circumference of 13.2 mm.

9. Find the area of a sector in a 7 cm radius circle if its central angle is 18°.

10. Find the area of a sector in an 8 cm radius circle if its central angle is 54°.

C 11. Find the diameter of a circle if a sector with a central angle of 104° has an area of 21.23 in^2.

C 12. Find the diameter of a circle if a sector with a central angle of 200° has an area of 16.73 cm^2.

13. Find the central angle (in radians) of a sector if the area of the sector is 54π cm^2 and the radius of the circle is 36 cm.

14. Find the central angle (in radians) of a sector if the area of the sector is $15\pi/2$ cm^2 and the radius is 3 cm.

15. Find the area of the shaded segment if ∠AOB = 60° and the radius is 30 cm.

16. Find the area of the shaded segment if ∠AOB = 90° and the radius is 20 cm.

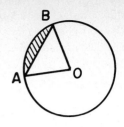

C 17. Find the area of the shaded segment if SR = 36 in. and ST = 77 in.

C 18. Find the area of the shaded segment if SR = 54 in. and ST = 26 in.

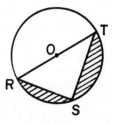

19. Find the area of the shaded segment if OA = 3 cm and AB = 4 cm.

20. Find the area of the shaded segment if OB = 8 cm and AB = 3 cm.

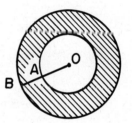

21. Find the area of the shaded segment if OA = 2 in., OB = 5 in., and ∠BOC = 108°.

22. Find the area of the shaded segment if OA = 3 in., OB = 4 in., and ∠BOC = $7\pi/12$ rad.

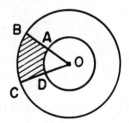

For Problems 23 to 40, find the area of the given figure. You may assume that line segments that appear to be perpendicular or parallel are. If the figure is shaded, find only the area of the shaded portion. Dots indicate the centers of circles or arcs.

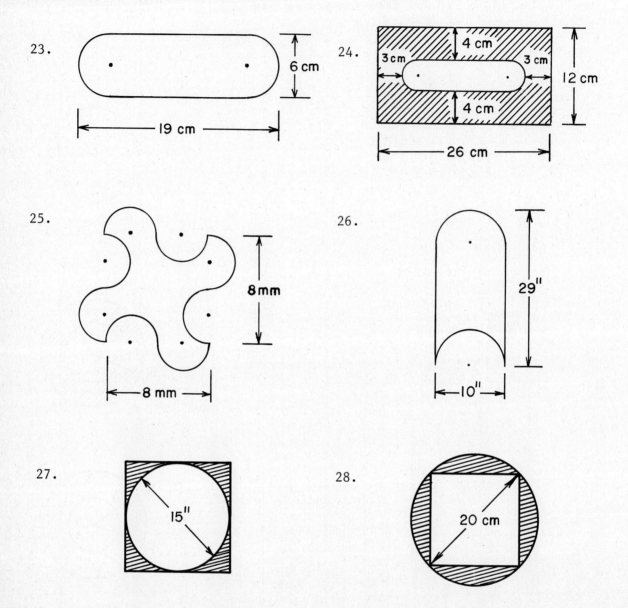

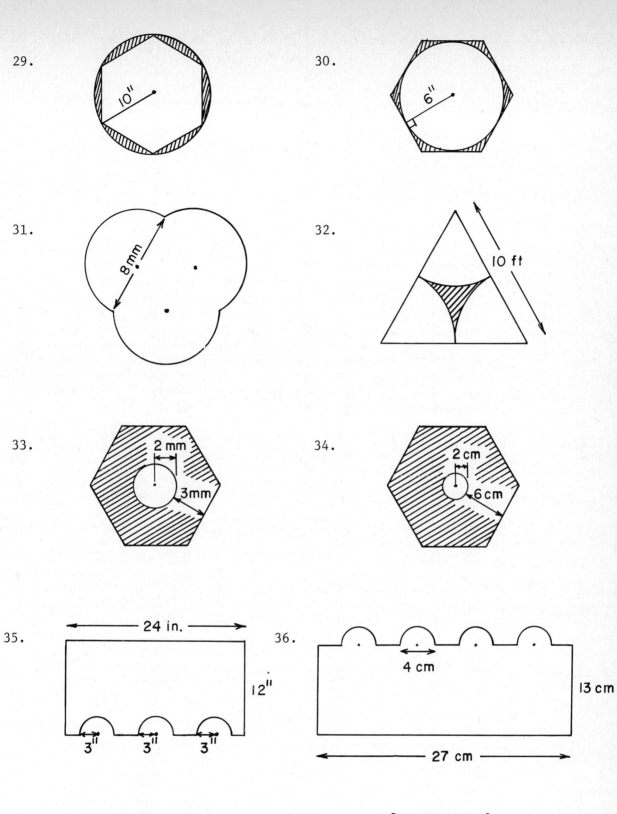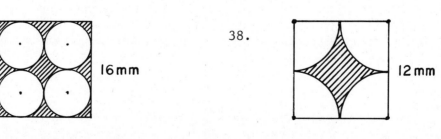

39.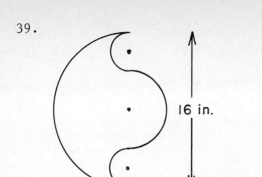

40.

41. Two concentric circles have the property that every chord of the larger circle drawn tangent to the smaller circle has length of 6 cm. Find the area of the annulus between the two circles.

42. Two concentric circles have the property that every chord of the larger circle drawn tangent to the smaller circle has length of 10 cm. Find the area of the annulus between the two circles.

**43. A Norman window consists of a rectangle topped by a semicircle. If the perimeter is P, given the expression for the total length of the window in terms of P and w, the width of the window.

**44. Give the expression for the area for the Norman window as described in Problem 41.

6.4 CONSTRUCTIONS

We can use the compass and straightedge for constructions involving circles. We know that the perpendicular bisector of a chord passes through the center of the circle. Therefore, we can locate the center of the circle by constructing the perpendicular bisectors of two distinct chords.

> To locate the center of a given circle:
>
> 1. Mark any three distinct points, A, B, and C, on the circle and draw the segment $\overline{AB}$ and the segment $\overline{BC}$.
>
> 2. Construct the perpendicular bisectors of $\overline{AB}$ and of $\overline{BC}$.
>
> 3. The intersection of the perpendicular bisectors is the center of the circle.

EXAMPLE 1   Locate the center of the given circle.

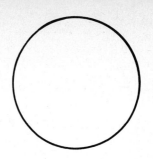

Solution   Step 1:   Mark three distinct points, A, B, and C, on the circle and draw the segments $\overline{AB}$ and $\overline{BC}$.

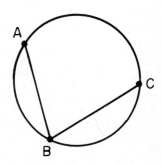

Step 2:   Construct the perpendicular bisectors of $\overline{AB}$ and $\overline{BC}$.

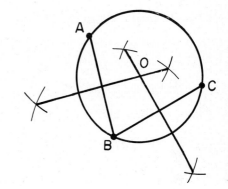

Step 3:   The intersection of the perpendicular bisectors, point O, is the center of the circle.

Because a tangent to a circle is perpendicular to a radius passing through the point of tangency, we can construct a tangent at a given point on a given circle.

> To construct a tangent to a given circle through a given point
> P on that circle:
>
> 1. Draw a secant containing the center of the circle and the
>   given point P. (Either the center is given, or it can be
>   located using the construction above.)
>
> 2. At the point P, construct a perpendicular to the secant.
>
> 3. The perpendicular constructed is the tangent to the
>   circle at the point P.

EXAMPLE 2   Construct a tangent to the given circle
with center O at the point P.

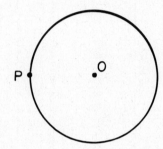

Solution   Step 1:   Draw a secant containing the center of the circle and the point P.

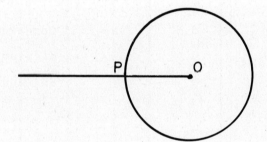

Step 2:   At the point P, construct a perpendicular to the secant.

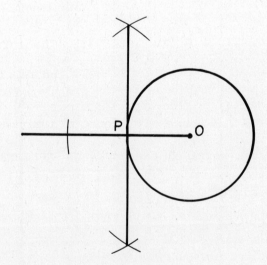

310

Step 3: The perpendicular constructed is the tangent to the circle at P.

We can also construct the tangent to a circle through a given point outside the circle.

To construct a tangent to a given circle through a given point P outside the circle.

1. Draw the segment from P to the center of the circle O.

2. Bisect $\overline{OP}$. Call the midpoint M.

3. Draw a circle at M with radius $\overline{OM}$. ($\overline{OP}$ will be a diameter of that circle.) Let Q and R be the points of intersection of the original circle with the circle just constructed.

4. Draw $\overleftrightarrow{QP}$ (or $\overleftrightarrow{RP}$). This line is tangent to the given circle at Q (or R). [∠OQP (or ∠ORP) is a right angle because it is inscribed in a semicircle.]

EXAMPLE 3   Construct a tangent to the given circle through P.

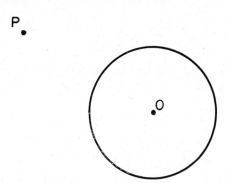

Solution   Step 1:   Draw the segment from P to the center of the circle

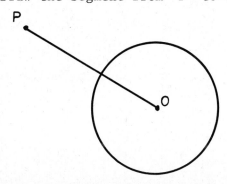

Step 2: Bisect $\overline{OP}$. Call the midpoint M.

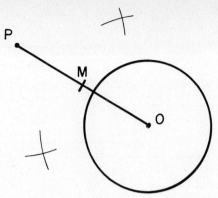

Step 3: Draw a circle with center at M and radius $\overline{OM}$. Let Q and R be the points of intersection of the original circle with the circle just constructed.

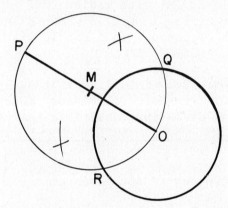

Step 4: Draw $\overleftrightarrow{QP}$. This line is tangent to the given circle at Q.

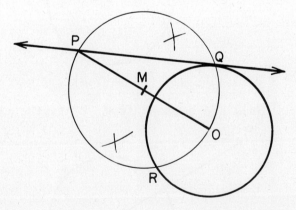

Let us examine why this construction guarantees a right angle at Q.
By construction we know PM = MQ, therefore △PMQ is an isosceles triangle with ∠QPM ≅ ∠PQM. We also know MQ = MO, so △MQO is also an isosceles triangle with ∠MQO ≅ ∠MOQ. Because the sum of the angles in a triangle is 180°, we know:

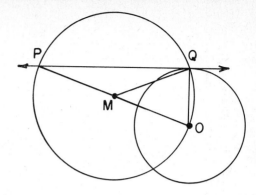

$$180° - (\angle MPQ + \angle PQM) = \angle \alpha$$
$$180° - 2 \cdot \angle PQM = \angle \alpha$$

We also know:

$$\angle \alpha = 180° - \angle QMO = 180° - [180° - (\angle MQO + \angle MOQ)]$$
$$= \angle MQO + \angle MOQ$$
$$\angle \alpha = 2 \cdot \angle MQO$$

Substituting for ∠α gives:

$$180° - 2 \cdot \angle PQM = 2 \cdot \angle MQO$$
$$180° = 2 \cdot \angle MQO + 2 \cdot \angle PQM$$
$$90° = \angle MQO + \angle PQM$$
$$90° = \angle PQO$$

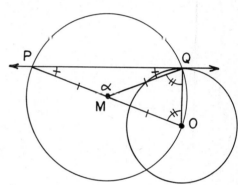

Thus we have a right angle at Q and PQ is a tangent.

A circle circumscribed about a triangle will have the sides of the triangle as chords. The perpendicular bisectors of the sides (chords) will pass through the center of the circle. Thus the perpendicular bisectors of the sides of a triangle will intersect in a point, the CIRCUMCENTER of the triangle, which is the center of the circumscribed circle.

> To circumscribe a circle about a triangle:
>
> 1. Construct the perpendicular bisector of each side of the triangle.
>
> 2. The point of intersection of the perpendicular bisectors (the circumcenter) is the center of the circumscribed circle. Set the compass to the radius, the distance from the center to a vertex of the triangle. Place the spike of the compass on the circumcenter and draw a circle about the triangle.
>
> 3. The circle constructed is circumscribed about the given triangle.

EXAMPLE 4   Circumscribe a circle about the given triangle.

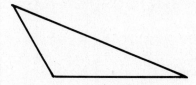

Solution   Step 1:   Construct the perpendicular bisector of each side of the triangle.

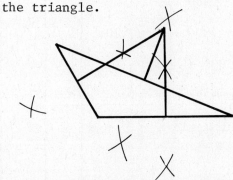

Step 2: The point of intersection of the perpendicular bisectors (the circumcenter) is the center of the circumscribed circle. Set the compass to the radius, the distance from the center to a vertex of the triangle. Place the spike of the compass on the circumcenter and draw a circle about the triangle.

Step 3: The circle constructed is circumscribed about the given triangle.

This construction allows us to construct a circle about a triangle because the intersection of the perpendicular bisectors of the sides is a point equidistant from each of the vertices (why ?). That distance then becomes the radius of the circle to be circumscribed about the triangle.

> To construct a circle containing three given noncollinear points, form the triangle with the three given points as vertices and proceed with the construction above.

A circle inscribed in a triangle will have the sides of the triangle as tangents. The tangents from the same point outside the circle to the circle are equal and each is perpendicular to a radius at the point of tangency. The line segment from the vertex of the triangle to the center of the circle is the common side of the two smaller triangles and these two triangles are congruent. Therefore the bisector of the vertex angle of a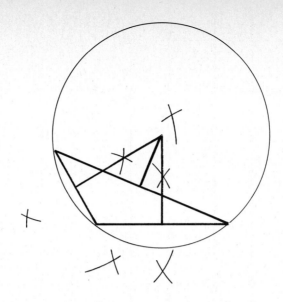

triangle passes through the center of the inscribed circle. The angle bisectors of a triangle will intersect in a point, the INCENTER of the triangle, which is the center of the inscribed circle.

> To inscribe a circle in a triangle:
>
> 1. Bisect each angle of the triangle.
>
> 2. The point of intersection of the bisectors (the incenter) is the center of the inscribed circle. Set the compass to the radius, the perpendicular distance from the center to a side of the triangle. Place the spike of the compass on the incenter and draw a circle.
>
> 3. The circle constructed is inscribed in the given triangle.

EXAMPLE 5   Inscribe a circle in the given triangle.

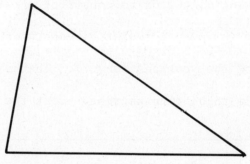

Solution   Step 1:   Bisect each angle of the triangle.

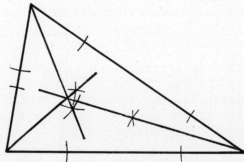

Step 2:   The point of intersection of the bisectors (the incenter) is the center of the inscribed circle. Set the compass to the radius, the perpendicular distance from the center to a side of the triangle. Place the spike of the compass on the incenter and draw a circle.

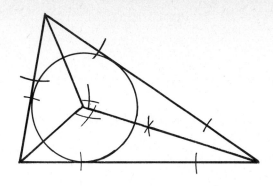

Step 3: The circle constructed is inscribed in the given triangle.

This construction allows us to inscribe a circle in a triangle because the intersection of the angle bisectors is a point equidistant from each of the sides of the triangle (why ?). That distance becomes the radius of the circle to be inscribed in the triangle.

We have seen that the perpendicular bisectors of the sides of a triangle intersect in a point (the circumcenter) and that the angle bisectors intersect in a point (the incenter). The altitudes of a triangle also intersect in a point (the ORTHOCENTER) but this point possesses less useful properties. A line segment connecting the midpoint of a side of a triangle with the opposite vertex is the MEDIAN of the triangle. The medians of a triangle also intersect in a point, the CENTROID of the triangle. If a triangle were constructed of uniform material, the centroid would be the center of gravity. To test this property, draw a triangle on cardboard, locate the centroid, then carefully cut out the triangle. If the point of a pin is placed at the centroid, the triangle will balance on it.

> To locate the centroid of a triangle:
>
> 1. Construct the medians of the triangle by connecting the midpoint of each side with the opposite vertex.
>
> 2. The point of intersection of the medians is the centroid of the triangle.

EXAMPLE 6   Locate the centroid of the given triangle.

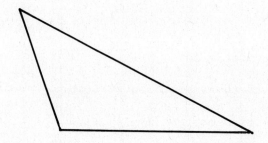

Solution   Step 1:   Construct the medians of the triangle.

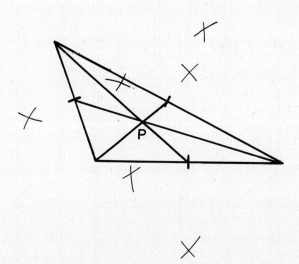

Step 2:   The point of intersection of the medians, P, is the centroid of the triangle.

We can also inscribe polygons in a circle. A central angle of $\frac{\pi}{3}$ rad will subtend a chord of length r, the radius of the circle. (The chord and two radii form an equilateral triangle.) Because there are six $\frac{\pi}{3}$ rad in 2π rad, six chords of length r can be marked off on a circle with radius r. We can use this fact to inscribe a regular hexagon in a circle.

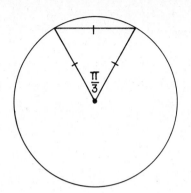

To inscribe a regular hexagon in a circle:

1. Draw a circle with radius r.

2. Without changing the setting of the compass, place the spike of the compass on the circle and mark off a distance r on the circle.

3. Place the spike on the intersection of the circle and the mark made in Step 2 and make another mark on the circle.

4. Repeat Step 3 for each new mark until a mark coincides with the point on the circle where the spike was first placed. You should have six evenly spaced marks on the circle. (If the last point does not coincide with the first spike mark, you should check the accuracy of your previous marks and the accuracy of the compass setting.)

5. Draw the chords connecting adjacent marks. These six chords form the sides of the inscribed hexagon.

EXAMPLE 7   Inscribe a hexagon in the given circle.

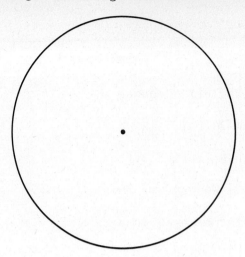

Solution   Steps 1 & 2:   Set the compass to the radius of the circle, place the spike on the circle and mark off a distance r

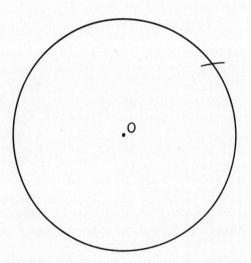

Steps 3 & 4:   Place the spike on the intersection of the circle and the mark made in Step 2 and make another mark. Repeat Step 3 for each new mark until a mark coincides with the point on the circle where the spike was first placed.

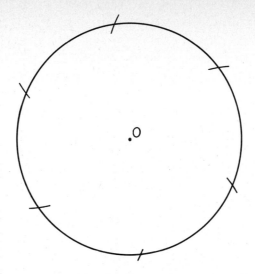

Step 5: Draw the chords connecting adjacent marks. These six chords form the sides of the inscribed hexagon.

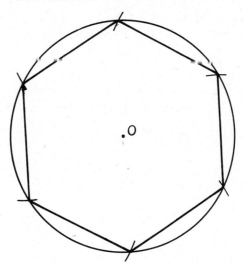

We can use much the same procedure to inscribe an equilateral triangle in a circle.

To inscribe an equilateral triangle in a circle:

1. Follow Steps 1 through 4 for inscribing a hexagon in a circle.

2. Connect every other mark. The three chords will form the sides of the inscribed equilateral triangle.

EXAMPLE 8 Inscribe an equilateral triangle in the given circle.

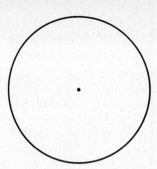

Solution   Step 1:   Set the compass to the radius of the circle and mark off that distance on the circle.

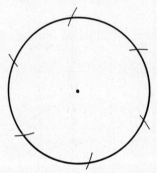

Step 2:   Connect every other mark. The three chords will form the sides of the inscribed equilateral triangle.

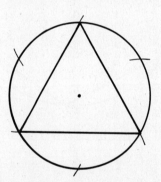

To inscribe a square in a circle:

1. Draw a diameter of the circle.

2. Construct the perpendicular bisector of the diameter (which will given a diameter perpendicular to the first).

3. Draw the chords from the endpoints of one diameter to the endpoints of the other.

4. The four chords will be the sides of the inscribed square.

EXAMPLE 9   Inscribe a square in the given circle.

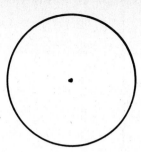

Solution   Step 1:   Draw a diameter of the circle.

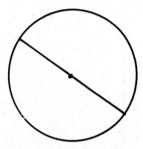

Step 2:   Construct a perpendicular bisector of the diameter.

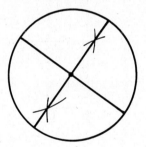

Step 3:   Draw the chords from the endpoints of one diameter to the endpoints of the other.

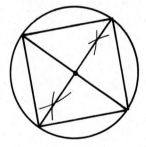

Step 4:   The four chords are the sides of the inscribed square.

We know that if an equilateral quadrilateral has one right angle, then the figure is a square. We can use this fact to circumscribe a square about a circle.

> To circumscribe a square about a circle:
>
> 1. Draw a diameter of the circle.
>
> 2. Construct the perpendicular bisector of the diameter. (This results in a diameter perpendicular to the first diameter, giving us the needed right angle.)
>
> 3. Set the compass to the radius of the circle. Place the spike at the endpoint of a diameter and draw a circle.
>
> 4. Repeat Step 3 for each endpoint of a diameter.
>
> 5. The four points of intersection of the circles from Steps 3 and 4 are the vertices of the circumscribed square. Connect them.
>
> 6. The resulting figure is a square circumscribed about the circle.

EXAMPLE 10   Circumscribe a square about the given circle.

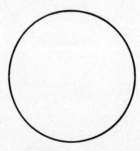

Solution   Step 1:   Draw a diameter of the circle.

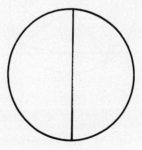

Step 2: Construct the perpendicular bisector of the diameter.

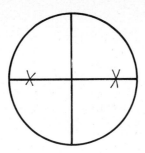

Step 3: Set the compass to the radius of the circle. Place the spike at the endpoint of a diameter and draw a circle.

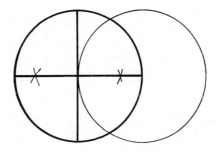

Step 4: Repeat Step 3 for each endpoint of a diameter.

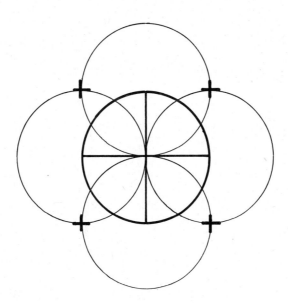

Step 5: The four points of intersection of the circles from Steps 3 and 4 are the vertices of the circumscribed square. Connect them.

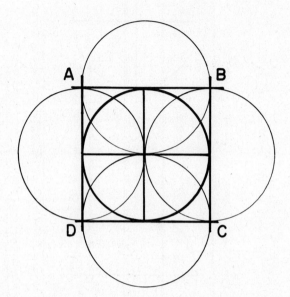

Step 6: The resulting figure ABCD is a square circumscribed about the given circle.

EXERCISE 6.4

Use only a straightedge and compass for the following problems.

1. Locate the center of the circle given in figure A.

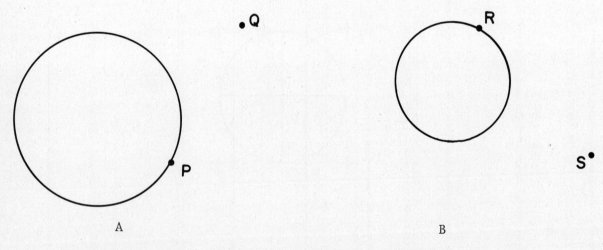

2. Locate the center of the circle given in figure B.

3. Construct a tangent to the circle in figure A at the point P.

4. Construct a tangent to the circle in figure B at the point R.

5. Construct a tangent to the circle in figure A through the point Q.

6. Construct a tangent to the circle in figure B through the point S.

7. Circumscribe a circle about the triangle in figure C.

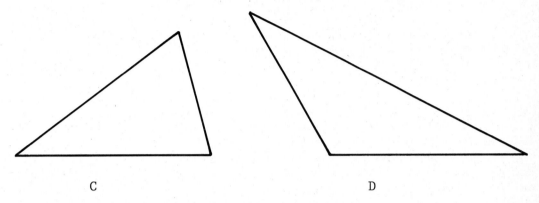

C   D

8. Circumscribe a circle about the triangle in figure D.

9. Inscribe a circle in the triangle in figure D.

10. Inscribe a circle in the triangle in figure C.

11. Locate the centroid of the triangle given in figure C.

12. Locate the centroid of the triangle given in figure D.

13. Circumscribe a circle about the triangle given in figure E.

14. Circumscribe a circle about the triangle given in figure F.

15. Choose three noncollinear points and construct a circle containing them.

16. Choose three noncollinear points and construct a circle containing them.

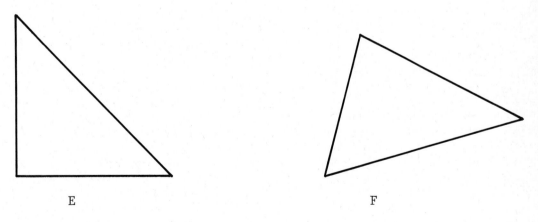

E   F

17. Inscribe a circle in the triangle given in figure F.

18. Inscribe a circle in the triangle given in figure E.

19. Locate the centroid of the triangle given in figure E.

20. Locate the centroid of the triangle given in figure F.

21. Inscribe a regular hexagon in the circle given in figure G.

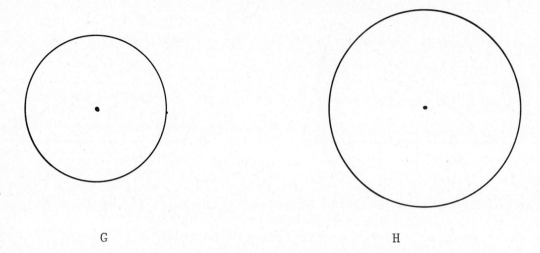

G H

22. Inscribe a regular hexagon in the circle given in figure H.

23. Inscribe an equilateral triangle in the circle given in figure H.

24. Inscribe an equilateral triangle in the circle given in figure G.

25. Inscribe a square in the circle given in figure G.

26. Inscribe a square in the circle given in figure H.

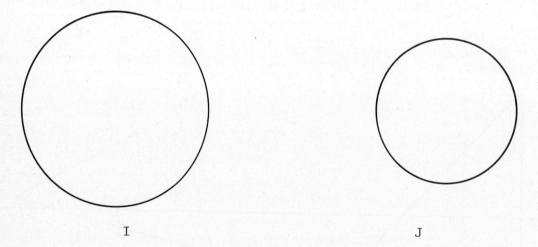

I J

27. Circumscribe a square about the circle given in figure I.

28. Circumscribe a square about the circle given in figure J.

29. Inscribe a regular octagon in the circle given in figure J.

30. Inscribe a regular octagon in the circle given in figure I.

31. Circumscribe an equilateral triangle about the circle given in figure I.

32. Circumscribe an equilateral triangle about the circle given in figure J.

33. Locate the orthocenter of the triangle given in figure E.

34. Locate the orthocenter of the triangle given in figure F.

6.5 PROOFS

We can use the definitions of circle, radius, diameter, chord, and tangent, and the relationship between central angles and inscribed angles in a circle to justify steps in a proof.

EXAMPLE 1   Given: Circle O with chord $\overline{AB}$

Prove: A radius bisecting the chord is perpendicular to the chord.

Solution   First we need a drawing to represent the facts given. The midpoint of $\overline{AB}$, we will label M. We can show that the radius bisecting chord $\overline{AB}$ is perpendicular to $\overline{AB}$ if we can show that ∠AMO is congruent to ∠BMO.

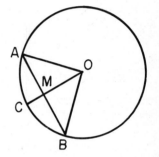

These angles are supplementary; if they are also congruent, each must be a right angle. In order to prove these angles congruent, we can prove that two triangles containing these angles are congruent. We can draw radii $\overline{AO}$ and $\overline{OB}$ to give △OMA and △OMB. We can show these two triangles congruent by an SSS correspondence.

	Statement	Reason
1.	$\overline{AB}$ is a chord in circle O.	Given
2.	M is the midpoint of $\overline{AB}$	Given
S 3.	$\overline{AM} \cong \overline{BM}$	2 & Def. of midpoint
S 4.	$\overline{OM} \cong \overline{OM}$	Common side
5.	$\overline{AO}$ and $\overline{BO}$ are radii	Def. of radii
S 6.	$\overline{AO} \cong \overline{BO}$	5 & Def. of radii
7.	$\triangle OMA \cong \triangle OMB$	3, 4, 6, & SSS
8.	$\angle OMA \cong \angle OMB$	7 & CPCTC
9.	$\overline{AB}$ is a line segment	Given
10.	$\angle OMA$ & $\angle OMB$ are supplementary	9 & Def. of supplementary
11.	$\angle OMA$ & $\angle OMB$ are right $\angle$s	8, 10 & angles that are both congruent and supplementary are right angles
12.	$\overline{OM}$ is $\perp$ to $\overline{AB}$	11 & Def. of $\perp$
13.	Radius $\overline{OC}$ is $\perp$ to $\overline{AB}$	Restatement of 12

EXERCISE 6.5

Use the definitions of circle, tangent, radius, etc., and the facts known about arcs and angles in the following proofs.

1. Given: $\triangle AOB$ with $\angle AOB$ a central angle

   $\angle AOB = \frac{\pi}{3}$

   Prove: $\triangle AOB$ is equilateral

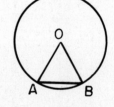

2. Given: $\overline{QR} \parallel \overline{ST}$

   Prove: $\overparen{QRT} = \overparen{SQR}$

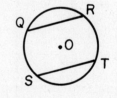

3. Given: $\overarc{AB} = \overarc{DC}$

   Prove: △AED is isosceles

4. Given: $\overline{EC} \cong \overline{EB}$

   Prove: $\overline{DC} \cong \overline{AB}$

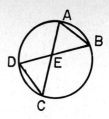

5. Given: $\overarc{RST} = 180°$

   $\overline{QR} \parallel \overline{ST}$

   Prove: QRST is a rectangle

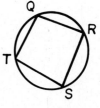

6. Given: $\overarc{AB} = \overarc{CD}$

   Chords $\overline{AC}$ and $\overline{BD}$

   Prove: △ABC ≅ △DCB

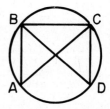

7. Prove that the point of intersection of the perpendicular bisectors of the sides of a triangle is equidistant from each vertex of the triangle.

8. Using the definition of tangent, prove that the tangents from a point outside the circle to the circle are the same length.

CHAPTER 6 REVIEW

1. Match each item in Column I with the appropriate item(s) from Column II. Items in Column II refer to the drawing to the right and can be used once, more than once, or not at all.

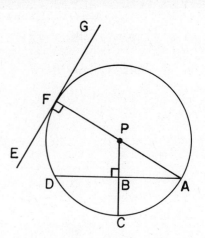

Column I	Column II
a) Point of tangency	$\overline{AB}$
b) An inscribed angle	B
c) A minor arc	$\overset{\frown}{FD}$
d) A diameter	F
e) A chord	$\overset{\frown}{FAC}$
f) Arc intercepted by ∠FAD	D
g) A major arc	∠FAD
h) A central angle	$\overline{AD}$
i) A radius	$\overline{PC}$
j) A tangent	∠CPF
k) The measure of one is twice the other	$\overset{\frown}{FA}$
	∠ABC
	$\overline{DA}$
	$\overline{FE}$
	$\overline{FA}$
	$\overset{\frown}{FC}$

2. If $\widehat{AB}$ = 1.83 rad and $\angle ABC$ = 0.92 find the number of radians in $\widehat{BC}$.

3. If $\widehat{ABC}$ = 294°, find the number of degrees in $\angle APC$.

4. If AP = 6 in. and $\angle APC = \frac{\pi}{3}$ rad, find the length of $\widehat{ABC}$. (Give an exact answer.)

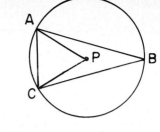

C 5. If $\widehat{AC}$ = 10 cm and $\angle ABC$ = 56°, find the length of $\widehat{ABC}$. (Give answer correct to hundredths.)

6. If $\widehat{ABC} = \frac{7\pi}{5}$ rad and $\widehat{CAB} = \frac{9\pi}{5}$ rad, find $\widehat{CB}$.

C 7. If $\angle ABO$ = 27° and OB = 15 cm, find the length of $\widehat{ACB}$.

8. If $\widehat{EFD}$ = 4.02 rad, find the number of radians in $\angle BAO$.

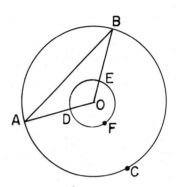

9. If $\angle ABD$ = 52°, find the number of degrees in $\angle BDC$.

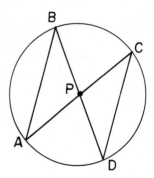

Give an exact answer for the remaining problems, unless a "C" indicates otherwise. For the problem marked with "C", give answer to nearest hundredth.

C 10. If $\stackrel{\frown}{RS}$ = 0.87 rad, $\stackrel{\frown}{TU}$ = 1.46 rad, $\stackrel{\frown}{UR}$ = 0.98 rad, find the number of radians in $\angle RUT$.

11. If $\stackrel{\frown}{RS}$ = x - 4°, $\stackrel{\frown}{ST}$ = 3x - 33°, $\stackrel{\frown}{TU}$ = 2x + 7°, and $\stackrel{\frown}{UR}$ = x + 26°, find the number of degrees in $\angle RST$.

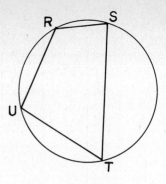

12. If AC = 2.8 cm and AB = 9.6 cm, find the circumference of the circle.

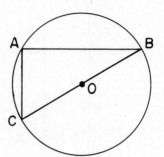

C 13. If $\overline{TS}$ is tangent to the circle at T, TP = 2.4 cm and RS = 5 cm, length of $\overline{TS}$.

*14. If $\overline{TS}$ is tangent to the circle at T, TS = 26 cm, and RS = 14 cm, find the area of the circle.

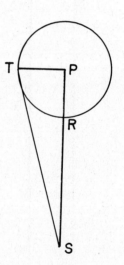

15. Find the circumference of a circle with area of 81π in^2.

16. If the diameter is 14 in., find the length of $\stackrel{\frown}{MN}$.

C 17. If the radius is 10 in., find the area shaded.

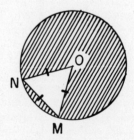

18. A 6 cm pendulum swings through an arc of 72°. Find the length of the arc.

19. A pendulum swings through an arc of 0.785 rad and the length of the arc is 15.7 cm; find the length of the pendulum.

20. If the diameter of the circle is 10 mm, find the length of AB.

21. If AB = 6 cm, find the area of the circle.

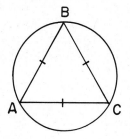

C 22. If the radius of the circle is 20 in., find the area shaded (to the nearest tenth).

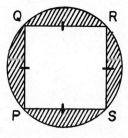

C 23. If BD = 2 mm and the radius of the circle is 7 mm, find the length of AB.

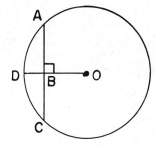

24. Given quadrilateral PQRS circumscribed about circle O, QR = 13.7 in., and PS = 9.6 in., find the perimeter of PQRS.

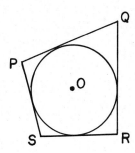

25. Given a circle with area 36π ft^2, find the dimensions of a square circumscribed about it. (Give an exact answer.)

26. Given $\triangle ACE$ circumscribed about circle P, AC = 24 mm, CE = 11 mm, and AE = 18 mm, and B, D, and F the points of tangency, find the length of AB, BC, and DE.

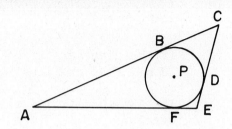

27. A regular hexagon is inscribed in a 20 inch radius circle. Find the area of the hexagon.

28. A $\frac{\pi}{6} - \frac{\pi}{3} - \frac{\pi}{2}$ rad triangle is inscribed in a circle. If the shorter leg is 7 cm, find the area of the circle.

C 29. A 45°-45°-90° triangle is circumscribed about a circle. If one leg is 32 in., find the radius of the circle.

C 30. Find the area of a sector in an 11 cm radius circle if its central angle is 54°.

C 31. Find the radius of a circle if a sector with central angle of 1.34 rad has area of 200 cm^2.

32. Find the central angle (in degrees) of a sector if the area of the sector is 129.6π and the radius is 18 in.

C 33. Find the area shaded if $\angle AOB = 120°$ and the radius is 22 mm.

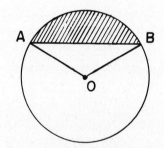

C 34. Find the area shaded if ED = 2.8 in. and EF = 9.6 in.

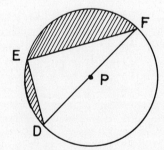

35. Find the area shaded if RQ = 4 m and QP = 5 m.

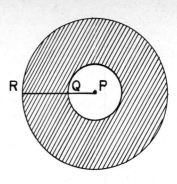

36. Find the area shaded if AO = 9 in., AB = 6 in., and ∠AOE = 108°.

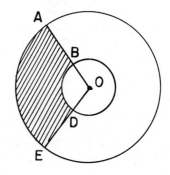

Find the area for each of the given figures for Problems 37 to 42. You may assume that line segments that appear perpendicular or parallel are. If the figure is shaded, find only the area of the shaded portion. Dots indicate the centers of circles or arcs.

37.

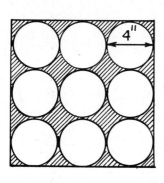

38.

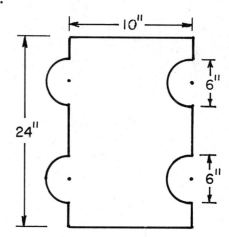

39.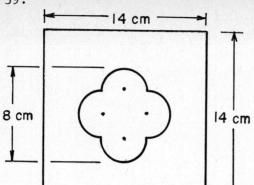

40.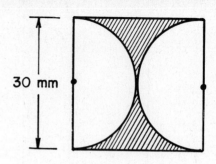

41.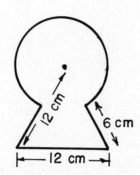

42.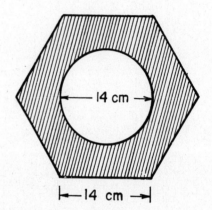

43. Locate the center of the circle to the right.

44. Construct a tangent to the circle at the right at the point S.

45. Construct a tangent to the circle at the right through the point R.

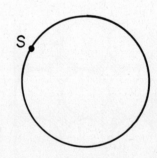

46. Inscribe a circle in △ABC.

47. Circumscribe a circle about △ABC.

48. Locate the centroid of △ABC.

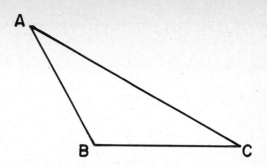

49. Choose three noncollinear points and construct a circle containing them.

50. Inscribe a regular hexagon in circle O.

51. Circumscribe an equilateral triangle about circle O.

52. Circumscribe a square about circle O.

53. Inscribe a regular octagon in circle O.

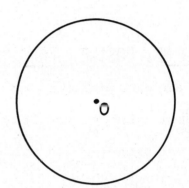

54. Locate the orthocenter of △ABC.

55. Given: A circle and a diameter perpendicular to a chord.

    Prove: The diameter bisects the chord.

56. Given: A circle with two chords the same distance from the center.

    Prove: The length of the chords is the same.

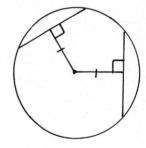

CHAPTER 7
INTRODUCTION TO TRIGONOMETRY

7.1 SINE, COSINE, AND TANGENT FOR ACUTE ANGLES

In our work with similar triangles, we use the fact that the sides in similar triangles are proportional. That is the fundamental principle upon which trigonometry is built. For instance, we have frequently used the fact that all 30°-60°-90° triangles have sides in the ratio: $1:\sqrt{3}:2$. We can state the same relationship by saying:

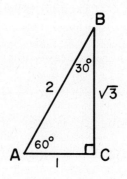

$$\frac{\text{leg opposite 60° angle}}{\text{hypotenuse}} = \frac{\sqrt{3}}{2}$$

$$\frac{\text{leg adjacent to 60° angle}}{\text{hypotenuse}} = \frac{1}{2}$$

$$\frac{\text{leg opposite 60° angle}}{\text{leg adjacent to 60° angle}} = \frac{\sqrt{3}}{1}$$

These relationships will be true for all 30°-60°-90° triangles. Because these basic relationships are used so frequently, mathematicians defined functions that stand for these ratios in right triangles.

$$\text{sine } \theta = \frac{\text{leg opposite } \theta}{\text{hypotenuse}}$$

$$\text{cosine } \theta = \frac{\text{leg adjacent to } \theta}{\text{hypotenuse}}$$

$$\text{tangent } \theta = \frac{\text{leg opposite } \theta}{\text{leg adjacent to } \theta}$$

where θ (theta) is one acute angle in a right triangle.

We usually write sin θ for sine θ, cos θ for cosine θ, and tan θ for tangent θ. For a 30°-60°-90° triangle, we have:

$$\sin 60° = \frac{\sqrt{3}}{2} \qquad \sin 30° = \frac{1}{2}$$

$$\cos 60° = \frac{1}{2} \qquad \cos 30° = \frac{\sqrt{3}}{2}$$

$$\tan 60° = \frac{\sqrt{3}}{1} = \sqrt{3} \qquad \tan 30° = \frac{1}{\sqrt{3}} = \frac{\sqrt{3}}{3}$$

Note that the trigonometric functions are numbers *without* units. They are formed from a ratio of like units, millimeters: millimeters, or inches: inches, for instance, so the units are cancelled.

To simplify the notation, we will adopt the following conventions.

Degrees will be given with the degree symbol, e.g., sin 47°.

Radians will be given without the "rad", unless it is needed for clarity, e.g., tan π/4 or tan 1.231.

In trigonometry, Greek letters are frequently used to represent angles. The three most commonly used letters are:

α alpha

β beta

θ theta

Let us now determine the value of these three functions using a $\frac{\pi}{4} - \frac{\pi}{4} - \frac{\pi}{2}$ triangle.

$$\sin \theta = \frac{\text{leg opposite } \theta}{\text{hypotenuse}}$$

$$\sin \frac{\pi}{4} = \frac{1}{\sqrt{2}} = \frac{\sqrt{2}}{2}$$

$$\cos \theta = \frac{\text{leg adjacent } \theta}{\text{hypotenuse}}$$

$$\cos \frac{\pi}{4} = \frac{1}{\sqrt{2}} = \frac{\sqrt{2}}{2}$$

$$\tan \theta = \frac{\text{leg opposite } \theta}{\text{leg adjacent to } \theta}$$

$$\tan \frac{\pi}{4} = \frac{1}{1} = 1$$

We have $\sin \frac{\pi}{4} = \frac{\sqrt{2}}{2}$, $\cos \frac{\pi}{4} = \frac{\sqrt{2}}{2}$, $\tan \frac{\pi}{4} = 1$.

Because we have seen how frequently these triangles are used, we can anticipate the importance of the trigonometric functions for these angles. Therefore, you should MEMORIZE both the definition for each of the functions and its particular value for these triangles.

$\theta =$	$\frac{\pi}{6} = 30°$	$\frac{\pi}{4} = 45°$	$\frac{\pi}{3} = 60°$
$\sin \theta$	$\frac{1}{2}$	$\frac{\sqrt{2}}{2}$	$\frac{\sqrt{3}}{2}$
$\cos \theta$	$\frac{\sqrt{3}}{2}$	$\frac{\sqrt{2}}{2}$	$\frac{1}{2}$
$\tan \theta$	$\frac{\sqrt{3}}{3}$	1	$\sqrt{3}$

Students first learning trigonometry usually do not realize that $\sin \frac{\pi}{3}$ is a single unit that specifically does NOT mean "sin times $\frac{\pi}{3}$". Rather "$\sin \frac{\pi}{3}$" is read "sin of $\frac{\pi}{3}$". Another common error students make is to think that $2\sin \theta = \sin (2\theta)$. This is WRONG. "$2\sin \theta$" means "2 times sin θ" or $2(\sin \theta)$. Let us look at a specific example.

$$\sin 30° = \frac{1}{2}$$

$$2\sin 30° = 2 \cdot \frac{1}{2} = 1$$

$$\sin 60° = \frac{\sqrt{3}}{2} \neq 1$$

Thus $2 \sin 30° \neq \sin 60°$.

We can calculate the trigonometric functions for angles of 30°, 45°, and 60° ($\frac{\pi}{6}$, $\frac{\pi}{4}$, and $\frac{\pi}{2}$) but we must use either a table or a calculator to determine the values for the trigonometric functions of other angles. Tables of the functions in degree and radian measures are shown in Appendix A. If the problem is given in degrees, be sure to use the degree table or be sure that your calculator is in the degree mode. If the problem is given in radians, then use the radian table or change your calculator to radian mode. READ the instruction manual for your calculator to determine how to change back and forth from degree to radian modes. If your answers are not correct, CHECK YOUR MODE.

Let us determine the value of $\sin 26°$ and $\tan \frac{\pi}{7}$. The first problem is in degrees, so we use the degree mode or table. Figure 7.1 illustrates the appropriate table to use. Because we want the sine of 26°, we go down to the row for 26° (marked by a horizontal arrow) and over to the column for the sine function (marked by a vertical arrow). The entry for $\sin 26°$ is circled and is the intersection of the desired row and column. Thus, $\sin 26° = 0.4384$.

For $\tan \frac{\pi}{7}$, we must first convert $\frac{\pi}{7}$ to a decimal approximation, giving 0.4488 (to four decimal places). Using a calculator, $\tan 0.4488 = 0.4816$ (to four decimal places). Using the table, we must round to 0.45, giving $\tan 0.45 = 0.4831$. The rounding caused the value from the table to be somewhat larger. Until you are sure you understand how to use your calculator, you may wish to check the calculator values against those from the table.

θ	sin θ	cos θ	tan θ
0°	.0000	1.0000	.0000
1°	.0175	.9998	.0175
2°	.0349	.9994	.0349
3°	.0523	.9986	.0524
4°	.0698	.9976	.0699
5°	.0872	.9962	.0875
6°	.1045	.9945	.1051
7°	.1219	.9925	.1228
8°	.1392	.9903	.1405
9°	.1564	.9877	.1584
10°	.1736	.9848	.1763
11°	.1908	.9816	.1944
12°	.2079	.9781	.2126
13°	.2250	.9744	.2309
14°	.2419	.9703	.2493
15°	.2588	.9659	.2679
16°	.2756	.9613	.2867
17°	.2924	.9563	.3057
18°	.3090	.9511	.3249
19°	.3256	.9455	.3443
20°	.3420	.9397	.3640
21°	.3584	.9336	.3839
22°	.3746	.9272	.4040
23°	.3907	.9205	.4245
24°	.4067	.9135	.4452
25°	.4226	.9063	.4663
26°	(.4384)	.8988	.4877
27°	.4540	.8910	.5095
28°	.4695	.8829	.5317
29°	.4848	.8746	.5543
30°	.5000	.8660	.5774

FIGURE 7.1

Evaluating a trigonometric function results in an answer that is a number only; there are NO units associated with it.

Once we know the value of a trigonometric function, we can determine the angle. For instance, if $\sin \theta = \frac{1}{2}$, we know that $\theta = 30°$ or $\frac{\pi}{6}$ because $\sin 30° = \sin \frac{\pi}{6} = \frac{1}{2}$. We can also use a calculator or the trigonometric tables to determine the angle if we are given the value of a function. We must judge from the context of the problem whether it is preferable to answer in degrees or radians and use that table or mode. To use the tables to find θ in radians when $\tan \theta = 0.1409$, we must first scan the tables to locate that portion of the table with tangent values that include 0.1409. Figure 7.2 gives the

t	$\sin t$	$\cos t$	↓ $\tan t$
.00	.0000	1.0000	.0000
.01	.0100	1.0000	.0100
.02	.0200	.9998	.0200
.03	.0300	.9996	.0300
.04	.0400	.9992	.0400
.05	.0500	.9988	.0500
.06	.0600	.9982	.0601
.07	.0699	.9976	.0701
.08	.0799	.9968	.0802
.09	.0899	.9960	.0902
.10	.0998	.9950	.1003
.11	.1098	.9940	.1104
.12	.1197	.9928	.1206
.13	.1296	.9916	.1307
→ .14	.1395	.9902	(.1409)
.15	.1494	.9888	.1511
.16	.1593	.9872	.1614
.17	.1692	.9856	.1717
.18	.1790	.9838	.1820
.19	.1889	.9820	.1923
.20	.1987	.9801	.2027
.21	.2085	.9780	.2131
.22	.2182	.9759	.2236
.23	.2280	.9737	.2341
.24	.2377	.9713	.2447
.25	.2474	.9689	.2553
.26	.2571	.9664	.2660
.27	.2667	.9638	.2768
.28	.2764	.9611	.2876
.29	.2860	.9582	.2984
.30	.2955	.9553	.3093
.31	.3051	.9523	.3203
.32	.3146	.9492	.3314
.33	.3240	.9460	.3425
.34	.3335	.9428	.3537
.35	.3429	.9394	.3650
.36	.3523	.9359	.3764
.37	.3616	.9323	.3879
.38	.3709	.9287	.3994
.39	.3802	.9249	.4111
.40	.3894	.9211	.4228
.41	.3986	.9171	.4346
.42	.4078	.9131	.4466

FIGURE 7.2

section that contains the tangent value of 0.1409. The value is circled. Once we find the desired value in the tangent column, we then read the radian entry for that row (0.14). Therefore, when $\tan \theta = 0.1409$, $\theta = 0.14$ rad.

What we have done is to take the inverse of the tangent function. Recall if $\sqrt{x} = 3$, then $x = 9$, because $(\sqrt{x})^2 = 3^2$. Squaring is the inverse of taking the square root, and taking the square root is the inverse of squaring. That is, the inverse of a function "undoes" what the function does. Because we want to "undo" the tangent function, we need the inverse of the tangent function, called the ARC TANGENT or the $\tan^{-1}$. On a calculator, the inverse of a trigonometric function may be computed by using an INV (inverse) key and the appropriate trigonometric key or by using the $\sin^{-1}$, $\cos^{-1}$, or $\tan^{-1}$ key. Until you are sure how to use your calculator, check your values against those in the tables. Note that if your calculator uses the "$\sin^{-1}$" notation, this DOES NOT MEAN $\frac{1}{\sin}$; it is a shortcut to writing inverse of the sine, or "arc sine."

Taking the inverse of a trigonometric function results in an answer in degrees or radians.

EXAMPLE 1   If $\sin \theta = 0.8746$, determine the value of θ.

Solution   Using the two tables, we get:

$$\sin \theta = 0.8746$$

$$\sin^{-1}(\sin \theta) = \sin^{-1} 0.8746$$

$$\theta = 61° \text{ or between } 1.06 \text{ and } 1.07 \text{ rad.}$$

Using a calculator, we get:

$$\sin \theta = 0.8746$$

$$\sin^{-1}(\sin \theta) = \sin^{-1} 0.8746$$

$$\theta = 60.9977° \text{ or } 1.0646 \text{ rad}$$

(to four decimal places)

The rest of the examples in this chapter will be worked using a calculator. Those using the tables in the Appendix will have slightly different answers due to rounding error.

We can use our knowledge of the trigonometric functions to determine sizes of angles or the length of a side in a right triangle. Because we have used the 3:4:5 right triangle so frequently, let us determine the angles in it.

EXAMPLE 2   Determine the acute angles in a 3:4:5 right triangle.

Solution   First we need a drawing.
We know that

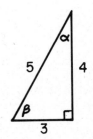

$$\sin \beta = \frac{4}{5} = 0.8$$

$$\sin^{-1}(\sin \beta) = \sin^{-1} 0.8$$

$$\beta = 53.13° \text{ or } 0.9273 \text{ rad}$$

$$\sin \alpha = \frac{3}{5} = 0.6$$

$$\sin^{-1}(\sin \alpha) = \sin^{-1} 0.6$$

$$\alpha = 36.87° \text{ or } 0.6435 \text{ rad}$$

We check to see that the sum is 90° or 1.5708 rad. The answers check, so we have β = 53.13° or 0.9273 rad and α = 36.87° or 0.6435 rad.

EXAMPLE 3   If $\angle$CAB = 29° and AC = 7 cm, find the length of $\overline{BC}$.

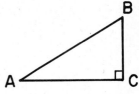

Solution   By entering the given information on the drawing, we see that we are concerned with the side opposite and the side adjacent to the 29° angle. Thus we should use the tangent function:

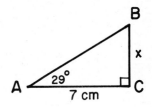

347

$$\tan \theta = \frac{\text{leg opposite}}{\text{leg adjacent}}$$

$$\tan 29° = \frac{x}{7 \text{ cm}}$$

$$7 \text{ cm} (\tan 29°) = x$$

$$3.88 \text{ cm} = x$$

BC is 3.9 cm (to the nearest tenth).

EXAMPLE 4 If AB = 19.3 in. and CB = 14.2 in., determine the number of radians in A.

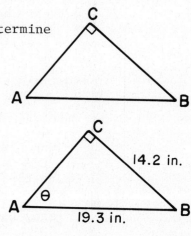

Solution By entering the given information on the drawing, we see that we should use the sine function.

$$\sin \theta = \frac{\text{leg opposite}}{\text{hypotenuse}}$$

$$\sin \theta = \frac{14.2}{19.3}$$

$$\sin \theta = 0.7358$$

$$\theta = 0.8268 \text{ rad}$$

The angle contains 0.8268 rad.

EXAMPLE 5 Determine an altitude of △PQR.

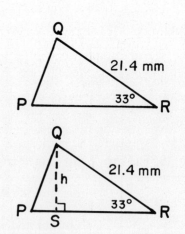

Solution △PQR is not a right triangle, but the altitude will form a right triangle, △QRS. We can use the sine function to determine h.

$$\sin \theta = \frac{\text{leg opposite}}{\text{hypotenuse}}$$

$$\sin 33° = \frac{h}{21.4 \text{ mm}}$$

$$21.4 \text{ mm } (\sin 33°) = h$$

$$11.6553 \text{ mm} = h$$

The altitude of the triangle is 11.66 mm (to the nearest hundredth).

EXAMPLE 6 Given $\cos \theta = \frac{15}{17}$, find the other trigonometric functions for θ. (Give an exact answer.)

Solution One could determine a value of θ and then calculate the values for the other two functions. Each, however, would involve two approximations: one determining the angle and the other the function value. We can, instead, use the definitions of the functions. For a right triangle, we know that

$$\cos \theta = \frac{\text{leg adjacent}}{\text{hypotenuse}}$$

and $\cos \theta = \frac{15}{17}$,

so we can let:

$$\text{leg adjacent} = 15$$

$$\text{hypotenuse} = 17$$

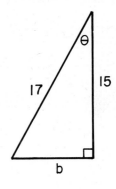

We can then use the Pythagorean theorem to determine the length of the other side.

$$a^2 + b^2 = c^2$$

$$15^2 + b^2 = 17^2$$

$$225 + b^2 = 289$$

$$b^2 = 64$$

$$b = \pm 8$$

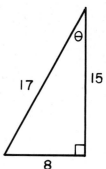

Because b represents a length, it must be a positive value: + 8. We can now find the values for the other trigonometric functions for θ.

$$\sin \theta = \frac{\text{leg opposite}}{\text{hypotenuse}}$$

$$\sin \theta = \frac{8}{17}$$

$$\tan \theta = \frac{\text{leg opposite}}{\text{leg adjacent}}$$

$$\tan \theta = \frac{8}{15}$$

Therefore, when $\cos \theta = \frac{15}{17}$, $\sin \theta = \frac{8}{17}$, and $\tan \theta = \frac{8}{15}$.

EXERCISE 7.1

A calculator may be used for all problems that do <u>not</u> have NC before the problem number. Give answers to the nearest hundredth.

1. Determine a) $\sin 59°$ b) $\cos 0.914$ c) $4 \cdot \tan \frac{\pi}{9}$.

2. Determine a) $\sin 0.34$ b) $\cos \frac{\pi}{17}$ c) $5 \cdot \tan 87°$.

3. Determine a) $\sin \frac{3\pi}{8}$ b) $3 \cdot \cos 11°$ c) $\tan 1.06$.

4. Determine a) $\sin 3°$ b) $2 \cdot \cos 1.52$ c) $\tan \frac{2\pi}{9}$.

5. Find θ (in radians) if

   a) $\sin \theta = 0.342$ b) $\cos \theta = 0.2079$ c) $\tan \theta = 5.1446$.

6. Find θ (in radians) if

   a) $\sin \theta = 0.6009$ b) $\cos \theta = 0.3584$ c) $\tan \theta = 1.2799$.

7. Find β (in degrees) if

   a) $\sin \beta = 0.7547$ b) $\cos \beta = 0.9816$ c) $\tan \beta = 0.0875$.

8. Find α (in degrees) if

   a) $\sin \alpha = 0.0872$ b) $\cos \alpha = 0.1219$ c) $\tan \alpha = 57.2906$.

9. Find the number of degrees in each acute angle of a 5:12:13 triangle.

10. Find the number of radians in each acute angle of a 5:12:13 triangle.

11. If ∠R = 69° and RT = 14.2 in., find the length of ST.

12. If ∠R = 17° and ST = 21.9 in., find the length of RS.

13. If ∠S = 0.95 and ST = 31.6 cm, find the length of RS.

14. If ∠S = 0.51 and ST = 19.3 cm, find the length of RT.

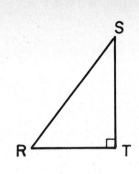

Use the drawing to the right for Problems 15 to 22.

15. If AC = 9.34 cm and CB = 14.21 cm, find the number of degrees in α.

16. If AC = 16.31 cm and AB = 20.01 cm, find the number of degrees in α.

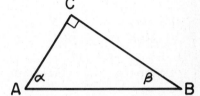

17. If AC = 2.3 in. and AB = 5.2 in., find the number of radians in β.

18. If AC = 7.004 in. and CB = 3.214 in., find the number of radians in β.

19. If α = $\frac{\pi}{7}$ and CB = 31.2 mm, find the length of CA.

20. If β = 1.01 and CB = 19.71 mm, find the length of AC.

21. If β = 0.11 and AB = 73.2 in., find the length of CA.

22. If α = $\frac{2\pi}{5}$ and AB = 1.213 mm, find the length of AC.

23. If β = 29° and ML = 16.3 cm, find an altitude of the triangle.

24. If α = 55° and KM = 37 in., find an altitude of the triangle.

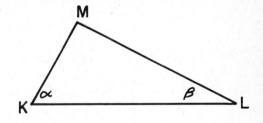

Use the drawing below for Problems 25 to 28.

25. If α = 0.94 and TS = 29 cm, find an altitude of the triangle.

26. If β = 0.63 and RT = 16 cm, find an altitude of the triangle.

27. If β = 0.49 and RT = 21 in., find an altitude of the triangle.

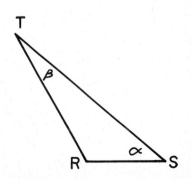

28. If α = 0.75 and TS = 31 in., find an altitude of the triangle.

NC 29. If $\cos\theta = \frac{4}{5}$, find the value of the other trigonometric functions of θ.

NC 30. If $\tan\theta = \frac{4}{3}$, find the value of the other trigonometric functions of θ.

NC 31. If $\sin\beta = \frac{24}{25}$, find the value of the other trigonometric functions of β.

NC 32. If $\cos\alpha = \frac{8}{17}$, find the value of the other trigonometric functions of α.

7.2 APPLICATIONS

Trigonometry allows us to work many more problems than were shown in Section 7.1. We have seen how trigonometry can be used to determine the altitude of a triangle. We can use a similar process to determine altitudes of other figures.

EXAMPLE 1 Calculate the area of trapezoid ABCD, if α = 53°.

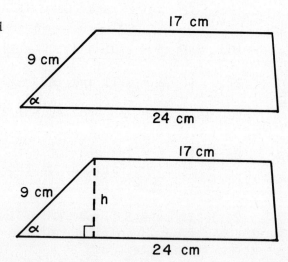

Solution In order to calculate the area of the trapezoid, we must have the altitude, h. We can calculate h using trigonometry.

$$\sin\alpha = \frac{\text{opposite side}}{\text{hypotenuse}}$$

$$\sin 53° = \frac{h}{9 \text{ cm}}$$

$$9 \text{ cm} \cdot (\sin 53°) = h$$

$$9 \text{ cm} \cdot (0.7986) = h$$

$$7.1877 \text{ cm} = h$$

We can now calculate the area of the trapezoid.

$$A_{trapezoid} = \frac{1}{2} h \cdot (b_1 + b_2)$$

$$= \frac{1}{2}(7.1877)(17 + 24)$$

$$= 147.3478 \text{ cm}^2 = 147.3 \text{ cm}^2$$
(to the nearest tenth)

The area of the trapezoid is 147.3 cm^2 (to the nearest tenth).

EXAMPLE 2   If the side of a regular pentagon is 7 cm, find the area of the pentagon.

Solution   In order to calculate the area of the pentagon, we need the perimeter, which is 5(7) = 35 cm, and the apothem. We can use trigonometry, once we know the value of α. We do know that α is one half of an interior angle. Each interior angle of a regular polygon $= \frac{(n-2)(180°)}{n}$,

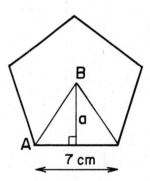

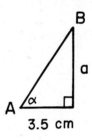

where n = the number of sides.

For a pentagon, n = 5, so we have:

Each interior angle $= \frac{(5-2)(180°)}{5}$

Each interior angle = 108°

Because each interior angle of a pentagon is 108°, $\alpha = 54°$.

We now can determine the apothem.

$$\tan \alpha = \frac{\text{opposite side}}{\text{adjacent side}}$$

$$\tan 54° = \frac{a}{3.5 \text{ cm}}$$

$$3.5 \text{ cm} \cdot (\tan 54°) = a$$

$$3.5 \text{ cm} \cdot (1.3764) = a$$

$$4.817 \text{ cm} = a$$

The apothem is 4.817 cm (to three places).

$$A_{pentagon} = \frac{1}{2} a \cdot p \quad \text{where } a \text{ is the apothem}$$
$$\text{and } p \text{ is the perimeter}$$

$$A_{pentagon} = \frac{1}{2} (4.817)(35)$$

$$A_{pentagon} = 84.2975 \text{ cm}^2$$

$$= 84 \text{ cm}^2 \text{ (to the nearest unit)}$$

The area of the pentagon is 84 cm^2.

The approach used to work Example 2 is just one of several ways of thinking about the problem. We could, for instance, have determined the central angle and taken one half of it to determine the apothem using the tangent function. We also could have calculated the area of one triangle composing the pentagon and multiplied that area by 5 rather than using the $\frac{1}{2}$ a·p formula. You should choose the method that you find easiest.

There are two terms that are used frequently in word problems. The ANGLE OF ELEVATION of an object is the angle measured from the horizontal <u>up</u> to the given object. The ANGLE OF DEPRESSION of an object is the angle measured from the horizontal <u>down</u> to the given object. In figure A, α is the angle of

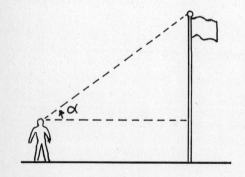

A

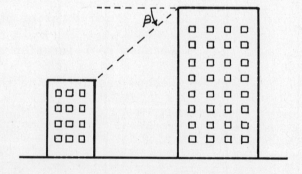

B

elevation from the observer to the top of the flagpole. In figure B, β represents the angle of depression from the roof of the taller building to the roof of the shorter building. Remember that both the angle of elevation and the angle of depression are measured from the horizontal; the angle of elevation goes up and the angle of depression goes down.

EXAMPLE 3 Determine the height of the flagpole in figure A if the angle of elevation, α, is 0.65 rad and the 6 ft tall person is 30 ft from the base of the flagpole.

Solution If h represents the height of the flagpole, then h - 6 represents the length of the side opposite the 0.65 rad angle in the right triangle. Because we are given the length of the side adjacent to the 0.65 rad angle, we will use the tangent function.

$$\tan \theta = \frac{\text{opposite side}}{\text{adjacent side}}$$

$$\tan 0.65 = \frac{h - 6}{30}$$

$$30 \cdot \tan 0.65 = h - 6$$

$$30 \cdot \tan 0.65 + 6 = h$$

$$30(0.7602) + 6 = h$$

$$28.8061 \text{ ft} = h$$

The flagpole is 28.8 ft tall (to the nearest tenth).

EXAMPLE 4 The angle of depression from the roof of one building to the roof of another building is 0.82 rad. The buildings have vertical walls and are 39 ft apart. If the shorter building is 62 ft tall, how tall is the other building?

Solution First we need a drawing to represent the conditions of the problem. Because the angle of depression is 0.82 rad, β must also be 0.82 rad (alternate interior angles of parallel lines). We let h represent the difference between the heights of the two buildings.

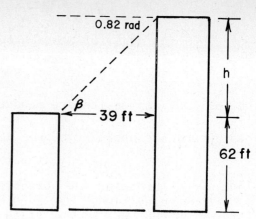

$$\tan \beta = \frac{\text{opposite side}}{\text{adjacent side}}$$

$$\tan 0.82 = \frac{h}{39}$$

$$39 \cdot \tan 0.82 = h$$

$$39(1.0717) = h$$

$$41.7968 = h$$

42 ft (to nearest foot) = h

The other building is 62 ft + 42 ft = 104 ft tall.

EXAMPLE 5 From a window in a lighthouse the keeper observes that the angles of depression of the bows of two boats following one another are 49° and 37°, respectively. If the window is 78 ft above the water, how far apart are the bows of the boats when they were observed?

Solution First we need a drawing to represent the conditions of the problem. Let d_1 be the distance the first boat is from the lighthouse and d_2 be the distance for the second boat. We can first determine d_1. We know that $\beta = 90° - 49° = 41°$.

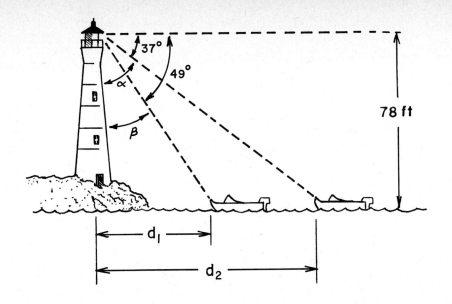

$$\tan \beta = \frac{\text{opposite side}}{\text{adjacent side}}$$

$$\tan 41° = \frac{d_1}{78}$$

$$78 \cdot \tan 41° = d_1$$

$$78(0.8693) = d_1$$

$$67.8054 = d_1$$

67.8 ft (to the nearest tenth) = d_1

We can now solve for d_2. We know that $\alpha = 90° - 37° = 53°$.

$$\tan \alpha = \frac{\text{opposite side}}{\text{adjacent side}}$$

$$\tan 53° = \frac{d_2}{78}$$

$$78 \cdot \tan 53° = d_2$$

$$78(1.3270) = d_2$$

$$103.5095 = d_2$$

103.5 ft (to the nearest tenth) = d_2

The distance between the bows of the boats is 103.5 ft − 67.8 ft = 35.7 ft.

EXERCISE 7.2

A calculator is recommended for all problems. Give answers to the nearest hundredth.

1. If ∠B = 1.34, AB = 17 cm, and BC = 31 cm, find the area of the triangle.

2. If ∠B = 0.35, AB = 6.2 in., and BC = 13.4 in., find the area of the triangle.

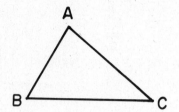

3. A parallelogram has sides of 22 in. and 7 in. If one angle is 0.43 rad, find the area of the parallelogram.

4. A parallelogram has sides of 10 in. and 26 in. If one angle is 0.189 rad, find the area of the parallelogram.

5. An isosceles trapezoid has a shorter base of 20 cm and legs of 8 cm. If the legs form a 62° angle with the longer base, find the area of the trapezoid.

6. An isosceles trapezoid has a shorter base of 18 cm and legs of 6 cm. If the legs form a 79° angle with the longer base, find the area of the trapezoid.

7. If a vertex angle of a rhombus is $\frac{3\pi}{7}$ rad and one side is 2.39 mm, find the diagonals.

8. If a vertex angle of a rhombus is $\frac{2\pi}{5}$ rad and one side is 7.09 mm, find the diagonals.

9. Calculate the area of a regular pentagon with 9 in. sides.

10. Calculate the area of a regular pentagon with 11 cm sides.

11. Calculate the area of a regular octagon with 4.2 mm sides.

12. Calculate the area of a regular decagon with 3.4 in. sides.

13. A surveyor wishes to determine the width of a river. He stands on one bank directly opposite a tree on the far bank of the river and drives in a stake. From the stake he walks 35 ft up the river and determines that the angle from the stake to the tree on the opposite bank is 52°. How far apart are the tree and the stake? (How wide is the river at that point?)

14. A surveyor must determine the width of a gorge so that a bridge can be built across it. On the far side she locates a large rock on the edge. She stands directly opposite the rock and marks the spot with a stake. From the stake she walks 40 ft along the gorge. There she find that the angle from the stake to the rock is 68°. How wide is the gorge? (What is the distance from the stake to the rock?)

15. When the angle of elevation of the sun is 69°, the shadow of a flag-pole is 12.3 ft. What is the height of the flagpole?

16. When the angle of elevation of the sun is 24°, the shadow of a telephone pole is 33.7 ft. How tall is the pole?

17. A 6 ft tall man standing 100 ft away from the base of a monument determines that the angle of elevation of the top of the monument is 0.48 rad. What is the height of the monument?

18. A 1.58 m tall woman standing 30 m away from the base of a monument measures the angle of elevation of the top of the monument and gets 0.59 rad. What is the height of the monument?

19. A helicopter pilot is hovering over a capsized boat. She must let down 56 ft of rope to reach the people in the water. If the angle of depression to the beach is 27°, how far must she travel to be over land? (Assume she maintains the same altitude.)

20. A balloonist wishes to determine his altitude. As he passes over a particular intersection, he notes that the angle of depression to his house is 19°. If his house is 750 ft away from the intersection, what is his altitude?

21. A lighthouse keeper spots a small boat headed directly toward the lighthouse and notes that the angle of depression to the boat is 5°. Five minutes later, he notices that the boat is still headed directly toward him, and the angle of depression now is 8°. If the keeper is 250 ft above the water level, how fast is the boat traveling (in mi/hr)? (1 mile = 5280 ft)

22. From the observation deck of the Tower of the Americas, 610 ft above street level, an observer sees a car on a street headed directly toward the tower, and determines the angle of depression to be 10°. One minute later, the angle of depression for the same car is 16°. How fast is the car traveling (in mi/hr)? (1 mile = 5280 ft)

*23. A balloonist wishes to determine his altitude. As he passes over a particular intersection, he measures the angle of depression to the top of the State Monument and gets 37°. The angle of depression to the base of the monument is 44°. If the State Monument is 75 ft tall, what is the balloonist's altitude?

*24. A pilot of a small plane is out at sea and needs to determine his altitude. He spots a ship headed in the same direction and determines that it is a very large crude carrier, which would be about 1200 ft long. The angle of depression to the stern is 42° and to the bow is 31°. From these figures he calculates his altitude. What does he estimate his altitude to be?

25. A 20 ft ladder is leaning up against a wall with the foot of the ladder 4 ft from the base of the wall. The base of the ladder is pulled away from the wall at a rate of 4 ft per second. What is the angle that the ladder makes with the wall after two seconds?

26. A girl is flying a kite at noon; the sun is directly overhead. She holds the string 3 ft above the ground and the wind holds the kite at a constant altitude of 53 ft. How much string does she have out when the shadow of the kite is 100 ft away from her? What is the angle of elevation of the kite? If she then lets more string out at a rate of 6 ft per second, what is the angle of elevation after three seconds?

**27. In Problems 25, what is the angle the ladder makes with the wall after t seconds?

**28. In Problem 26, what is the angle of elevation after t seconds?

29. A balloon rises vertically from a point 200 ft from an observer. If the balloon is rising at a constant rate of 5 ft per second, what is the angle of elevation of the balloon after 15 seconds?

30. A boat is pulled toward a dock by a rope passed through a ring on the dock 3 ft higher than the bow of the boat (where the other end of the rope is attached). What is the angle of depression (from the horizontal) of the rope when 12 ft of rope is out? If the rope is pulled in at a rate of 2 ft per second, what is the angle of depression of the rope after two seconds?

**31. In Problem 29, what is the angle of elevation of the balloon after t seconds?

**32. In Problem 30, what is the angle of depression of the rope after t seconds?

7.3 OTHER TRIGONOMETRIC FUNCTIONS

In addition to the sine, cosine, and tangent functions, there are three other trigonometric functions: the cosecant, secant, and cotangent.

$$\text{cosecant } \theta = \frac{1}{\sin \theta} = \frac{\text{hypotenuse}}{\text{leg opposite } \theta}$$

$$\text{secant } \theta = \frac{1}{\cos \theta} = \frac{\text{hypotenuse}}{\text{leg adjacent to } \theta}$$

$$\text{cotangent } \theta = \frac{1}{\tan \theta} = \frac{\text{leg adjacent to } \theta}{\text{leg opposite } \theta}$$

We usually write csc θ for cosecant θ, sec θ for secant θ, and ctn θ or cot θ for cotangent θ. For a 30° angle, we have:

$$\csc 30° = \frac{1}{\sin 30°} = \frac{1}{\frac{1}{2}} = 2$$

$$\sec 30° = \frac{1}{\cos 30°} = \frac{1}{\frac{\sqrt{3}}{2}} = \frac{2}{\sqrt{3}} = \frac{2\sqrt{3}}{3}$$

$$\cot 30° = \frac{1}{\tan 30°} = \frac{1}{\frac{1}{\sqrt{3}}} = \sqrt{3}$$

We can expand the table for the basic angles to include these functions.

$\theta =$	$\frac{\pi}{6} = 30°$	$\frac{\pi}{4} = 45°$	$\frac{\pi}{3} = 60°$
$\sin \theta$	$\frac{1}{2}$	$\frac{\sqrt{2}}{2}$	$\frac{\sqrt{3}}{2}$
$\cos \theta$	$\frac{\sqrt{3}}{2}$	$\frac{\sqrt{2}}{2}$	$\frac{1}{2}$
$\tan \theta$	$\frac{\sqrt{3}}{3}$	1	$\sqrt{3}$
$\csc \theta$	2	$\sqrt{2}$	$\frac{2\sqrt{3}}{3}$
$\sec \theta$	$\frac{2\sqrt{3}}{3}$	$\sqrt{2}$	2
$\cot \theta$	$\sqrt{3}$	1	$\frac{\sqrt{3}}{3}$

To evaluate cosecant, secant, or cotangent for a particular value of θ, evaluate the reciprocal function for the given value of θ, and take the reciprocal.

EXAMPLE 1   Find the cotangent of 1.41.

Solution   Because $\cot \theta = \dfrac{1}{\tan \theta}$, we determine the value of tan 1.41 and take the reciprocal:

$$\cot 1.41 = \frac{1}{\tan 1.41} = \frac{1}{6.1654} = 0.1622$$

When using a calculator, first find the value of tan 1.41, then press the $\dfrac{1}{x}$ button to take the reciprocal.

Cot 1.41 = 0.1622.

EXERCISE 7.3

A calculator may be used for all problems.

1. Determine the value of the cosecant, secant, and cotangent for a 60° angle.

2. Determine the value of the cosecant, secant, and cotangent for a $\dfrac{\pi}{4}$ angle.

For Problems 3 through 20, find the value of the given function for the given angle. Give answers to nearest hundredths.

3. cot 57°   4. sec $\dfrac{\pi}{8}$ 5. csc 1.63   6. csc 1.26

7. sec $\dfrac{\pi}{5}$ 8. cot 13°   9. sin $\dfrac{\pi}{9}$ 10. cot 71°

11. sec 0.67   12. cos 38°   13. csc $\dfrac{2\pi}{7}$ 14. tan 0.23

15. cot 0.05   16. cos 0.73   17. csc 7°   18. tan $\dfrac{3\pi}{7}$

19. sec 27°   20. sin $\dfrac{2\pi}{5}$

7.4 TRIGONOMETRIC FUNCTIONS FOR $90° \leq \theta \leq 360°$

The previous discussions of the trigonometric functions have assumed that θ is an angle in a right triangle. We may also think of θ in relation to a coordinate system. As shown below, the vertex of θ is the origin, $(0,0)$. One side coincides with the x-axis, and is called the initial side. The other

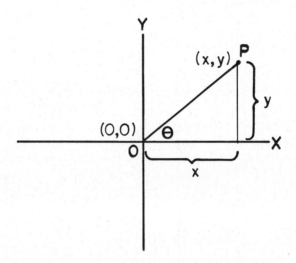

side, the terminal side of θ, we will call $\overline{OP}$. The point P has coordinates (x,y). $\overline{OP}$ then has length $r = \sqrt{x^2 + y^2}$. We can now use the coordinates of P to define the principal trigonometric functions of θ:

$$\sin \theta = \frac{y}{r}$$

$$\cos \theta = \frac{x}{r}$$

$$\tan \theta = \frac{y}{x}$$

A

B

C

Figure A illustrates a 30° angle where $r = 1$. We have:

$$\sin 30° = \frac{y}{r} = \frac{\frac{1}{2}}{1} = \frac{1}{2}$$

$$\cos 30° = \frac{x}{r} = \frac{\frac{\sqrt{3}}{2}}{1} = \frac{\sqrt{3}}{2}$$

$$\tan 30° = \frac{y}{x} = \frac{\frac{1}{2}}{\frac{\sqrt{3}}{2}} = \frac{\sqrt{3}}{3}$$

Figure B illustrates a $\frac{\pi}{4}$ angle where $r = 1$. We have:

$$\sin \frac{\pi}{4} = \frac{y}{r} = \frac{\frac{\sqrt{2}}{2}}{1} = \frac{\sqrt{2}}{2}$$

$$\cos \frac{\pi}{4} = \frac{x}{r} = \frac{\frac{\sqrt{2}}{2}}{1} = \frac{\sqrt{2}}{2}$$

$$\tan \frac{\pi}{4} = \frac{y}{x} = \frac{\frac{\sqrt{2}}{2}}{\frac{\sqrt{2}}{2}} = 1$$

Figure C illustrates a 60° angle where $r = 1$. We have:

$$\sin 60° = \frac{y}{r} = \frac{\frac{\sqrt{3}}{2}}{1} = \frac{\sqrt{3}}{2}$$

$$\cos 60° = \frac{x}{r} = \frac{\frac{1}{2}}{1} = \frac{1}{2}$$

$$\tan 60° = \frac{y}{x} = \frac{\frac{\sqrt{3}}{2}}{\frac{1}{2}} = \sqrt{3}$$

We see that we obtain the same values for the trigonometric functions as we did when we used the right triangle to determine them.

Using the same definitions for the functions, let us determine the values of the functions when $\theta = 0°$, as shown in figure D.

$$\sin 0° = \frac{y}{r} = \frac{0}{1} = 0$$

$$\cos 0° = \frac{x}{r} = \frac{1}{1} = 1$$

$$\tan 0° = \frac{y}{x} = \frac{0}{1} = 0$$

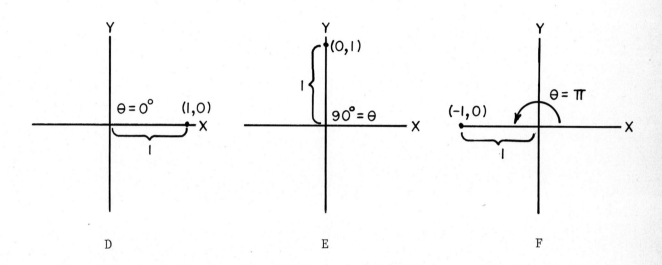

D   E   F

The principal functions for $\theta = 90°$, illustrated in figure E, are:

$$\sin 90° = \frac{y}{r} = \frac{1}{1} = 1$$

$$\cos 90° = \frac{x}{r} = \frac{0}{1} = 0$$

$$\tan 90° = \frac{y}{x} = \frac{1}{0} = \text{undefined} \quad \text{(Remember, division by 0 is not defined.)}$$

When $\theta = \pi$, as illustrated in figure F, we obtain the values:

$$\sin \pi = \frac{y}{r} = \frac{0}{1} = 0$$

$$\cos \pi = \frac{x}{r} = \frac{-1}{1} = -1$$

$$\tan \pi = \frac{y}{x} = \frac{0}{-1} = 0$$

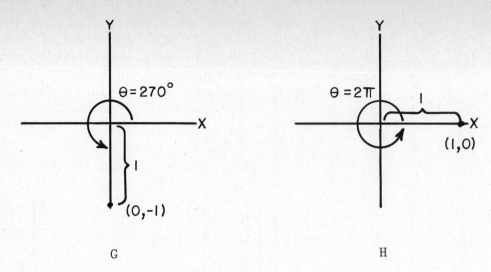

G   H

When θ = 270°, as illustrated in figure G, we have:

$$\sin 270° = \frac{y}{r} = \frac{-1}{1} = -1$$

$$\cos 270° = \frac{x}{r} = \frac{0}{1} = 0$$

$$\tan 270° = \frac{y}{x} = \frac{-1}{0} = \text{undefined}$$

Figure H illustrates θ = 2π. We have:

$$\sin 2\pi = \frac{y}{r} = \frac{0}{1} = 0$$

$$\cos 2\pi = \frac{x}{r} = \frac{1}{1} = 1$$

$$\tan 2\pi = \frac{y}{x} = \frac{0}{1} = 0$$

Notice that the values obtained for 2π are identical to those obtained for 0°. Because an angle of 2π is indistinguishable from an angle of 0°, one would expect that the trigonometric functions would have the same value for each.

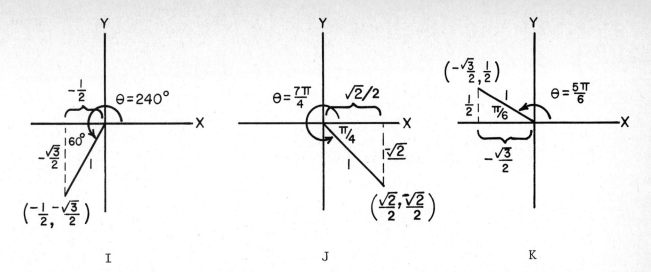

I J K

Let us determine the value of the trigonometric functions for 240°, as shown in figure I.

$$\sin 240° = \frac{y}{r} = \frac{-\frac{\sqrt{3}}{2}}{1} = -\frac{\sqrt{3}}{2}$$

$$\cos 240° = \frac{x}{r} = \frac{-\frac{1}{2}}{1} = -\frac{1}{2}$$

$$\tan 240° = \frac{y}{x} = \frac{-\frac{\sqrt{3}}{2}}{-\frac{1}{2}} = \sqrt{3}$$

For $\frac{7\pi}{4}$, as illustrated in figure J, we have:

$$\sin \frac{7\pi}{4} = \frac{y}{r} = \frac{-\frac{\sqrt{2}}{2}}{1} = -\frac{\sqrt{2}}{2}$$

$$\cos \frac{7\pi}{4} = \frac{x}{r} = \frac{\frac{\sqrt{2}}{2}}{1} = \frac{\sqrt{2}}{2}$$

$$\tan \frac{7\pi}{4} = \frac{y}{x} = \frac{-\frac{\sqrt{2}}{2}}{\frac{\sqrt{2}}{2}} = -1$$

The values of the functions when $\theta = \dfrac{5\pi}{6}$, as given in figure K, are:

$$\sin \frac{5\pi}{6} = \frac{y}{r} = \frac{\frac{1}{2}}{1} = \frac{1}{2}$$

$$\cos \frac{5\pi}{6} = \frac{x}{r} = \frac{-\frac{\sqrt{3}}{2}}{1} = -\frac{\sqrt{3}}{2}$$

$$\tan \frac{5\pi}{6} = \frac{y}{x} = \frac{\frac{1}{2}}{-\frac{\sqrt{3}}{2}} = -\frac{\sqrt{3}}{3}$$

We can thus calculate the value of the trigonometric functions for angles from 0° to 360° using the coordinate system. When evaluating a trigonometric function, sketch the angle on a coordinate system to determine the correct sign for the function.

Problems 13 through 18 in the Exercises ask for graphs of the functions. You need to become familiar with all three of the ways of thinking about the trigonometric functions: right triangle, unit circle on a coordinate system, and the graphs of the functions themselves.

EXERCISE 7.4

For Problems 1 through 12, complete the appropriate row of the tables below. A calculator may be used. Compute to nearest hundredth.

	θ in degrees	0°	30°	45°	60°	90°	120°	135°	150°	180°
	θ in radians	0	$\frac{\pi}{6}$	$\frac{\pi}{4}$	$\frac{\pi}{3}$	$\frac{\pi}{2}$	$\frac{2\pi}{3}$	$\frac{3\pi}{4}$	$\frac{5\pi}{6}$	π
1.	$\sin \theta$									
2.	$\cos \theta$									
3.	$\tan \theta$									
4.	$\csc \theta$									
5.	$\sec \theta$									
6.	$\cot \theta$									

	θ in degrees	180°	210°	225°	240°	270°	300°	315°	330°	360°
	θ in radians	π	$\frac{7\pi}{6}$	$\frac{5\pi}{4}$	$\frac{4\pi}{3}$	$\frac{3\pi}{2}$	$\frac{5\pi}{3}$	$\frac{7\pi}{4}$	$\frac{11\pi}{6}$	2π
7.	$\sin \theta$									
8.	$\cos \theta$									
9.	$\tan \theta$									
10.	$\csc \theta$									
11.	$\sec \theta$									
12.	$\cot \theta$									

For Problems 13 through 18, use degrees (or radians) for values on the x axis. Use the function values for values on the y axis. An example of a coordinate system to graph sin θ is given below.

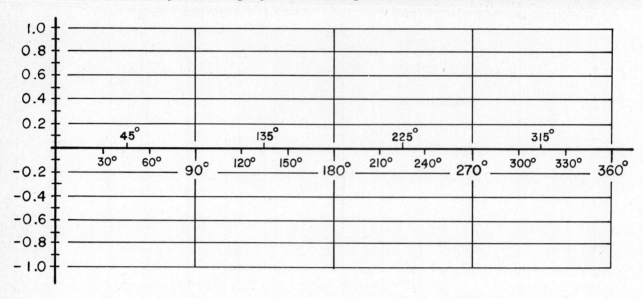

13. Using the values found in Problems 1 and 7, graph sin θ.

14. Using the values found in Problems 2 and 8, graph cos θ.

15. Using the values found in Problems 3 and 9, graph tan θ.

16. Using the values found in Problems 4 and 10, graph csc θ.

17. Using the values found in Problems 5 and 11, graph sec θ.

18. Using the values found in Problems 6 and 12, graph cot θ.

19. For what values of θ is sin θ negative?

20. For what values of θ is cos θ negative?

21. For what values of θ is tan θ negative?

22. For what values of θ is csc θ negative?

23. For what values of θ is sec θ negative?

24. For what values of θ is cot θ negative?

25. For what values of θ is tan θ undefined?

26. For what values of θ is csc θ undefined?

27. For what values of θ is sec θ undefined?

28. For what values of θ is cot θ undefined?

CHAPTER 7 REVIEW

1. Match each item in Column I with the appropriate item(s) from Column II. Items in Column II can be used once, more than once, or not at all.

   Column I

   a) $\sin 45°$

   b) angle measured from the horizontal down

   c) $\cos^{-1} \frac{\sqrt{2}}{2}$

   d) $\frac{\text{leg adjacent}}{\text{hypotenuse}}$

   e) $\frac{1}{\sin \theta}$

   f) $2 \cdot \cos \frac{\pi}{6}$

   Column II

   A. $30°$

   B. angle of elevation

   C. $\csc \theta$

   D. $\frac{1}{2}$

   E. $\frac{\sqrt{2}}{2}$

   F. $\cos \theta$

   G. $\sqrt{3}$

   H. $\sec \theta$

   I. angle of depression

   J. $\frac{\sqrt{3}}{2}$

   K. $\sin \theta$

   L. 1

   M. $\frac{\pi}{4}$

A calculator is recommended for the following problems except those marked with NC. Give answers to nearest hundredth.

2. If $\angle R = 53°$, and $ST = 14.3$ cm, find the length of $\overline{RS}$.

3. If $\angle S = 1.04$ and $ST = 7.3$ in., find the length of $\overline{RT}$.

4. If $ST = 34$ mm and $RS = 53$ mm, find $\angle S$.

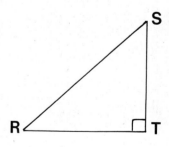

5. If ∠A = 1.3 and AB = 7.1 in., find an altitude of the triangle.

6. If ∠C = 34°, BC = 13 in., and AC = 27 in., find the area of the triangle.

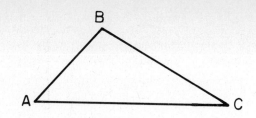

NC 7. If $\cos \theta = \frac{9}{41}$, find the value of the other trigonometric functions.

8. Calculate the area of a regular octagon with 5 mm sides.

9. A parallelogram has sides of 19 in. and 23 in. If one angle is 69°, find the area of the parallelogram.

10. A rhombus with 10 in. sides has a vertex angle of 50°. Find the length of the diagonals.

11. When the angle of elevation of the sun is 0.89 rad, the shadow of a building is 52 ft. How tall is the building?

12. A surveyor needs to determine the distance to an island in the middle of a river. He stands on a bank directly opposite the island and drives in a stake. He walks 50 feet up the river and determines that the angle from the stake to the island is 37°. What is the distance from the stake to the island?

13. From the top of a lighthouse 160 ft above sea level, the angle of depression to a boat is 35°. Find, to the nearest foot, the distance from the boat to the foot of the lighthouse.

14. The longer diagonal of rhombus is 24 ft and the shorter diagonal is 10 ft. Find, to the nearest degree, the angle which the longer diagonal makes with a side of the rhombus.

15. The diagonals of a rectangle are each 22 mm and intersect at an angle of 110°. Find, to the nearest whole number, the dimensions of the rectangle.

16. At A, a point 85 feet from the base of the building, the angle of elevation to the top of the building is 49°. At B, the angle of elevation to the top of the building is 26°.

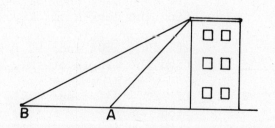

a) Find the height of the building to the nearest foot.

b) Find the distance from the top of the building to B, to the nearest foot.

372

17. $\overline{AB}$ and $\overline{CD}$ represent the heights of two cliffs on opposite sides of a river 120 ft wide. From B the angle of elevation to D is 20° and the angle of depression to C is 25°. Find, to the nearest foot,

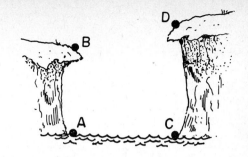

   a) the height of cliff, $\overline{AB}$.

   b) the height of cliff, $\overline{CD}$.

*18. In the diagram at the right, P represents a point 310 ft from the foot of a vertical cliff. A flagpole stands on the edge of the cliff. At P the angle of elevation to the base of the flagpole is 21° and to the top is 25°. Find, to the nearest foot:

   a) the distance from the foot of the cliff to the top of the flagpole.

   b) the height of the flagpole.

CHAPTER 8
THREE-DIMENSIONAL (SOLID) GEOMETRY

8.1 BASIC DEFINITIONS

In two-dimensional (plane) geometry, figures have dimensions of length and width. In three-dimensional (solid) geometry, the dimension of height (or thickness) is added. Circles and polygons are two-dimensional figures while spheres and polyhedra are three-dimensional.

A SPHERE is the set of all points in space equidistant from a fixed point, called the CENTER. The RADIUS of a sphere is the fixed distance. A RADIUS is also any line segment joining the center with any point on the sphere. In the sphere below, O is the center of the sphere and r is a radius. If two

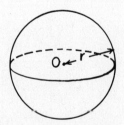

spheres have the same radius, they are called EQUAL SPHERES, or CONGRUENT SPHERES.

A POLYHEDRON (plural, polyhedra) is the union of a finite number of polygonal regions so that: (1) the interiors of any two regions do not intersect, and (2) every side of any of the polygons is the side of exactly one of the other polygons. Each of the polygonal regions is a FACE of the polyhedron. The intersection of any two faces of a polyhedron is an EDGE of the polyhedron. The intersection of any two edges is a VERTEX of the polyhedron. Thus, a sphere , a cone , and a cylinder , are not polyhedra; but a cube , a pyramid , and a prism are.

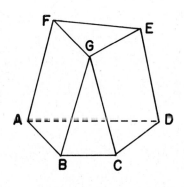

FIGURE 8.1

The polyhedron above has six faces: triangles GBC and FGE, and quadrilaterals ABGF, ADEF, CDEG, and ABCD. The eleven edges are $\overline{FE}$, $\overline{FG}$, $\overline{GE}$, $\overline{AF}$, $\overline{BG}$, $\overline{CG}$, $\overline{DE}$, $\overline{AB}$, $\overline{BC}$, $\overline{CD}$, and $\overline{AD}$. The seven vertices are A, B, C, D, E, F, and G. One possible pattern for the polyhedron in Figure 8.1 might be:

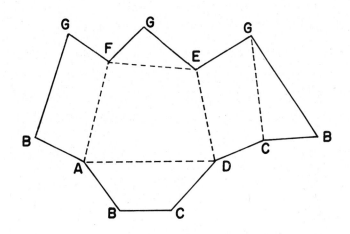

One way to name, or classify, polyhedra is by the number of faces. Some of these names are shown in Table 8.1.

Table 8.1 Names of polyedra according to number of faces.

No. of Faces	Name
4	Tetrahedron
5	Pentahedron
6	Hexahedron
7	Heptahedron
8	Octahedron
12	Dodecahedron
20	Icosahedron

The polyhedron in Figure 8.1 is a hexahedron. Note that a poly<u>hedron</u> is a three-dimensional figure, while a poly<u>gon</u> is only two-dimensional.

Another way to classify polyhedra is by the type of base, or bases, and faces. A PRISM is a polyhedron having two polygonal faces, called BASES, which are parallel and congruent and whose other faces, called LATERAL FACES, are parallelograms. The ALTITUDE of a prism is the perpendicular distance between the bases. A RIGHT PRISM has bases that are perpendicular to the lateral faces and all lateral faces are rectangles. An OBLIQUE PRISM has bases that are not perpendicular to the lateral faces and has parallelograms for all lateral faces. A RECTANGULAR PRISM has a rectangle for each base. A TRIANGULAR PRISM has a triangle for each base. Thus, the shape of the base is indicated in the name of the figure. A REGULAR PRISM is a right prism with bases that are regular polygons. Some of these polyhedra are shown in Figure 8.2.

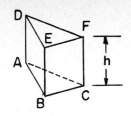 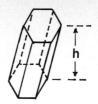

Right triangular prism   Oblique hexagonal prism   Regular square prism (cube)

FIGURE 8.2

In the right triangular prism shown in Figure 8.2, △ABC and △DEF are the bases. Rectangles ABED, BCFE, and CADF are the lateral faces, and h is the altitude.

A PARALLELEPIPED is a prism whose bases are parallelograms. A RECTANGULAR PARALLELEPIPED is a parallelepiped in which all faces and bases are rectangles. Thus, cubes and rectangular boxes are rectangular parallelepipeds. Specifically, a CUBE is a rectangular parallelepiped whose faces are all squares.

A PYRAMID is a polyhedron with one face, called the BASE, which is a polygon and with other faces, called LATERAL FACES, which are triangles that meet at a common point called the VERTEX. The intersections of pairs of lateral faces are called LATERAL EDGES. The ALTITUDE of a pyramid is the perpendicular distance from the vertex to the base. A REGULAR PYRAMID has a regular polygon as a base and the foot of the altitude is at the center of the base. The SLANT HEIGHT of a regular pyramid is the altitude of lateral faces. Figure 8.3 illustrates some types of pyramids.

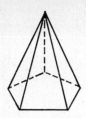 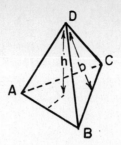 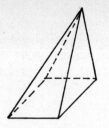

Regular pentagonal pyramid   Regular triangular pyramid   Oblique square pyramid

FIGURE 8.3

In the regular triangular pyramid shown in Figure 8.3, the base is △ABC; the lateral faces are △ABD, △ACD, and △BCD; the lateral edges are $\overline{AD}$, $\overline{BD}$, and $\overline{CD}$. The vertex is D; the altitude is h; and the slant height is b.

A CYLINDER is (1) a simple closed surface with parallel and congruent bases, each of which is a plane region bounded by a simple closed curve, plus (2) a curved surface connecting the two bases so that a straight line segment, called an ELEMENT, can be drawn from base to base and be contained in the curved surface. The curved surface is the LATERAL SURFACE. The cylinder is a CIRCULAR CYLINDER if the bases are circles; it is an ELLIPTICAL CYLINDER if the bases are ellipses. The ALTITUDE of a cylinder is the perpendicular distance between the bases. A RIGHT CYLINDER has elements that are perpendicular to the base. An OBLIQUE CYLINDER has elements that are not perpendicular to the base. The BOUNDARIES of a cylinder are the curves formed by the intersection of the bases with the lateral surface. Two types of cylinders are shown in Figure 8.4.

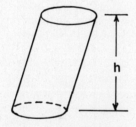 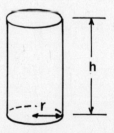

Oblique cylinder   Right circular cylinder

FIGURE 8.4

The right circular cylinder has two circular bases with radius r, a lateral surface with height h, and boundaries that are circles of radius r.

Given a simple closed curve in a plane and a point not in that plane, a CONE is the region enclosed by the simple closed curve, called the BASE, and the set of line segments, called ELEMENTS, that join the simple closed curve to the point not in the plane, the VERTEX. The perpendicular distance from the vertex to the base is the ALTITUDE of the cone. A CIRCULAR CONE has a circle for its base; an ELLIPTICAL CONE has an ellipse for its base. A RIGHT CIRCULAR CONE has a circle for its base and the center of the circle is the foot of the altitude of the cone. An OBLIQUE CIRCULAR CONE is a circular cone in which the line segment joining the center of the circle and the vertex is not perpendicular to the base. The SLANT HEIGHT of a right circular cone is the length of an element of the cone. The LATERAL SURFACE of a cone is the surface formed by the elements of the cone. The BOUNDARY of a cone is the curve formed by the intersection of the base and the lateral surface. Two types of cones are shown in Figure 8.5. In the right circular cone shown in Figure 8.5, the

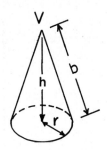   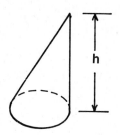

Right circular cone   Oblique elliptical cone

FIGURE 8.5

vertex is V, the altitude is h, the slant height is b, and the radius of the base is r.

When we name these three-dimensional figures, we have used three characteristics:

the angle formed by the side (or sides) and the base,
the type of base, and
the number of bases.

One way to outline these choices is given in Table 8.2. From each set of brackets we can choose one of the given words. We have a precise vocabulary for

Table 8.2   Outline of choices for naming three dimensional figures.

Angle with Base	Type of Base		Number of Bases
Right Oblique	Regular (Not regular)*	Triangular Quadrilateral Pentagonal ⋮ etc.	Prism (2)
Regular (Not regular)*	Regular (Not regular)*	Triangular Quadrilateral Pentagonal ⋮ etc.	Pyramid (1)
Right Oblique		Circular Elliptical (Not regular)*	Cylinder (2) Cone (1)

*The words in parentheses, (not regular), indicate what we assume if the word "regular" were not used.

many quadrilaterals so we can, for instance, replace the term "regular quadrilateral" by the word "square." Thus a "right regular quadrilateral prism" becomes a "right square prism."

EXERCISE 8.1

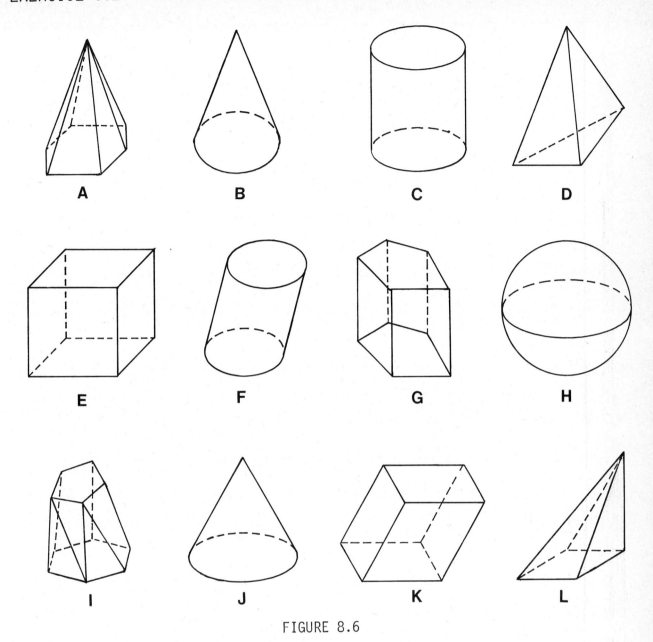

FIGURE 8.6

1. List the letters of the figures in Figure 8.6 that are polyhedra.

2. Name each of the figures in Figure 8.6 using three-word labels where possible (e.g., oblique octagonal prism). All bases are either polygons or circles.

3. Complete the following chart using the figures in Figure 8.6. If the entry does not apply to the figure, enter "NA" (not applicable). For instance, figure H has no edges, so NA has been entered in that space on the chart.

Figure	No. of Faces or Surfaces	No. of Bases	No. of Lateral Faces or Lateral Surfaces	No. of Edges or Boundaries	No. of Vertices
A					
B					
C				2	NA
D					
E					
F					NA
G					
H		0	1	NA	
I					
J					
K					
L					

4. For each of the figures below, name the line segments that represent the: a) slant height b) altitude.

 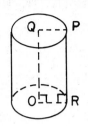 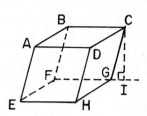

For Problems 5 and 6, name the line segments that represent the:

a) slant height
b) altitude
c) lateral edge
d) altitude of the base
e) altitude of a lateral face

5.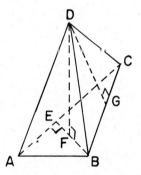

6.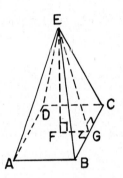

For Exercises 7 through 14, determine whether the given statement is true or false.

7. Every pyramid is a prism.

8. Every pyramid is a polyhedron.

9. Some polyhedra are prisms.

10. The bases of prisms may lie on planes perpendicular to one another.

11. A pyramid with eight faces has a hexagon as the border for its base.

12. If the sides of a prism are rectangles, then it is a right prism.

13. All prisms have rectangles for bases.

14. Some polyhedra are cylinders.

In the Exercises 15 through 24, draw a figure similar to the one in Figure 8.7 that can be used to make a model for the given figure. (The drawing does not have to be to scale.)

15. Rectangular right prism.

16. Triangular right prism.

17. Hexagonal right prism.

18. Pyramid with a square base.

19. Tetrahedron.

20. Octahedron.

21. Pyramid with an octagonal base.

22. Pyramid with a pentagonal base.

23. Right circular cone.

24. Right circular cylinder.

Pattern for model of a cube

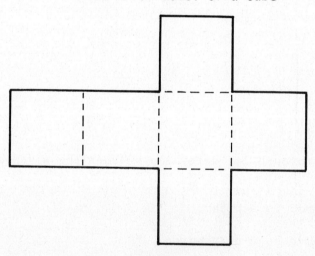

FIGURE 8.7

8.2 SURFACE AREA OF PRISMS AND PYRAMIDS

Now that we have the terminology for three-dimensional figures, we can develop methods for calculating the surface area for many of them. The SURFACE AREA of a three-dimensional figure is the sum of the measure of the areas of all the surfaces of the figure, and is symbolized SA. Thus, the

SURFACE AREA of a POLYHEDRON is equal to the sum of the areas of each polygonal face.

The TOTAL SURFACE AREA (TSA) of a prism consists of the area of both bases plus the area of the lateral faces, or the LATERAL SURFACE AREA (LSA) This can be symbolized:

$$TSA_{prism} = LSA + A_{bases}$$

A right rectangular prism and its surfaces are shown in Figure 8.8.

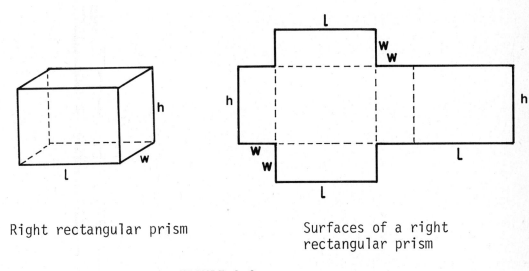

Right rectangular prism   Surfaces of a right rectangular prism

FIGURE 8.8

A right rectangular prism of length ℓ, width w, and height h, will have two congruent rectangular bases with dimensions ℓ by w, two rectangular lateral faces with dimensions ℓ by h, and two rectangular lateral faces with dimensions w by h. The total surface area would then be:

$$\begin{aligned}TSA_{right\ rectangular\ prism} &= LSA + A_{bases} \\ &= (2\ell h + 2wh) + 2\ell w \\ &= 2h\ell + 2wh + 2\ell w.\end{aligned}$$

EXAMPLE 1 What is the TSA of a right rectangular prism with 3" by 5" bases and an altitude of 7"?

Solution

$$A_{bases} = (2)(3)(5) = 30 \text{ in}^2$$

$$LSA = (2)(3)(7) + (2)(5)(7) = 42 + 70 = 112 \text{ in}^2$$

$$TSA = LSA + A_{bases}$$

$$= 112 + 30$$

$$= 142 \text{ in}^2$$

Thus the TSA is 142 in^2.

A right triangular prism (shown in Figure 8.9) of altitude h with bases having sides of lengths b, c, and d, would have two congruent triangular

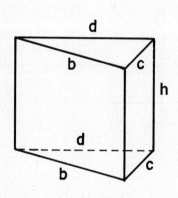
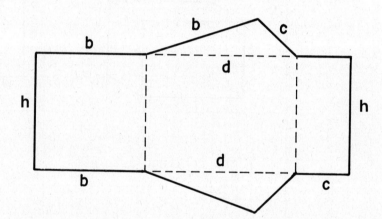

Right triangular prism   Surfaces of a right triangular prism

FIGURE 8.9

bases and three rectangular lateral faces with dimensions b by h, c by h, and d by h. The total surface area would then be:

$$TSA_{right\ triangular\ prism} = LSA + A_{bases}$$

$$= bh + ch + dh + A_{bases}$$

where the area of the triangular bases would be found using the specific facts given for the triangle.

EXAMPLE 2 A right triangular prism with an altitude of 9 cm has a right triangular base with legs of 6 cm and 8 cm. Find the total surface area.

Solution Because the base is a right triangle and the two legs given are in the ratio 3:4, we know we have a 3:4:5 right triangle; thus the hypotenuse is 10 cm.

$$A_{bases} = 2A_{triangle} = 2(\frac{1}{2}hb)$$
$$= 2(\frac{1}{2})(6)(8)$$
$$= 48 \text{ cm}^2$$

$$LSA = bh + ch + dh$$
$$= (6)(9) + (8)(9) + (10)(9)$$
$$= 216 \text{ cm}^2$$

$$TSA = LSA + A_{bases}$$
$$= 216 + 48$$
$$= 264 \text{ cm}^2$$

The TSA for the prism is 264 cm^2.

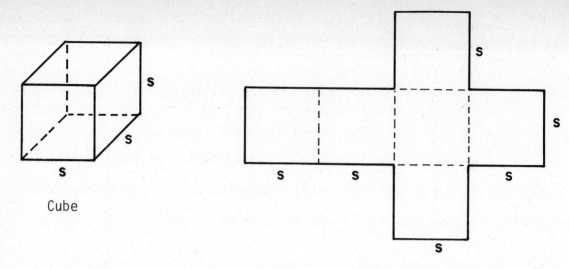

Cube

Surface area of a cube

FIGURE 8.10

A cube, as shown in Figure 8.10, with edges of length s has six square faces each with sides of length s; therefore, the total surface area of a cube would be

$$\boxed{\text{TSA}_{\text{cube}} = 6s^2.}$$

EXAMPLE 3   Find the total surface area for a cube with a 2 ft edge.

Solution

$$\text{TSA} = 6(2)^2$$
$$= 24 \text{ ft}^2$$

The surface area of the cube is 24 ft^2.

This method for determining total surface area can be used for many different types of prisms.

EXAMPLE 4   Determine the total surface area for a right regular hexagonal prism with an altitude of 11 cm if the base has an apothem (see Section 5.4) of $5\sqrt{3}$ cm.

Solution Because the base has an apothem of $5\sqrt{3}$ cm, each side of the hexagon must be 10 cm.

$$A_{bases} = 2(\tfrac{1}{2} \text{ pa})$$
$$= 2(\tfrac{1}{2})(6)(10)(5\sqrt{3})$$
$$= 300\sqrt{3} \text{ cm}^2$$

$$\text{LSA} = 6(10)(11)$$
$$= 660 \text{ cm}^2$$

$$\text{TSA} = \text{LSA} + A_{bases}$$
$$= 660 + 300\sqrt{3} \text{ cm}^2$$

The total surface area of the right prism is $660 + 300\sqrt{3}$ cm^2 or 1179.6 cm^2 (to the nearest tenth).

**EXAMPLE 5 Find the expression for the total surface area of a box if its length is 5 more than twice the width and the height is 12 more than the width.

Solution Let x = width
2x + 5 = length
12 + x = height

$$\text{TSA} = 2\ell h + 2wh + 2\ell w$$
$$= 2(2x + 5)(x + 12) + 2(x)(x + 12) + 2(2x + 5)(x)$$
$$= 4x^2 + 58x + 120 + 6x^2 + 34x$$
$$= 10x^2 + 92x + 120$$

The expression for the TSA is $10x^2 + 92x + 120$.

T EXAMPLE 6 Find the total surface area of a right regular octagonal prism with a height of 11 cm and base edges of 6 cm.

Solution The lateral surface area is easily computed, because there are 8 rectangular faces with a width of 6 cm and a length of 11 cm.

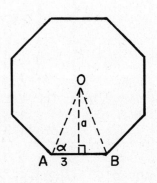

$$LSA_{prism} = 8(6)(11)$$

$$= 528 \text{ cm}^2$$

The area of the bases, however, requires more work. If we are to determine the area directly, we must use trigonometry to calculate the apothem. (See Section 7.2.) First let us calculate the number of degrees in each interior angle. (See Section 4.5.)

Each interior angle $= \dfrac{(n - 2)(180°)}{n}$

$$= \dfrac{6(180°)}{8}$$

$$= 135°$$

Because each interior angle contains 135°, the base angle of $\triangle OAB$, α, contains 67.5°. We can form a right triangle, $\triangle ACO$ and use trigonometry to determine the apothem, a.

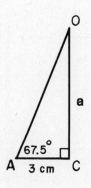

390

$$\tan 67.5° = \frac{a}{3}$$

$$3 \tan 67.5° = a$$

$$3(2.4142) = a$$

$$7.2426 \text{ cm} = a$$

We can now calculate the area of the octagon to find the area of one base.

$$A_{octagon} = \frac{1}{2} p \cdot a$$

$$= \frac{1}{2}(8)(6)(7.2426)$$

$$= 173.8224$$

$$A_{octagon} = 173.8 \text{ cm}^2$$

We now have:

$$TSA = LSA + A_{bases}$$

$$= 528 \text{ cm}^2 + 2(173.8 \text{ cm}^2)$$

$$= 875.6 \text{ cm}^2$$

The total surface area is 876 cm^2 (to nearest whole unit).

The total surface area for a pyramid consists of the area of the base plus the area of the lateral faces, which are triangles. If we are given the

$$\boxed{TSA_{pyramid} = LSA + A_{base}.}$$

slant height of a regular pyramid and the sides of the base, we can easily compute the lateral surface area. We would need more information to calculate the area of the base.

For a regular triangular pyramid with an equilateral triangle with side s for a base, we would have the following surfaces (illustrated in Figure 8.11): three lateral faces, each of which is a triangle with altitude ℓ, the

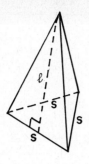
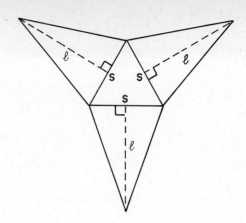

Regular equilateral triangular pyramid

Surfaces of a regular triangular pyramid

FIGURE 8.11

slant height of the pyramid, and base s, a side of the base. The base itself would have area of $\frac{s^2}{4}\sqrt{3}$ (see Section 5.2). The total surface area would then be:

$$TSA_{regular\ triangular\ pyramid} = LSA + A_{base}$$

$$= 3(\frac{1}{2}s\ell) + \frac{s^2\sqrt{3}}{4}$$

$$= \frac{3}{2}s\ell + \frac{s^2\sqrt{3}}{4}$$

where s is the edge of the base and ℓ is the slant height of the pyramid.

It is best not to memorize many different formulas for surface area; rather you should think about the shape of each face and how you may determine the area of that figure. In general, then you would only need to remember the area of a triangle and of a rectangle.

EXAMPLE 7   Find the total surface area for a regular triangular pyramid with a slant height of 15 in. and a base with a 6 in. side.

Solution   We know that the base is an equilateral triangle because the figure is a regular triangular pyramid. The area of an equilateral triangle with a 6 in. side gives the area of the base.

$$A_{base} = \frac{1}{4}(6)^2(\sqrt{3})$$

$$= 9\sqrt{3} \text{ in}^2$$

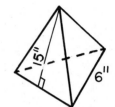

$$LSA = 3(\frac{1}{2})(15)(6)$$

$$= 135 \text{ in}^2$$

$$TSA = LSA + A_{base}$$

$$= 135 + 9\sqrt{3} \text{ in}^2$$

The TSA of the pyramid is $135 + 9\sqrt{3}$ in^2 or 150.6 in^2 (to the nearest tenth).

Instead of the slant height, however, we could be given the altitude of the pyramid. In that case, we would need to compute the slant height. The altitude is one leg of a right triangle, the slant height is the hypotenuse, and the other leg is the line segment in the base that joins the slant height to the altitude. Thus we can use the Pythagorean theorem to compute the slant height.

EXAMPLE 8   Determine the total surface area for a regular square pyramid with altitude of 14 cm if the base has 8 cm sides.

Solution The altitude, one leg of the right triangle ABC, is 14 cm. The hypotenuse $\overline{AB}$ is the slant height. Because the vertex A of the pyramid is centered over the base, C lies in the center of the square; therefore, $\overline{CB}$ is one half the length of a side of the square, or 4 cm. The slant height then is:

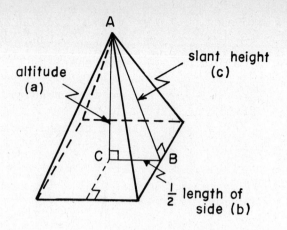

$$c^2 = a^2 + b^2$$

$$(AB)^2 = (AC)^2 + (CB)^2$$

$$= 14^2 + 4^2$$

$$= 212$$

$$AB = 2\sqrt{53} \text{ cm}$$

$$A_{base} = 8^2$$

$$= 64 \text{ cm}^2$$

$$LSA = 4(\tfrac{1}{2})(cb)$$

$$= 4(\tfrac{1}{2} \cdot 8 \cdot 2\sqrt{53})$$

$$= 32\sqrt{53} \text{ cm}^2$$

$$TSA = 64 + 32\sqrt{53} \text{ cm}^2$$

The TSA of the pyramid is $64 + 32\sqrt{53}$ cm^2 or 297.0 cm^2 (to nearest tenth).

Another type of problem is that in which the length of a lateral edge is given instead of the slant height. As in the previous example, the Pythagorean theorem is applied, but now the slant height forms one leg of the right triangle, the lateral edge is the hypotenuse, and the other leg is one half a side of the base of the pyramid.

EXAMPLE 9 Determine the total surface area of a regular hexagonal pyramid with a lateral edge of 6 in. if the base has 4 in. sides.

Solution Because the base is a regular hexagon with 4 in. sides, the apothem is $2\sqrt{3}$ in. (See Section 5.4.)

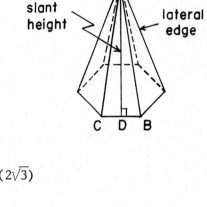

$$A_{base} = \frac{1}{2} pa$$

$$= \frac{1}{2} \cdot (6)(4)(2\sqrt{3})$$

$$= 24\sqrt{3} \text{ in}^2$$

In order to calculate the lateral surface area, we must first determine the slant height. Now the slant height is also the altitude of a lateral face, which is an isosceles triangle, and the altitude of an isosceles triangle is the perpendicular bisector of the base of the triangle. Therefore, $\overline{AD}$, the altitude of $\triangle ABC$, bisects $\overline{CB}$ and is perpendicular to it. Thus $\triangle ABD$ is a right triangle with legs $\overline{AD}$ and $\overline{DB}$, and hypotenuse $\overline{AB}$. Now $\overline{AB}$ is a lateral edge of the pyramid and is 6 in. long; $\overline{DB}$ is 2 in. long. We can find the length of $\overline{AD}$ using the Pythagorean theorem:

$$c^2 = a^2 + b^2$$

$$(AB)^2 = (AD)^2 + (DB)^2$$

$$6^2 = (AD)^2 + 2^2$$

$$(AD)^2 = 36 - 4$$

$$= 32$$

$$AD = 4\sqrt{2} \text{ in.}$$

$$LSA = 6(\tfrac{1}{2})(\ell b)$$

$$= 6(\tfrac{1}{2})(4\sqrt{2})(4)$$

$$= 48\sqrt{2} \text{ in}^2$$

$$TSA = LSA + A_{base}$$

$$= 48\sqrt{2} + 24\sqrt{3} \text{ in}^2$$

The TSA for the pyramid is $48\sqrt{2} + 24\sqrt{3}$ in^2 or 109.5 in^2 (to nearest tenth).

By examining the preceding three examples, we see the importance of distinguishing the problems by carefully reading and labeling a drawing. The altitude of the prism, the lateral edge, and the altitude of a face (which is the slant height of the pyramid) are all _different_ dimensions of the figure.

EXERCISE 8.2

(Note: The use of a calculator will speed many of the calculations necessary for this problem set.) Give an exact answer unless rounding is indicated. Find the surface area of the right rectangular prisms with dimensions:

1. ℓ = 6 cm, w = 5 cm, h = 7 cm

2. ℓ = 10 cm, w = 10 cm, h = 10 cm

3. ℓ = 5.7 in., w = 8.9 in., h = 13.7 in. (to the nearest tenth)

4. $\ell = 5\frac{1}{2}$ yd, w = 4 yd, h = $10\frac{1}{2}$ yd

Find the surface area of the following cubes (s = length of an edge).

5. s = 12 cm

6. s = 1.05 cm (to the nearest tenth)

7. s = $\frac{1}{2}$ yd

8. s = 0.15 in. (to the nearest tenth)

9. Rob wants to make a canvas tent with a canvas floor and ends, like the one shown at the right. How many square yards of canvas will he use? (Ignore waste.)

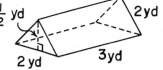

Find the surface area of each regular pyramid with the given dimension (ℓ is slant height).

10.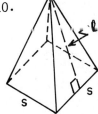

s = 7"
ℓ = 9"

11.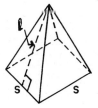

s = $3\frac{1}{2}$"
ℓ = $4\frac{1}{2}$"

12.

ℓ = 5 yd
s = 3 yd

*13. The base of a right prism is a rhombus with 10 in. sides; one diagonal of the base is 16 in. The altitude of the prism is 15 in. Find its surface area.

*14. Find the surface area of a regular pyramid whose base is a hexagon with a perimeter of 60 cm and a slant height of 20 cm.

15. The diagonal of one face of a cube is 8". Find the surface area of the cube.

16. Find the surface area of a right prism that has as the base an isosceles right triangle with a 10 in. hypotenuse and a 12 in. lateral edge.

17. A storage vault has a rectangular floor 72 ft by 48 ft and vertical walls that are 15 ft high. Find the total area of floors, wall, and ceiling.

18. Find the surface area of a regular pyramid having an altitude of 12 in. and a square base 10 in. on each side.

*T 19. Given a right prism with a base as pictured and a height of 8 ft, find its lateral area and total surface area.

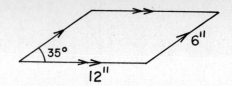

*T 20. Find the total area of a regular pyramid having a lateral edge of 40 in. and having a regular octagon with 30 in. sides as its base.

8.3 SURFACE AREA OF CYLINDERS, CONES, AND COMPOSITE FIGURES

The total surface area of a cylinder is given by adding the area of the two bases and the lateral surface area. For a right circular cylinder, the bases are two congruent circles. The lateral surface, obtained by making a vertical cut and unrolling the cylinder, is a rectangle with a width of the circumference of the base and a height of the height (altitude) of the cylinder (as shown in Figure 8.12).

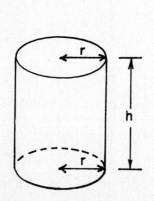

Right circular cylinder

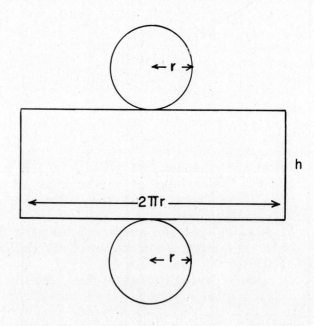

Surfaces of a right circular cylinder

FIGURE 8.12

The total surface area for a right circular cylinder of altitude h and base with radius r is given by:

$$TSA_{cylinder} = LSA + A_{bases}$$
$$= h(2\pi r) + 2\pi r^2$$
$$= 2\pi hr + 2\pi r^2$$

EXAMPLE 1   What is the total surface area of a right circular cylinder 6 ft high and 12 ft across?

Solution
$$TSA = LSA + A_{bases}$$
$$= 2\pi(6)(6) + 2\pi(6)^2$$
$$= 72\pi + 72\pi$$
$$= 144\pi \text{ ft}^2$$

The total surface area of the cylinder is 144π ft^2 or 452.39 ft^2 (to nearest hundredth).

The total surface area of a right circular cone can be found by adding the lateral surface area to the area of the base. The lateral surface of a cone can be cut vertically from the edge of the base to the vertex and then unrolled, forming a sector of a circle. A right circular cone with slant height ℓ and base with radius r has a lateral surface area that is a sector of a circle with radius ℓ and an arc length the circumference of the base of the cone.

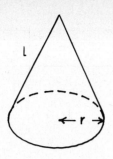

Right circular cone

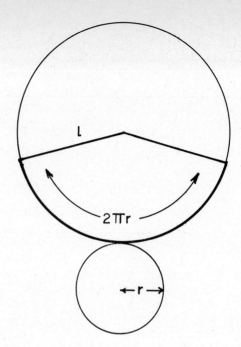

Surfaces of a right circular cone

FIGURE 8.13

The area of the base of the cone in Figure 8.13 is the area of a circle with radius r. The lateral surface is a sector of a circle with radius ℓ,

$$\frac{\text{Area of sector}}{\pi \ell^2} = \frac{2\pi r}{2\pi \ell}$$

$$\text{Area of sector} = \pi r \ell$$

$$\text{LSA}_{\text{cone}} = \pi r \ell = \frac{1}{2} C\ell,$$

Therefore, if C is the circumference of the base:

$$\boxed{\begin{aligned} \text{TSA}_{\text{cone}} &= \text{LSA} + A_{\text{base}} \\ &= \frac{1}{2} C\ell + \pi r^2 \end{aligned}}$$

EXAMPLE 2 Find the total surface area for a right circular cone with 16 ft slant height and base with an 8 ft radius.

Solution

$$A_{base} = \pi 8^2$$

$$= 64\pi \text{ ft}^2$$

$$LSA = \frac{1}{2}(2\pi 8)(16)$$

$$= 128\pi \text{ ft}^2$$

$$TSA = LSA + A_{base}$$

$$= 128\pi + 64\pi$$

$$= 192\pi \text{ ft}^2$$

The TSA of the cone is 192π ft^2 or 603.19 ft^2 (to nearest hundredth).

As was the case for pyramids, we can be given the altitude of the cone and will have to calculate the slant height. Because the slant height is the hypotenuse of a right triangle with the altitude of the cone as one leg and the radius of the base as the other, we can use the Pythagorean theorem.

EXAMPLE 3 What is the total surface area of a right circular cone of altitude 12 cm and base of radius 5 cm?

Solution The slant height, $\overline{AB}$, is the hypotenuse of right triangle AOB; the altitude $\overline{OA}$, 12 cm, and the radius of the base $\overline{OB}$, 5 cm, are the two legs of the triangle. Using the Pythagorean theorem,

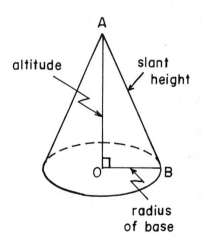

$$c^2 = a^2 + b^2$$

$$(AB)^2 = (AO)^2 + (OB)^2$$

$$= (12)^2 + 5^2$$

$$= 169$$

$$AB = 13 \text{ cm}$$

$$TSA = LSA + A_{base}$$

$$= \frac{1}{2}(2\pi)(5)(13) + \pi(5)^2$$

$$= 65\pi + 25\pi$$

$$= 90\pi \text{ cm}^2$$

The TSA of the cone is 90π cm^2 or 282.74 cm^2 (to nearest hundredth).

EXAMPLE 4   If the slant height is four times the radius of the base in a cone with a total surface area of 45π m^2, find the dimensions of the cone.

Solution   Let   x = radius of base
4x = slant height

$$TSA = LSA + A_{base}$$

$$= \frac{1}{2}(2\pi r)(\ell) + \pi r^2$$

$$45\pi = \pi x(4x) + \pi x^2$$

$$45\pi = 5\pi x^2$$

$$9 = x^2$$

$$\pm 3 = x$$

We use only the positive root because we are measuring distance, so the radius of base is 3 m, and the slant height is 12 m.

The surface area of a sphere of radius r is:

$$SA_{sphere} = 4\pi r^2$$

EXAMPLE 5   What is the surface area of a 12 in. diameter sphere?

Solution   If $d = 12$ in., then $r = 6$ in., and:

$$SA = 4\pi r^2$$
$$= 4\pi 6^2$$
$$= 144\pi \text{ in}^2$$

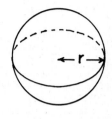

The surface area of the sphere is 144π in^2 or 452.39 in^2 (to nearest hundredth).

We can now find the surface area of more complicated figures by finding the area for each separate piece and then taking the total of the areas found.

EXAMPLE 6   Find the surface area of the mushroon shaped figure. Heavy dots indicate centers of arcs, circles, and spheres.

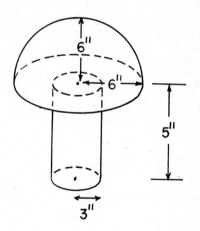

Solution   The figure consists of a cylinder topped by a hemisphere. The surfaces of the figure are: the surface of the hemisphere, the outer area of the circle that is the bottom of the hemisphere (note, the part of the circle that tops the cylinder is NOT part of the surface area of the mushroom), the lateral surface of the cylinder, and the base of the cylinder. We can now use our given formulas to find each of these areas.

403

$$SA_{hemisphere} = \frac{1}{2} A_{sphere} = \frac{1}{2}(4\pi r^2)$$
$$= 72\pi \text{ in}^2$$

$$SA_{outer\ portion\ of\ circle} = A_{outer} - A_{inner}$$
$$= \pi 6^2 - \pi 3^2$$
$$= 27\pi \text{ in}^2$$
$$LSA_{cylinder} = 2\pi rh$$
$$= 30\pi \text{ in}^2$$

$$A_{base\ cylinder} = \pi r^2$$
$$= 9\pi \text{ in}^2$$

Taking the total of the above areas gives:

$$TSA_{mushroom} = 138\pi \text{ in}^2 \text{ or } 433.54 \text{ in}^2 \text{ (to the nearest hundredth).}$$

EXAMPLE 7 Find the total surface area for a cone formed by revolving an 8 in.-15 in.-17 in. right triangle about its longer leg.

Solution First we must visualize how the cone is generated. Because the triangle is revolved about its longer leg, we draw the triangle with the 15 in. side vertical and the 8 in. side as the base. The cone formed then will have a base with radius of 8 in., an altitude of 15 in., and a slant height of 17 in. We can now calculate the surface area.

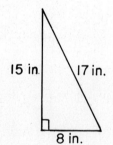

$$A_{base} = \pi r^2$$
$$= \pi(8)^2$$
$$= 64\pi \text{ in}^2$$

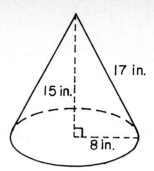

$$LSA_{cone} = \frac{1}{2} C\ell$$
$$= \frac{1}{2} 2\pi(8)(17)$$
$$= 136\pi \text{ in}^2$$

$$TSA_{cone} = LSA_{cone} + A_{base}$$
$$= 64\pi + 136\pi \text{ in}^2$$
$$= 200\pi \text{ in}^2 \text{ or } 628.3 \text{ in}^2 \text{ (to nearest tenth)}$$

EXERCISE 8.3

A calculator may be used. Give answers correct to hundredths.
Find the surface area of the following right circular cylinders:

1. $r = 3\,m$, $h = 0.5\,m$ (to nearest whole unit)

2. $r = 0.80$ in., $h = 1.25$ in. (to the nearest tenth)

Find the surface area of the following right circular cones:

3. $r = 1.5$ cm, $h = 4$ cm

4. $r = 2\frac{1}{2}$ ft, $h = 7$ ft

Find the surface area of the following spheres:

5. $r = 4$ ft

6. $r = \frac{1}{2}$ in.

7. How many square inches of aluminum are used to make a beer can with a diameter of 2 1/2" and a height of 4 3/4"?

8. How many square centimeters of paper are needed to make a label for a can of grapefruit juice if the can is 5 cm in diameter and 17 cm high?

9. Frozen orange juice cans have tops and bottoms of aluminum and sides of cardboard. For a can 4 1/2" high with a diameter of 2 1/2", find the amount of aluminum in each can; find the amount of cardboard needed for each.

10. Find the surface area of a right circular cone with radius 5 in. and altitude 12 in.

11. A sphere with an 8 in. diameter just fits inside (is inscribed in) a right circular cylinder. Find the lateral area of the cylinder.

12. An oil drum's height is three times its radius. If the surface area is 96π ft, what are the drum's dimensions?

13. The radius of one cylinder is twice that of a second cylinder, and they are the same height. What is the ratio of the surface area of the larger cylinder to the surface area of the smaller cylinder?

14. Find the radius of a sphere with surface area 144π m^2.

Find the lateral and total area for each solid in Problems 15 to 18.

15. A cylinder formed by revolving rectangle ABCD about AB; AB = 6", BC = 3".

16. A cone formed by revolving a 3-4-5 cm right triangle about its longer leg.

17. A cone formed by revolving a 5-12-13 cm right triangle about its its shorter leg.

18. A cylinder formed by revolving rectangle FGHI about FG; FG = 8", GH = 5".

19. How many square yards are in the outside surface of a cylindrical smokestack 50' high the exterior diameter of which is 6'?

C 20. A steam engine contains 120 cylindrical pipes called flues, each with an inside diameter of 1.5" and 12 ft long, through which the water passes to be heated. What is the total internal heating surface of these flues?

21. A cylindrical pail is 6 in. deep and 7 in. in diameter. Find the amount of tin needed for its construction. (Ignore waste.)

C 22. Assuming the earth to be a sphere 7960 mi in diameter, what is the area of its surface?

23. If the surface area of a cube is 96 cm^2, what is the length of an edge?

24. Given that the total surface area of a right circular cylinder is 168π cm^2 and the height is 5 cm more than three times the radius, find the dimensions of the cylinder.

*25. What are the dimensions of a right circular cone with total surface area of 500π cm^2 if the radius is 30 cm less than the slant height?

26. Find the dimensions of a right rectangular prism if the length is 5 in. greater than twice the width, the height is 4 in. less than the length, and the lateral surface area is 110 in^2.

27. Find the total surface area of the figure to the right. Include the lateral surface area of the cylinder, the area of the base, and the area of the hemisphere.

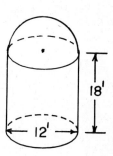

C 28. Find the surface area of a regular pentagonal pyramid with a height of 10 in. and a base with sides of 3 in. and an apothem of 0.92 in.

29. Find the area of the exposed outside surface of a house with dimensions as given in the drawing. (Assume that the floor is rectangular.)

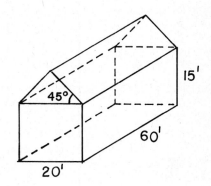

30. Find the total surface area for the figure to the right. Heavy dots indicate the centers of circles.

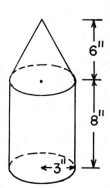

C 31. A cone with the dimensions given in the drawing increases its temperature at the rate of 1° per 500 in^2 of surface area per minute. How much does the temperature increase after 10 minutes?

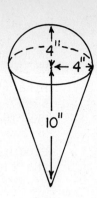

C 32. Find the total surface area of the given solid block of metal that has two holes bored through it.

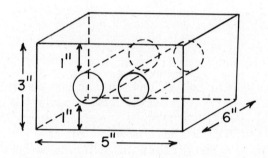

33. A right circular cylinder is three times as tall as it is across. If the total surface area is 504π mm^2, find the dimensions of the cylinder.

*34. A right rectangular prism has total surface area 562 in^2. If the width is 1 in. more than the height and the length is 1 in. less than twice the width, find the dimensions of the prism.

**35. A rectangular piece of cardboard 12 in. by 18 in. is to be made into an open box by cutting a small square from each corner and bending up the sides. If x is the side of the square cut out, give the expression for the surface area of the box.

**36. A rectangular sheet of tin b in. wide and 2b in. long is to be used to make an open box by cutting a small square of tin from each corner and bending up the sides. If x is the side of the square cut out, give the expression for the surface area of the box.

**37. Find the lateral surface area of a right circular cylinder with radius $\sqrt{x}$ and altitude Δx.

**38. Find the lateral surface area of a right circular cylinder with radius $\sqrt[3]{y}$ and altitude Δy.

**39. A wire 18 cm long is cut into two pieces, one bent to form a square and the other to form an equilateral triangle. If x is the length of wire used to form the triangle, give the expression for the sum of the areas of the two figures.

**40. A wire 20 cm long is cut into two pieces, one bent to form a square and the other to form a circle. If x is the length of wire used to form the circle, give the expression for the sum of the areas of the two figures.

8.4 VOLUME

Fortunately, volumes are usually easier to calculate than are surface areas. Volume measures the number of cubic units that the figure contains. If each block is a one unit cube, then figure A, shown below, contains six

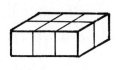

A

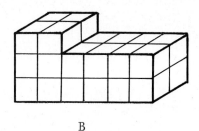

B

cubic units. B contains four stacks of three blocks each and eight stacks of two blocks each, giving a total of (4)(3) + (8)(2) = 28 cubic units.

To find the volume of any prism or cylinder, multiply the area of the base times the altitude. Thus, we have the formula:

$$\text{Volume}_{\text{cylinder or prism}} = A_{\text{base}} \cdot h,$$

where h is the altitude.

EXAMPLE 1 What is the volume of an 8 in. tall right circular cylinder with a 6 in. diameter base?

Solution
$$A_{\text{base}} = \pi r^2$$
$$= 9\pi \text{ in}^2$$

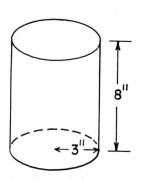

409

$$V = A_{base} \cdot h$$
$$= 9\pi \cdot 8$$
$$= 72\pi \text{ in}^3$$

The volume of the cylinder is 72π in^3 or 226.2 in^3 (to nearest tenth).

EXAMPLE 2   Determine the volume of a right triangular prism of altitude 9 cm with a right triangular base with legs 7 cm and 8 cm long.

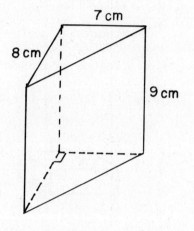

Solution
$$A_{base} = \frac{1}{2} hb$$
$$= \frac{1}{2}(7)(8)$$
$$= 28 \text{ cm}^2$$

$$V = A_{base} \cdot h$$
$$= 28(9)$$
$$= 252 \text{ cm}^3$$

The volume of the cylinder is 252 cm^3.

Two cylinders or prisms with the same bases and same altitudes will have the same volume even though one may be oblique and the other right. That means that figure A and B, below, will have the same volume as long as $r_1 = r_2$ and $h_1 = h_2$, despite the fact that A is oblique and B is right. Therefore it is important to remember that the <u>altitude</u>, <u>not</u> the <u>slant height</u>,

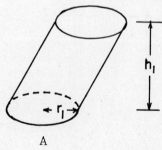

A

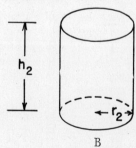

B

is the measurement used in calculating volume.

EXAMPLE 3   Determine the volume of an oblique circular cylinder with a base of diameter 12 cm and an altitude of 11 cm.

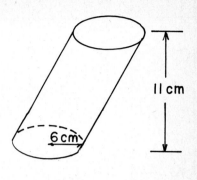

Solution

$$A_{base} = \pi r^2$$
$$= \pi(6)^2$$
$$= 36\pi \text{ cm}^2$$

$$V = A_{base} \cdot h$$
$$= 36\pi \cdot (11)$$
$$= 396\pi \text{ cm}^3$$

The volume of the cylinder is 396π cm^3 or 1244.1 cm^3 (to nearest tenth).

EXAMPLE 4   What is the volume of an oblique rectangular prism if it has a 15" by 18" base, a slant height of 12", and one side forms a 60° angle with the 18" edge of the base?

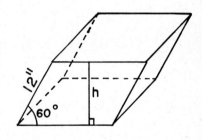

Solution   The altitude must be calculated. Because the base and one side form a 60° angle, we will have a 30°-60°-90° right triangle with a hypotenuse of 12" and the altitude as the leg opposite the 60° angle. (See Section 3.3.) We can use the proportion for the 30°-60°-90° triangle:

$$\frac{\text{opposite}}{\text{hypotenuse}} = \frac{\sqrt{3}}{2}$$

$$\frac{\text{altitude}}{12} = \frac{\sqrt{3}}{2}$$

$$\text{altitude} = 6\sqrt{3}''$$

$$A_{\text{base}} = \ell w$$

$$= 15(18)$$

$$= 270 \text{ in}^2$$

$$V = A_{\text{base}} \cdot h$$

$$= 270 \cdot 6\sqrt{3}$$

$$= 1620\sqrt{3} \text{ in}^3$$

The volume of the prism is $1620\sqrt{3}$ in^3 or 2805.9 in^3 (to nearest tenth).

EXAMPLE 5 Determine the volume of a 10 in. high right circular cylinder if the total surface area of the cylinder is 150π in^2.

Solution We must first determine the value of r:

$$\text{TSA}_{\text{cylinder}} = \text{LSA} + A_{\text{bases}}$$

$$= 2\pi hr + 2\pi r^2$$

$$150\pi = 2\pi \cdot 10r + 2\pi r^2$$

$$150\pi = 20\pi r + 2\pi r^2$$

$$0 = 2\pi r^2 + 20\pi r - 150\pi$$

$$= 2\pi(r^2 + 10r - 75)$$

$$= 2\pi(r - 5)(r + 15)$$

$$r = 5'' \text{ or } r = -15''$$

Because we are measuring distance, we only consider the positive value. We can now find the volume:

$$A_{base} = \pi r^2$$
$$= 25\pi \text{ in}^2$$

$$V = A_{base} \cdot h$$
$$= 25\pi \cdot 10$$
$$= 250\pi \text{ in}^3$$

The volume of the cylinder is 250π in^3 or 785.40 in^3 (to nearest hundredth).

EXAMPLE 6 Given a right rectangular prism with the length of the base twice the width and altitude five more than three times the length, write the expression representing the volume of the prism.

Solution
$$w = x$$
$$\ell = 2x$$
$$h = 3(2x) + 5$$

$$V = A_{base} \cdot h$$
$$= x(2x)(6x + 5)$$
$$= 12x^3 + 10x^2$$

The expression for the volume is $12x^3 + 10x^2$.

EXAMPLE 7 Given a right circular prism where the radius of the base is given by $x^2 + 3x$ and the altitude is Δx (delta x), write the expression representing the volume.

Solution
$$A_{base} = \pi r^2$$
$$= \pi(x^2 + 3x)^2$$

$$V = A_{base} \cdot h$$

$$= \pi(x^2 + 3x)^2 \cdot \Delta x$$

The expression for the volume is $\pi(x^2 + 3x)^2 \cdot \Delta x$.

To find the volume of any cone or pyramid, multiply one third the area of the base by the altitude. Thus we have the formula:

$$V_{cone \text{ or } pyramid} = \frac{1}{3} A_{base} \cdot h,$$

where h is the altitude.

EXAMPLE 8   Determine the volume of a right circular cone with a 7 m radius base and an altitude of 15 m.

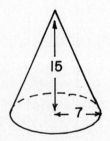

Solution

$$A_{base} = \pi r^2$$

$$= 49\pi \text{ m}^2$$

$$V = \frac{1}{3} A_{base} \cdot h$$

$$= \frac{1}{3}(49\pi)(15)$$

$$= 245\pi \text{ m}^3$$

The volume of the cone is 245π m^3 or 769.69 m^3 (to nearest hundredth).

EXAMPLE 9   What is the volume of a regular hexagonal pyramid with an altitude of 21 ft and a base with an edge of 8 ft?

Solution   Apothem = $4\sqrt{3}$ ft (Why?)

$$A_{base} = \frac{1}{2} ap$$

$$= \frac{1}{2} \cdot 4\sqrt{3} \cdot 6 \cdot 8$$

$$= 96\sqrt{3} \text{ ft}^2$$

$$V = \frac{1}{3} A_{base} \cdot h$$

$$= \frac{1}{3}(96\sqrt{3})(21)$$

$$= 672\sqrt{3} \text{ ft}^3$$

The volume of the pyramid is $672\sqrt{3}$ ft^3 or 1163.94 ft^3 (to nearest hundredth).

As was the case with cylinders and prisms, two pyramids or cones with the same bases and same altitudes will have the same volume even though one may be oblique and the other not. That means that figures A and B, shown below, will have the same volume as long as $r_1 = r_2$ and $h_1 = h_2$ despite the fact that

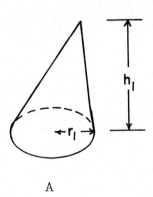

A

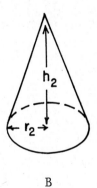

B

A is oblique and B is not. Once again it is necessary to remember that the altitude, not the slant height, is the measurement used in calculating volume.

EXAMPLE 10 What is the volume of an oblique square pyramid with a 13 in. base and an altitude of 21 in.?

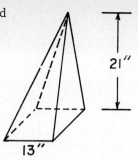

Solution

$$A_{base} = 169 \text{ in}^2$$

$$V = \frac{1}{3} A_{base} \cdot h$$

$$= \frac{1}{3}(169)(21)$$

$$= 1183 \text{ in}^3$$

The volume of the pyramid is 1183 in^3.

EXAMPLE 11 Determine the volume of a right circular cone if the base has a diameter of 24 in. and elements of length 30 in.

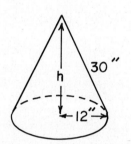

Solution Because we are not given the altitude, we must use the Pythagorean theorem where the radius of the base is one leg of the right triangle and the element (slant height) is the hypotenuse. The altitude will be given by:

$$a^2 = c^2 - b^2$$

$$(\text{altitude})^2 = 30^2 - 12^2$$

$$= (2 \cdot 3 \cdot 5)^2 - (2^2 \cdot 3)^2$$

$$= 2^2 \cdot 3^2 \cdot 5^2 - 2^4 \cdot 3^2$$

$$= 2^2 \cdot 3^2 (5^2 - 2^2)$$

$$\text{altitude} = 2 \cdot 3 \sqrt{25 - 4}$$

$$= 6\sqrt{21} \text{ in.}$$

$$A_{base} = \pi r^2$$
$$= 144\pi \text{ in}^2$$

$$V = \frac{1}{3} A_{base} \cdot h$$
$$= \frac{1}{3}(144\pi)(6\sqrt{21})$$
$$= 288\pi\sqrt{21} \text{ in}^3$$

The volume of the cone is $288\pi\sqrt{21}$ in^3 or 4146.2 in^3 (to nearest tenth).

EXAMPLE 12 Determine the equation that represents the volume of a cone generated by revolving the line $y = 8x$ about the x axis (that is, if the cone has altitude x, then it will have base with radius $8x$).

Solution
$$A_{base} = \pi r^2$$
$$= 64\pi x^2$$

$$V = \frac{1}{3} A_{base} \cdot h$$
$$= \frac{1}{3}(64\pi x^2)(x)$$
$$= \frac{64}{3} \pi x^3$$

The desired equation is $V = \frac{64}{3} \pi x^3$.

There are many applications in calculus for these formulas for volume.

EXAMPLE 13 A town is laying new sewer lines and is digging a trench 4 ft deep by 2 ft wide. If a property owner has a lot 80 ft wide, how many cubic ft of dirt will be removed when the trench is dug? If the sewer pipe has a 12" outside diameter, how many cubic ft of dirt will remain after the pipe is laid?

Solution   The trench is a prism 2 ft by 4 ft by 80 ft, so the volume is:

$$V = A_{base} \cdot h$$

$$= 2 \cdot 4 \cdot 80$$

$$= 640 \text{ ft}^3$$

640 ft^3 of dirt will be removed from the trench.

The sewer pipe is an 80 ft cylinder with radius $\frac{1}{2}$ ft.

$$V = A_{base} \cdot h$$

$$= \frac{1}{4} \pi \cdot 80$$

$$= 20\pi \text{ ft}^3$$

The dirt remaining will equal the volume of the pipe, and thus is 20π ft^3 or 62.8 ft^3 (to nearest tenth).

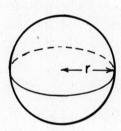

The volume of a sphere can be derived with calculus.

$$V_{sphere} = \frac{4}{3} \pi r^3,$$

where r is the radius.

EXAMPLE 14   Calculate the volume of a 15 in. radius sphere.

Solution

$$V = \frac{4}{3} \pi r^3$$

$$= \frac{4}{3} \pi (15)^3$$

$$= 4500\pi \text{ in}^3$$

The volume of the sphere is 4500π in^3 or 14137 in^3 (to nearest unit).

EXAMPLE 15 What is the volume of a sphere with surface area of 484 in^2?

Solution We first must determine the radius of the sphere:

$$SA = 4\pi r^2$$

$$484\pi = 4\pi r^2$$

$$121 = r^2$$

$$11" = r$$

$$V = \frac{4}{3}\pi r^3$$

$$= \frac{4}{3}\pi(11)^3$$

$$= \frac{5324}{3}\cdot \pi \text{ in}^3$$

The volume of the sphere is $\frac{5324}{3}\cdot \pi$ in^3 or 5,575 in^3 (to nearest unit).

We can now find the volume of more complicated figures by finding the volume for separate portions of the figure and then taking the total of the volumes found.

EXAMPLE 16 Find the volume of the figure to the right if it has a 6" square base.

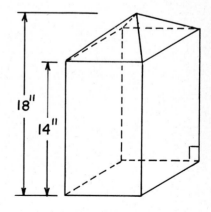

Solution Because the figure is composed of a regular square pyramid on top of a square prism, we will find the two separate volumes then add them.

419

$$V_{pyramid} = \frac{1}{3} \cdot A_{base} \cdot h$$

$$= \frac{1}{3}(36)(4)$$

$$= 48 \text{ in}^3$$

$$V_{prism} = A_{base} \cdot h$$

$$= (36)(14)$$

$$= 504 \text{ in}^3$$

Taking the total of the above volumes gives:

$$V_{figure} = 552 \text{ in}^3$$

Note that it would not be difficult to compute the surface area for the figure in Example 16. We would have to compute the slant height for the pyramid before we could compute the surface area for that portion. In general, either we are given all of the necessary dimensions for the volume and must make a calculation in order to compute surface area, or we are given the necessary dimensions for the surface area and must make a calculation to determine the volume.

EXAMPLE 17   Find the volume and total surface area for a right circular cone with an altitude of 20 cm and a base of radius 8 cm.

Solution   We are given the necessary dimensions to determine the volume.

We must find the slant height in order to compute the surface area. Using the Pythagorean theorem, we find:

$$c^2 = a^2 + b^2$$

$$(\text{Slant height})^2 = (20)^2 + 8^2$$

$$= 464$$

$$\text{Slant height} = 4\sqrt{29} \text{ cm}$$

$$V_{cone} = \frac{1}{3} A_{base} \cdot h$$

$$= \frac{1280\pi}{3} \text{ cm}^3$$

$$TSA_{cone} = LSA + A_{base}$$

$$= \frac{1}{2}(16\pi)(4\sqrt{29}) + \pi 8^2$$

$$= 32\sqrt{29}\,\pi + 64\pi \text{ cm}^2$$

The TSA of the cone is $32\sqrt{29}\,\pi + 64\pi$ cm^2 or 742.44 cm^2 (to nearest hundredth).

EXERCISE 8.4

Note: The use of a calculator will speed many of the calculations necessary for this problem set. Give answers to nearest tenth.

For Problems 1 to 4, find the volumes of the prism with the given dimensions (B = area of base).

1. B = 88 mm^2, h = 3.25 mm
2. B = 15.8 m^2, h = 1.04 m
3. ℓ = 9 in., w = 7 in., h = 11 in. (rectangular prism)
4. ℓ = 11 cm, w = 4 cm, h = 7 cm (rectangular prism)

Find the volumes of the circular cylinders with the given dimensions.

5. B = 31.4 in^2, h = 5.2 in.
6. r = 18 cm, h = 18 cm
7. r = 6 mm, h = 10 mm
8. d = 5.2 in., h = 12.8 in.

9. What is the volume of a right square prism whose base measures 5 in. by 5 in. and the altitude of which is 10 in.?

10. What is the volume of a cylinder with a diameter of 14 in. and an altitude of 7 in.?

11. The area of the base of a triangular prism is 75 mm^2 and the height is 4 mm. Find the volume.

12. The area of the base of an octagonal prism is 2.4 yd^2 and the height is 0.75 yd. To the nearest tenth of a cubic yd, what is the volume?

13. Find (to the nearest hundredth) the volume of a cube that measures 2.54 cm on each edge.

14. Find the volume of a cube that measure 1 in. on each edge.

15. Using the answers to Problems 13 and 14 and the fact that 1 in. is equivalent to 2.54 cm, make the following sentence true. The number of cubic centimeters in a cubic inch is _____, to the nearest hundredth of a cubic centimeter.

16. The volume of a prism is 100 in^3. If the area of the base is 25 in^2, what is the height of the prism?

17. Rob has two cylindrical containers for gasoline. In one, the inside diameter of the base is 10 in. and the height is 12 in. In the other, the inside diameter of the base is 12 in. and the height is 10 in. Which container holds the greater amount of gasoline, the taller one or the one with the greater diameter? What is the difference in volume?

18. Two cylindrical cans each have an inside height of 8 cm. The first can has an inside radius of 3 cm, while the second has an inside radius of 6 cm. The second can will hold _____ times as much as the first.

In Exercises 19 to 24, round your answers to the nearest whole cubic unit indicated by the dimensions. (B = area of base.)

Find the volumes of the pyramids with these dimensions.

19. B = 30.7 cm^2, h = 12.1 cm   20. B = 18 cm, h = 18 cm

Find the volumes of the rectangular pyramids with these dimensions.

21. ℓ = 22.3 mm, w = 18.9 mm, h = 34.1 mm

22. ℓ = 1.7 ft, w = 1.5 ft, h = 2.5 ft

Find the volumes of the circular cones with these dimensions.

23. d = 12 ft, h = 2 yd,   24. r = 2.6 m, h = 110 cm

25. What is the volume of a circular cone if the area of the base is 55 in^2 and the height is 12 in.?

26. What is the volume of a rectangular pyramid whose base measures 4 in. by 5 in. and whose altitude is 12 in.?

*T 27. The base of a regular pentagonal pyramid has 6 cm sides. The pyramid's height is 10 cm. What is its volume?

*T 28. To the nearest tenth of a cubic in., what is the volume of a circular cone with a diameter of 4 in. if the elements (slant height) form a 65° angle with the radius?

29. What is the relationship between the volume of a prism and the volume of a pyramid that have the same altitude and base area?

30. What is the relationship between the volume of a cylinder and the volume of a cone that have the same base area and altitude?

Find the volume <u>and</u> the surface area for a sphere with the given radius.

31. 2.5 ft
32. 3.1 cm

33. How many cubic in. of rubber (to the nearest whole unit) are there in a solid rubber ball 10 in. in diameter?

34. How many marbles $\frac{3}{4}$ in. in diameter can be made from 100 in^3 of glass, if there is no waste in melting?

C 35. A hollow rubber ball is 2 in. in diameter and the rubber is $\frac{5}{16}$ in. thick. How much rubber would be used in the manufacture of 1000 such balls?

36. If a sphere of iron weighs 999 lb, determine the weight of another sphere of iron with one half the diameter of the first.

37. If the radius of one sphere is 6 in. and the radius of another is 3 in., what is the ratio of the volume of the first sphere to the volume of the second?

38. Suppose a sphere fits exactly into a cube. If the radius of the sphere is 1 cm, what part of the volume of the cube is occupied by the sphere?

C *39. Find the surface area and volume for a right hexagonal prism with a 3 in. edge base and an altitude of 15 in.

*40. Given a right hexagonal pyramid with slant height of 9 in. and a base with edges of 6 in., find the volume and surface area.

41. Find the volume of iron in a 10 ft length of cylindrical pipe with an inside diameter of 10 in. and 1/2 in. thick walls.

In Problems 42 and 43, assume large dots are centers of spheres, circles, or arcs.

42. Find the volume and total surface area for the figure to the right. Assume the base is rectangular.

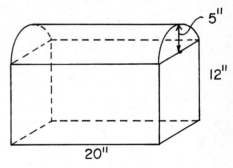

423

43. Find the volume and total surface area for the figure to the right.

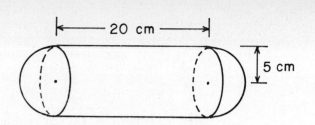

44. The volume of a rectangular prism is 48 cm^3. If the length is three times the width and the height is twice the width, find the dimensions of the prism.

*45. The volume of a right circular cylinder is 108π ft^3. If the altitude is four times the radius, what are the dimensions of the cylinder?

**46. A square sheet of tin with sides of a in. is to be used to make an open-top box by cutting a small square of tin from each corner and bending up the sides. If x is the side of the square cut out, give the expression for the volume of the box.

**47. Give the expression for the volume of a right circular cylinder with a radius $\sqrt{x}$ and height Δx.

**48. Give the expression for the volume of a right circular cylinder with a radius x^2 and height Δx.

**49. A tin can is to be made with fixed volume V. Express the surface area of the can as a function of V and r.

**50. A rectangular box, open at the top, with a square base, is to have a volume of 4000 in^3. Express the surface area as a function of x, an edge of the base.

**51. An open rectangular box is to be made from a piece of cardboard 8 in. wide and 15 in. long by cutting a square with sides of length x, from each corner and bending up the sides. Give the equation for the volume of the box as a function of x.

**52. Sand is falling onto a conical pile. The radius of the base of the pile is always equal to one half of its altitude. Give the equation for the volume of the cone as a function of the height of the pile, x.

**53. Water is withdrawn from a conical reservoir 8 ft in diameter and 10 ft deep (vertex down). Let x be the depth of the water. Give the equation for the volume of water as a function of x.

CHAPTER 8 REVIEW

1. For the figure to the right, name the line segments that represent the:

   a) lateral edge

   b) slant height

   c) an apothem

   d) altitude of figure

   e) altitude of lateral face

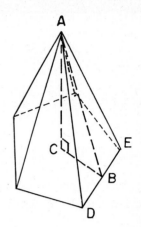

2. Find the total surface area and volume of a right rectangular prism with ℓ = 8 cm, w = 13 cm, and h = 10 cm.

3. Find the total surface area of a cube with $\frac{1}{4}$ in. sides.

4. Find the surface area and volume of a right circular cone with a 2 in. radius and a 5 in. altitude. (Give exact answers.)

5. Find the surface area and volume of a sphere with a 2 m radius. (Give an exact answer.)

6. Find the volume of a prism with $3\frac{1}{2}$ in. altitude and $6\frac{1}{2}$ in^2 base.

7. Find the volume of a pyramid with 7.5 m altitude and 25.25 m^2 base.

8. The volume of a prism is 200 in^3. If the area of the base is 15 in^2, what is the altitude of the prism?

9. Find the surface area and volume of a right circular cylinder with radius 4 mm and altitude 7 mm. (Give exact answer.)

C 10. How many square mm of metal are used to make a tuna can 45 mm high and 85 mm across? (Give answer to nearest tenth.)

11. A 10 ft by 12 ft storage room with an 8 ft ceiling needs to have the walls painted. The only opening is a 3 ft by 7 ft doorway. How much paint must be bought if a quart of paint covers 80 ft^2?

12. A sphere with a 6 in. diameter just fits inside a right circular cylinder. Find the volume of the air space between the two figures. (Give answer to nearest tenth.)

C 13. How many square inches of paper are in a right circular cone paper cup 4 in. deep with a 3 in. diameter. (Give answer to nearest tenth.)

*14. Find the dimensions of a right rectangular prism if the length is twice the width, the height is 2 m more than the width, and the total surface area is 208 m^2.

C 15. Find the volume of a regular square pyramid if the base is 10 in. on a side and the slant height is 12 in.

T 16. Find the surface area of a roof of a 12 ft by 30 ft building if the roof makes a 40° angle with the wall.

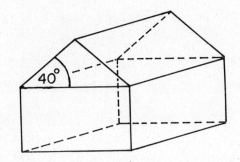

17. Find the volume of a cylinder formed by revolving a 6 in. by 8 in. rectangle about its longer side.

18. Find the volume of a cylinder formed by revolving a 6 in. by 8 in. rectangle about its shorter side.

19. Find the volume of a cone formed by revolving an equilateral triangle about its altitude if the sides are 12 in.

20. Find the lateral surface area of a cone formed by revolving an isosceles triangle about its altitude if the base is 12 cm and the equal sides are 10 cm.

C 21. How many rubber balls 1 in. in diameter can be made from 20 pounds of rubber if 1 in^3 of rubber weighs 0.033 lb?

T 22. Find the surface area of a regular square pyramid if the base has 8 in. sides and the lateral faces form a 70° angle with the base.

C 23. Find the volume of cement in a pipe 4 ft in diameter and 10 ft long if the wall is 2 in. thick. (Give answer to nearest tenth of a ft^3.)

T 24. Find the surface area of a right prism with a regular 11-ogon for a base if the altitude is 15 mm and the edge of the 11-ogon is 4 mm.

25. If the volume of a sphere is 108π in^3, find the surface area.

**26. An open rectangular box is to be made from a piece of tin that is twice as long as it is wide. A square with sides, x, is to be cut from each corner and the sides are to be bent up. If W is the width of the piece of tin, give the equation for the volume of the box as a function of x.

**27. Give the expression for the volume of a right circular cylinder with radius $x^4 + 3x^2 - 4$ and altitude of Δx.

**28. Gravel is falling onto a conical pile where the diameter is always twice the height. Give the equation of the volume of the gravel as a function of the height of the pile, x.

*TC 29. A water trough has been made by cutting a barrel in half vertically then laying one half on its side. The barrel is 6 ft long and has a 32 in. diameter. If the water is 6 in. deep in the trough, find the volume of the water to nearest ft^3.

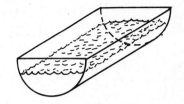

T**30. The water is x in. deep in the trough from Problem 29. Find the expression for the volume of water.

**31. A wire L cm long is cut into two pieces, one bent to form an equilateral triangle and the other to form a circle. If x is the length of the wire used to form the triangle, give the expression for the sum of the areas of the two figures.

For Problems 32 through 35, find the surface area and volume of the given figure. Heavy dots indicate the centers of circles or semicircles. You may use a calculator if you wish.

32.

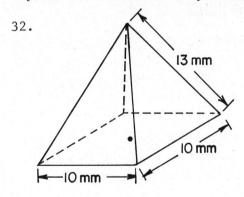

33.

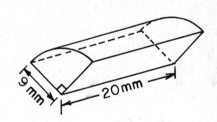

34.

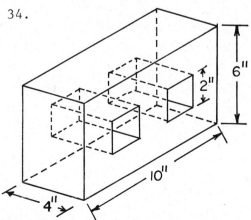

35.

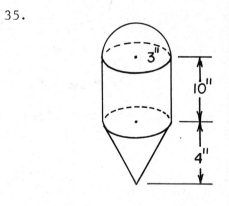

CHAPTER 9
LOGIC

Geometry, like all mathematics, depends on careful, logical thought. Frequently, however, mathematicians cannot agree on exactly what is meant by "logical thought." Therefore, we will take some time to study logic so that we will have more systematic means to determine whether or not an argument (a proof or a problem solution) is logical.

9.1 STATEMENTS; CONJUNCTION AND DISJUNCTION

We must first define the type of sentences we will consider in our discussion of logic:

> A STATEMENT is defined to be a declarative sentence that is either true <u>or</u> false but not both true <u>and</u> false at the same time.

EXAMPLE 1 $4 + 7 = 2$.

EXAMPLE 2   Where is my coat?

EXAMPLE 3   I will study geometry on Thursday night or Friday morning.

EXAMPLE 4   Mice and hamsters eat grain.

EXAMPLE 5   Come here.

EXAMPLE 6   If you do not do your homework, you will not understand the material.

EXAMPLE 7   Does 11 - 5 = 6?

EXAMPLE 8   All squares have five sides.

Examples 1, 3, 4, 6, and 8 <u>are</u> statements. Examples 2 and 7 are <u>not</u> statements because they are questions. Example 5 is <u>not</u> a statement because it is a command.

In order to simplify what we write and to help us to see the basic patterns, we will use a letter to represent a statement. For instance, we can let q represent the statement "4 + 7 = 2." We can then say "q is a false statement," and not have to repeat the words, numbers, or symbols that q represents.

CONJUNCTION

Examples 1 and 8 above are simple statements; Examples 3, 4, and 6 are compound statements because each contains two simple statements that are connected. The statements in Example 3 are connected by "or," those in Example 4 by "and," and those in Example 6 by "if." The simple statements that make up Example 4 are "Mice eat grain" and "hamsters eat grain." We can let q represent the first statement and r represent the second to give:

q: Mice eat grain.

r: Hamsters eat grain.

In order to write Example 4 in symbolic form we must define the symbol for "and," the conjunction.

> A statement that can be written in the form of "p and q" is defined to be the CONJUNCTION of p and q and is symbolized by $p \wedge q$.

Example 4 can be written in symbolic form as: $q \wedge r$.

For any two statements p and q, we need to know under what conditions we will consider a conjunction of p and q false and when we will consider it true. There are four possible combinations of conditions:

> Both statements are true.
> The first statement is true and the second statement is false.
> The first statement is false and the second statement is true.
> Both statements are false.

These possibilities can be presented in table form:

p	q
T	T
T	F
F	T
F	F

Certainly if the first statement is false we do not consider p ∧ q to be true; and if the second statement is false, p ∧ q will be false. If the first <u>and</u> second are true simultaneously, then p ∧ q will be true. We can enter these truth values for p ∧ q in a table (called a truth table) for the conjunction:

p	q	p ∧ q
T	T	T
T	F	F
F	T	F
F	F	F

Note that there is only <u>one</u> instance in which p ∧ q is true: when <u>both</u> p <u>and</u> q are true. For all other combinations p ∧ q is false.

DISJUNCTION

Example 3, "I will study geometry on Thursday night or Friday morning," may be rewritten as the two simple statements "I will study geometry on Thursday night" and "I will study geometry on Friday morning" joined by the word "or." To write this in symbolic form, we must define the symbol for "or," the disjunction.

> A statement that can be written in the form of "p or q" is
> defined to be the DISJUNCTION of p and q and is symbolized by
> p ∨ q.

Writing Example 3 in symbolic form gives:

    s: I will study geometry on Thursday night.

    t: I will study geometry on Friday morning.

$$s \vee t$$

For any two statements p and q we need to know under what conditions we will consider a disjunction of p and q false and when we will consider it true. If both statements are false, $p \vee q$ will be false. When one is true and the other is false, the disjunction will be true. The English language is ambiguous about the case when both statements are true; sometimes the disjunction is considered true and at other times false. To avoid this ambiguity, here we will consider the disjunction true if both statements are true. The truth table for the disjunction is:

p	q	p ∨ q
T	T	T
T	F	T
F	T	T
F	F	F

Note there is only one instance in which $p \vee q$ is false: when <u>both</u> p <u>and</u> q are false. For all other combinations, $p \vee q$ is true.

Let us now use these symbols to write the following sentences in symbolic form.

EXAMPLE 9   Algebra, geometry, and trigonometry must be taken prior to calculus.

   Solution   Let   r: Algebra must be taken prior to calculus.

                 s: Geometry must be taken prior to calculus.

t: Trigonometry must be taken prior to calculus.

The sentence can then be symbolized:

$$r \wedge s \wedge t$$

EXAMPLE 10   A quadratic equation may be solved by factoring, by completing the square, or by using the quadratic formula.

Solution   Let   s: A quadratic equation may be solved by factoring.

t: A quadratic equation may be solved by completing the square.

u: A quadratic equation may be solved by using the quadratic formula.

The sentence can then be symbolized by:

$$s \vee t \vee u$$

Note that in each of the above examples each letter represents a single, simple statement.

EXAMPLE 11   This figure is a square and has right angles, or it is a rhombus.

Solution   Let   p: This figure is a square.

q: This figure has right angles.

r: It is a rhombus.

The sentence can then be symbolized:

$$(p \wedge q) \vee r$$

Note that parenthesis must be used in the above example to symbolize it correctly.

Now that we have these facts about truth values for the conjunction and disjunction, we can combine several different kinds of statements and determine the truth value of the resulting statement.

EXAMPLE 12   Form the disjunction of the statements "27 is a prime number" and "$x^2 + 2x + 1$ is a linear equation." Determine the truth value of the compound statement.

Solution   The compound statement is "27 is a prime number, or $x^2 + 2x + 1$ is a linear equation." The statement is false because each simple statement is false and the disjunction of two false statements is a false statement. (See the fourth row in the truth table for the disjunction.)

If we are given the truth values for both the compound statement and one of its simple statements, we can sometimes determine the truth value of the other simple statement.

EXAMPLE 13   If $s \vee t$ is false and t is false, determine the truth value of s.

Solution   The truth table for $s \vee t$ is:

s	t	$s \vee t$
T	T	T
T	F	T
F	T	T
F	F	F

Only the fourth line corresponds to the conditions given: $s \vee t$ is false, and t is false. Because s is false for that line, the answer is "FALSE."

EXAMPLE 14   If $s \wedge t$ is false and s is false, determine the truth value of t.

Solution   The truth table for $s \wedge t$ is:

s	t	$s \wedge t$
T	T	T
T	F	F
F	T	F
F	F	F

Both the third and fourth lines satisfy the conditions given: $s \wedge t$ is false, and s is false. Because t is true in line three and false in line four, there is no one truth value for t. The answer is "CANNOT BE DETERMINED."

EXERCISE 9.1

Determine which of Sentences 1 through 10 are statements and which are not.

1. All rectangles are squares.

2. $3 + 5 = 10$.

3. Where is my geometry book?

4. Parallel lines do not intersect.

5. Do your homework.

6. If two lines intersect, they form an angle.

7. Prisms and pyramids are three-dimensional figures.

8. Why did he ask that question?

9. Either Sue registers and goes to class, or she gets a job and earns some money.

10. Turn down that stereo.

For each of Problems 11 through 14, there are two statements.
   a) Join these statements to form the disjunction.
   b) Join them to form the conjunction.
   c) Determine the truth value of the statements you formed in part (a).
   d) Determine the truth value of the statements you formed in part (b).

11. Three is an even number. Twenty-five is a perfect square.

12. The solution of the equation $2x + 3 = 5$ is -1. Eleven is divisible by 4.

13. $x^2 - 3x - 4 = (x - 4)(x + 1)$. $(15)^2 + (20)^2 = (25)^2$.

14. Eighty-seven is divisible by 3. The solution of the equation $6 - 4x = 12 - x$ is 2.

Each of Sentences 15 through 20 is a compound statement. For each sentence, write the simple statements and assign a letter to each. Write each sentence in symbolic form.

15. He is studying both English and geometry.

16. Squares and trapezoids are quadrilaterals.

17. He may go with me this summer, or he may get a job.

18. A number that is divisible by 2 ends in 0, 2, 4, 6, or 8.

19. Quadrilaterals, triangles, and circles are two-dimensional figures.

20. Either you stay awake and study, or you fail the test tomorrow.

For Problems 21 through 24, give an example of a sentence of the given form.

21. p ∧ (q ∨ r)
22. (p ∧ q) ∨ r
23. p ∨ q ∨ r
24. p ∧ q ∧ r

For Problems 25 through 29, p and q are statements. Using the given truth values, determine whether the truth value of p is true, false, or cannot be determined.

25. p ∨ q is false and q is false.

26. p ∧ q is true and q is true.

27. p ∧ q is false and q is true.

28. p ∨ q is true and q is false.

29. p ∨ q is true and q is true.

9.2 NEGATIONS AND QUANTIFIERS

At times we may wish to write a statement that has the opposite truth value to a given statement p. The new statement is called the negation of p. In order for a statement q to be the negation of p, q must be true whenever p is false, and whenever q is true p must be false. Likewise, q must be false whenever p is true, and whenever q is false p must be true. In short, q must always have the opposite truth value of p; there can be no circumstance in which q and p are simultaneously both true, nor can they both be false simultaneously. Further, the negation of the negation must be the original statement.

> The NEGATION of p is that statement "not p," symbolized by ~p. The truth value of ~p is always the opposite of the truth value of p. The negation of the negation of p is p; that is ~~p = p.

The truth table for the negation is:

p	$\sim p$
T	F
F	T

Sometimes two statements p and q will have exactly the same truth values. Whenever p is false, q is false, and whenever q is false, p is false. Likewise, whenever p is true, q is true, and whenever q is true, p is true. Such statements are equivalent.

A statement q that always has the same truth values as p is defined to be EQUIVALENT to p.

In the English language we would not write a sentence beginning with the word "not" or the phrase "it is false that." More commonly we would incorporate the negation into the sentence. Examples 1 through 6 demonstrate this practice.

	Statement	Negation
EXAMPLE 1	This course is difficult.	This course is not difficult.
EXAMPLE 2	Today is not Tuesday.	Today is Tuesday.
EXAMPLE 3	All rectangles are squares.	Some rectangles are not squares.
EXAMPLE 4	Some lines are parallel.	No lines are parallel.
EXAMPLE 5	No lines intersect.	Some lines intersect.
EXAMPLE 6	Some triangles do not have exactly three sides.	All triangles have exactly three sides.

Note that the negation of Example 3 is NOT "all rectangles are not squares." For the statement "all rectangles are squares" to be false, we need only one rectangle that is not a square, or "some rectangles are not squares."

To negate a statement that begins with "all," change "all" to "some" and negate the verb.

In mathematics the word "all" can be replaced by the word "any," "every," or "each," and the meaning of the sentence will not change. For example, all of the following sentences have the same meaning:

>All rectangles have four sides.
>Each rectangle has four sides.
>Every rectangle has four sides.
>Any rectangle has four sides.

Occasionally we use the article "a" (or "an") as a synonym for "all." For example:

>A rectangle has four right angles.
>An equilateral triangle is equiangular.

Note that in formal mathematics "a" is taken to mean "all" as in each of the examples above, whereas in common usuage, the article "a" can be ambiguous. For example:

A dog is digging a hole. (Here "a" means one particular dog.)
A man came to see you. (Here "a" means one particular man.)
A dog barks. (Here "a" can refer to all dogs.)

<u>In this text we will interpret the article "a" (or "an") to mean "all."</u>

The negation for Example 6 is NOT "all triangles do not have exactly three sides." In order to ensure that the statement "some triangles do not have exactly three sides" is false, there must not be <u>any</u> triangle that does not have exactly three sides; hence, "all triangles have exactly three sides."

To negate a statement that begins with "some," change the "some" to "no" or "none," <u>or</u> change "some" to "all" and negate the verb.

Negations should be written to avoid "double negatives" whenever possible.

EXAMPLE 7   Statement: Some dogs are not brown.

                Negation: No dogs are not brown.   (This statement is awkward with the double negative "no . . . not")

                               All dogs are brown.   (This statement is much easier to understand because it does not have the double negative.)

There are several ways of writing a "some" statement. For example, the following sentences all have the same meaning:

        Some angles are right angles.
        There are angles that are right angles.
        There exist angles that are right angles.
        Not all angles are not right angles.
        At least one angle is a right angle.
        There exists an angle that is a right angle.

Notice that "some" means "There exists <u>at least</u> one." There may be only one, several, many, or all with the given property.

What is the negation of a conjunction? For example, what sentence will have the opposite truth value to the sentence: "This dog barks and this cat purrs?"

   Let p: This dog barks.
         q: This cat purrs.

Now $p \wedge q$ symbolizes the sentence, and the truth table for $p \wedge q$ and its negation $\sim(p \wedge q)$ is:

p	q	$p \wedge q$	$\sim(p \wedge q)$
T	T	T	F
T	F	F	T
F	T	F	T
F	F	F	T

We want a statement with truth values <u>opposite</u> to those of $p \wedge q$. We can obtain the negation by using the statement "This dog does not bark or this cat does not purr." This can be symbolized as $\sim p \vee \sim q$, and in truth table form we have:

p	q	p ∧ q	~(p ∧ q) ①	~p	~q	~p ∨ ~q ②
T	T	T	F	F	F	F
T	F	F	T	F	T	T
F	T	F	T	T	F	T
F	F	F	T	T	T	T

Columns 1 and 2 have exactly the same entries (truth values), so the two statements are equivalent.

The **negation** of a conjunction, p ∧ q, is (~p) ∨ (~q).

Let us determine the negation of the disjunction "We will eat at MacDonald's, or we will eat at the Dairy Queen."

Let q: We will eat at MacDonald's.
r: We will eat at the Dairy Queen.

Then q ∨ r symbolizes the sentence, and the truth table for q ∨ r and its negation ~(q ∨ r) is:

q	r	q ∨ r	~(q ∨ r)
T	T	T	F
T	F	T	F
F	T	T	F
F	F	F	T

The negation will be a conjunction: "We will not eat at MacDonald's, and we will not eat at the Dairy Queen." This can be symbolized as (~p) ∧ (~r), and in truth table form we have:

q	r	q ∨ r	~(q ∨ r) ①	~q	~r	~q ∧ ~r ②
T	T	T	F	F	F	F
T	F	T	F	F	T	F
F	T	T	F	T	F	F
F	F	F	T	T	T	T

Columns 1 and 2 have exactly the same entries, so the two statements are equivalent.

The negation of a disjunction, $p \vee q$, is $(\sim p) \wedge (\sim q)$.

We can use the information from this section together with that from Section 9.1 to write the following statements in symbolic form.

EXAMPLE 8   Lines m and n are parallel, or ∠A and ∠B are not equal.

Let   r: Lines m and n are parallel.
      s: ∠A and ∠B are equal.

The sentence may be symbolized:

$$r \vee (\sim s)$$

EXAMPLE 9   It is false that all triangles and all squares are congruent.

Let   p: All triangles are congruent.
      q: All squares are congruent.

The sentence may be symbolized:

$$\sim(p \wedge q)$$

Because $\sim(p \wedge q)$ is equivalent to $(\sim p) \vee (\sim q)$, the statement "Some triangles are not congruent, or some squares are not congruent" is equivalent to the original statement.

EXAMPLE 10   Form the negation of the statement "All numbers divisible by 15 are divisible by 3 or by 7." The statement is a disjunction.

Let   r: All numbers divisible by 15 are divisible by 3.
      s: All numbers divisible by 15 are divisible by 7.

The statement may be symbolized as:

$$r \vee s$$

The negation of $r \vee s$ is $(\sim r) \wedge (\sim s)$.

Let ~r: Some numbers divisible by 15 are not divisible by 3.
~s: Some numbers divisible by 15 are not divisible by 7.

The negation of the original statement is "Some numbers divisible by 15 are not divisible by 3 and some numbers divisible by 15 are not divisible by 7." It also may be worded "Some numbers divisible by 15 are not divisible by 3 and by 7."

Sometimes we encounter sentences that use the words "neither . . . nor." For example, the sentence "Neither Clarissa nor Helga will go" means Clarissa will not go <u>and</u> Helga will not go. The original sentence can be symbolized as follows:

c: Clarissa will go.
h: Helga will go.

(~c) ∧ (~h), which can also be written: ~(c ∨ h)

In words the second expression is "It is false that Clarissa or Helga will go." Because this sentence is equivalent to the original sentence, we now have another way to express the negation of a disjunction.

EXAMPLE 11 Negate the following sentence: "Either the figure is a quadrilateral or a triangle."

Solution Because the given sentence is a disjunction, we may express the negation in two ways:

1. The figure is not a quadrilateral and not a triangle, <u>or</u>
2. The figure is neither a quadrilateral nor a triangle.

Now that we know how to form the negation for statements, we can examine the truth values of statements.

For the following examples, <u>accept each sentence as true</u>, and give the truth value (true, false, or cannot be determined) of the underlined part.

EXAMPLE 12   The solution of the equation $x^2 + x = 156$ is $+12$ or -12. -12 is not a solution to the equation.

Solution   Let   s: A solution of the equation $x^2 + x = 156$ is $+12$.
                 t: A solution of the equation $x^2 + x = 156$ is -12.

The first statement can be written as $s \vee t$, which has the following truth table:

s	t	$s \vee t$
T	T	T
T	F	T
F	T	T
F	F	F

The lines in which the disjunction is true are circled. Because we are given that the statement "The solution to the equation is not -12" (which is $\sim t$) is true, then t must be false. Of the circled lines, only the second meets this condition, so we must eliminate the first and third lines because t is true there. In the second line, s is true, so the answer to the problem is "TRUE."

EXAMPLE 13 $\angle ABC$ is congruent to $\angle FGH$ or $\angle ABC$ is the complement of $\angle FGH$.
$\angle ABC$ is congruent to $\angle FGH$.

Solution   Let   r: $\angle ABC$ is congruent to $\angle FGH$.
                 s: $\angle ABC$ is the complement of $\angle FGH$.

The first statement can be symbolized $r \vee s$. The second statement can be symbolized r.

The truth table is:

r	s	$r \vee s$
T	T	T
T	F	T
F	T	T
F	F	F

The first two lines agree with the given conditions: $r \vee s$ is true and r is true. Because s is true in line one and false in line two, there is no one truth value for s. The answer is "CANNOT BE DETERMINED."

EXERCISE 9.2

Form the negation for Sentences 1 through 22.

1. This book is heavy.

2. The phone is black.

3. This pen does not write.

4. $4 + 2 \neq 6$.

5. All vertical angles are equal.

6. Each triangle has three vertices.

7. Some numbers are divisible by 4.

*8. At least one triangle has sides of length 4 in., 5 in., and 10 in.

9. No circle is round.

10. No number is divisible by 15.

11. Some rectangles are not congruent.

12. Some lines are not parallel.

13. $p \wedge \sim r$

14. $s \vee \sim r$

15. $\sim p \wedge \sim t$

16. $\sim s \vee \sim r$

17. Squares and circles are two-dimensional figures.

18. Congruent figures are the same size and shape.

19. The weight of Bill can be measured in feet or pounds.

20. The quadratic equation or factoring are the usual techniques used to graph the equation $2x - 3y = 12$.

*21. Either squares are round or circles do not have a center.

*22. Circles are round and squares have five sides.

23 through 26. Determine the truth value for each of the original Sentences 19 through 22.

Let p be the statement "This class is small," and q be the statement "This class meets on Tuesdays." Write each of the Sentences 27 through 30 in symbolic form using p and q.

27. This class is large and meets on Tuesday.

28. This class does not meet on Tuesdays, or it is small.

29. It is not true that this class is small and meets on Tuesdays.

30. It is false that this class is large or meets on Tuesdays.

31 to 34. Define "small" as being a class with an enrollment of less than 10. Using this definition and your class, determine the truth value for Statements 27 through 30.

For Exercises 35 to 38 accept each entire sentence as true. That does NOT necessarily mean that each PART is true. Give the truth value (true, false, or cannot be determined) of the underlined parts.

35. This figure is a parallelogram and the opposite sides are parallel.
This figure is a parallelogram.

36. State University has a winning football team, or some of the football players are unhappy.
All of the football players are happy.

37. All triangles have three sides, or 11 is not an even number.
Eleven is not an even number.

38. These angles are congruent, or these lines are not parallel.
These angles are not congruent.

9.3 CONDITIONAL STATEMENTS

Many times statements such as "if I have the money, I will go" are made. This type of statement is called a conditional statement.

A statement that can be written in the form of "if p then q" is defined to be a CONDITIONAL statement or an IMPLICATION. The statement is symbolized by $p \rightarrow q$, where p is the PREMISE and q is the CONCLUSION of the conditional statement.

The premise, the statement associated with the "if," need not be the first part of the sentence. For each of the following conditional statements, the premise is underlined.

If <u>lines are parallel</u>, they do not intersect.

An equation is a quadratic equation if <u>it is a second degree equation</u>.

If <u>a number is divisible by four</u>, it is an even number.

The animal is a mammal if <u>it is a dog</u>.

A quadrilateral is a square if <u>the sides are all of equal length and the angles are all of equal measure</u>.

Sometimes conditional sentences do not use the "if . . . then" form. All of the following sentences have the same meaning.

1. If you passed the course, then you took the final.
2. You took the final if you passed the course.
3. You passed the course only if you took the final.
4. Passing the course implies that you took the final.
5. When you passed the course you took the final.
6. You took the final because you passed the course.

    Let p: You passed the course.
            q: You took the final.

Then the sentences above can be written:

1. $p \to q$
2. q if p
3. p only if q
4. p implies q
5. when p, q
6. q because p

Because all the forms of Sentences 1 to 6 convey the <u>same</u> meaning, <u>each</u> is symbolized: $p \to q$.

EXAMPLE 1   Write the following statement in symbolic form, indicating the meanings of each letter used:

He will cash a check if the bank is still open.

Solution   Let s: He will cash a check.
                      t: The bank is still open.

The statement is symbolized: $t \to s$

Sometimes it is convenient to express declarative sentences in the form, "if . . . then." This can be done by using the subject as the premise and the predicate as the conclusion.

EXAMPLE 2

Original statement	"If . . . then" form
A square has four sides.	If the figure is a square, then the figure has four sides.
Vertical angles are congruent.	If two angles are vertical angles, then the angles are congruent.
The sum of the measures of the angles in a triangle is 180°.	If the number of degrees is the sum of the measures of the angles in a triangle, then the number is 180°.

Let us examine the truth values for the conditional statement. Suppose your friend made the following statement to you: "If I get my check today, then I will take you out to eat." Under what circumstances would you say that the promise was broken? That is, when would you consider that statement made by your friend to be false? Let p represent "He got his check today," and q represent "He takes you out to eat." These, then, are the possibilities:

1. He got his check today, and he took you out to eat. (p true, q true)
2. He got his check today, and he did not take you out to eat. (p true, q false)
3. He did not get his check today, and he did take you out to eat. (p false, q true)
4. He did not get his check today, and he did not take you out to eat. (p false, q false)

Only in circumstance 2 would you be angry with your friend, saying that he had not kept his word. For <u>none</u> of the other three possibilities would you justifiably say that he was dishonest with you. Thus, the truth table for the conditional is:

p	q	p → q
T	T	T
T	F	F
F	T	T
F	F	T

> The <u>only</u> time a conditional statement is considered false is when the premise is true and the conclusion is false.

We can use what we have learned about the truth values of conjunctions, disjunctions, and implications to evaluate the truth value of given statements.

EXAMPLE 3 Assume that p is true, q is false, and s is false. Determine the truth value of each complete statement.

   a) $(\sim s) \rightarrow (p \wedge q)$
   b) $[(p \vee s) \wedge q] \rightarrow (\sim p)$

Solution

   a) First we substitute the given truth values into the statement:

$$(\sim F) \rightarrow (T \wedge F)$$

Just as one simplifies numerical and algebraic expressions, we begin by working within the parentheses first, giving:

$$(T) \rightarrow (F),$$

which has a truth value

$$F$$

Therefore, the complete statement is false for the given values of p, q, and s.

   b) Using the same procedure, we get:

$$[(T \vee F) \wedge F] \rightarrow (\sim T)$$
$$[(T) \wedge F] \rightarrow (F)$$
$$[F] \rightarrow F$$

The truth value here is

$$T$$

Therefore, the complete statement is true for the given values of p, q, and s.

The truth table for the conditional can be used to determine the truth values of components of compound sentences. For the following examples, accept each sentence as true, even though you may know in reality that the sentence is false. Given that each sentence is true, give the truth value (T, F, or cannot be determined) of the underlined parts.

EXAMPLE 4   If 5 × 3 = 15, then your answer is correct.
            5 × 3 = 15.

   Solution   Let p: 5 × 3 = 15.
                   q: Your answer is correct.

The first statement can be written as $p \to q$, which has the truth table:

p	q	$p \to q$
T	T	T
T	F	F
F	T	T
F	F	T

The lines in which the implication is true are circled. We are given that p is true; of the circled lines only the first meets this condition. We must eliminate the third and fourth lines because p is false there. In the first line q is true, so the answer to the problem is "TRUE."

EXAMPLE 5   You took all of the tests if you got an A in the course.
            You did not take all of the tests.

   Solution   Let r: You took all of the tests.
                   s: You got an A in the course.

The first sentence can be symbolized by $s \to r$, which has the truth table:

s	r	$s \to r$
T	T	T
T	F	F
F	T	T
F	F	T

The lines in which the implication is true are circled. We are given that r is false; only the last line circled meets this condition. We must eliminate the first and third lines because r is true there. In the last line s is false, so the answer to the problem is "FALSE."

EXAMPLE 6   If <u>two triangles are congruent</u>, then the corresponding angles of the triangles are congruent.
The corresponding angles of the triangles are congruent.

Solution   Let   q: Two triangles are congruent.
                r: The corresponding angles of the triangles are congruent.

Then the first statement is an implication, $q \to r$, which has the truth table:

q	r	q → r
T	T	T
T	F	F
F	T	T
F	F	T

The lines in which the implication is true are circled. We are given that r is true; both the first line and the third line of the truth table meet this condition. We must eliminate the fourth line because r is false there. For the first line, q is true; for the third line, q is false. Therefore we cannot determine the truth value of q, so the answer to the problem is "CANNOT BE DETERMINED."

EXAMPLE 7   If <u>12 is a prime number</u>, then 6 is not divisble by 2.
If <u>some numbers are prime</u>, <u>6 is divisible by 2</u>.
Twenty-five is a multiple of 5, and some numbers are prime.

Solution   Let   r: Twelve is a prime number.
                s: Six is divisible by 2.
                t: Some numbers are prime.
                u: Twenty-five is a multiple of 5.

Symbolizing the sentences gives:

$$r \rightarrow (\sim s)$$
$$t \rightarrow s$$
$$u \wedge t$$

Now we know that both elements of a conjunction must be true if the conjunction is true, so both u and t are true. Let us examine the truth table for t → s:

t	s	t → s
T	T	T
T	F	F
F	T	T
F	F	T

The table gives s true when t is true and t → s is true. The answer for s is "TRUE."

Now that we know that s is true, we can examine the truth table for r → (~s)

r	s	~s	r → (~s)
T	T	F	F
T	F	T	T
F	T	F	T
F	F	T	T

The lines in which the implication is true are circled. We are given that s is true, so ~s must be false. The third line corresponds to ~s false, which gives r false. The answer for r is "FALSE."

EXERCISE 9.3

Write Sentences 1 through 10 in the symbolic form p → q. Be sure to indicate the meaning of each letter that you use.

1. If the number is even, it is divisible by two.

2. A number is divisible by 10 if it ends in zero.

3. You will pass only if you take the final examination.

4. He dropped the course because he was ill.

5. $x^2 + 4 \neq (x + 2)^2$ because $(x + 2)^2 = x^2 + 4x + 4$.

6. When lines intersect, they form vertical angles.

7. If two lines form a right angle, then they are perpendicular.

8. Odd numbers are not divisible by two.

9. Squares have four equal sides.

10. I will drive only if he won't take me.

Using the indicated substitutions, write Sentences 11 to 16 in symbolic form.

    p: Dinosaurs can fly.

    q: Unicorns have wings.

    r: Rudolph has a red nose.

11. If dinosaurs can fly, then unicorns have wings.

12. If it is false that Rudolph has a red nose, then some unicorns don't have wings.

13. It is false that if dinosaurs can fly, then Rudolph doesn't have a red nose.

14. If it is false that some dinosaurs can't fly, then Rudolph has a red nose.

15. If unicorns have wings or Rudolph doesn't have a red nose, then dinosaurs can fly.

16. Either some dinosaurs can't fly, or if Rudolph has a red nose then unicorns have wings.

17. Given: r: Charlie is tall.

            s: Charlie is agile.

            t: Charlie is a basketball player.

            u: Charlie likes to read.

Write in correct English:

    a) $\sim t \wedge \sim r$ b) $\sim u \to t$ c) $(r \wedge s) \to t$

18. Given: p: Triangles are not circles.

            q: Squares are not polygons.

Write in correct English:

a) $\sim(p \land q)$ b) $p \to \sim q$ c) $\sim p \lor q$

19. Given: p: I like math.
    q: All polygons are beautiful.
    r: Some homework is difficult.

    Write in correct English:

    a) $p \land r$ b) $\sim(q \land r)$ c) $\sim r \to p$

20. Given: p: Fido has fleas.
    q: Morris does not eat Nine Lives.
    r: Squirrels eat nuts.

    Write in correct English:

    a) $\sim q \to (p \lor \sim r)$ b) $\sim(\sim p \land q)$ c) $(\sim q \lor r) \to p$

Using the <u>facts that you know</u>, label each of Statements 21 through 42 true or false.

21. $4 + 5 \neq 9$ 22. $\sim(4 + 5 = 8)$
23. $(4 + 5 = 9) \to (2 + 3 = 5)$ 24. $(4 + 5 = 9) \to (2 + 3 = 6)$
25. $(4 + 5 = 20) \to (2 + 3 = 5)$ 26. $(4 + 5 = 20) \lor (2 + 3 = 6)$
27. $(4 + 5 = 9) \lor (2 + 3 = 6)$ 28. $(4 + 5 = 9) \land \sim(2 + 3 = 6)$
29. $(4 + 5 = 20) \to (2 + 3 = 6)$ 30. $(4 + 5 = 20) \lor \sim(2 + 3 = 6)$
31. $(4 + 5 = 9) \land (2 + 3 = 5)$ 32. $[\sim(4 + 5 = 9)] \lor (2 + 3 = 6)$
33. $(4 + 5 = 20) \land (2 + 3 = 6)$ 34. $(4 + 5 = 20) \land (2 + 3 = 5)$
35. $[\sim(4 + 5 = 20)] \to (2 + 3 = 5)$ 36. $(2 + 3 = 5) \to \sim(4 + 5 = 9)$

37. If you have yellow eyes, then today is February 31.

38. If you don't have yellow eyes, then today is not February 31.

39. If you don't have yellow eyes, then today is February 31.

40. If you have yellow eyes, then today is not February 31.

41. An acorn has feathers only if a dog is an animal.

42. A pine tree has needles only if a dog is an animal.

For Exercises 43 to 47, assume that p is false, q is true, and r is false. Indicate the truth value of each of the problems.

43. $p \to [q \wedge (\sim p \vee q)]$

44. $(p \wedge q) \vee r$

45. $(p \to q) \wedge r$

46. $(r \vee \sim p) \wedge (p \to q)$

47. $(q \to p) \wedge [(\sim p) \vee q]$

For Exercises 48 through 60 accept each sentence as true and give the truth value (T, F, or cannot be determined) of the underlined parts.

48. If <u>all elephants sing</u>, then all monkeys hold their ears.
    It is false that all monkeys hold their ears.

49. If each student fails the course, <u>they will fire the teacher</u>.
    Each student fails the course.

50. If <u>Mr. Williams works late</u>, Mrs. Williams goes to bed early.
    Mrs. Williams goes to bed early.

51. All birds sing when <u>the sun rises</u>.
    Some birds are not singing.

52. If Mr. Smith wins the election, <u>he will go to Washington</u>.
    Mr. Smith loses the election.

53. If $\triangle ABC \cong \triangle DEF$ then <u>$\angle D \cong \angle A$</u>.
    $\triangle ABC \cong \triangle DEF$.

54. If <u>$\triangle ABC \cong \triangle DEF$</u> then <u>$\angle D \cong \angle A$</u>.
    $\angle D \cong \angle A$.

55. If <u>you eat a snack at 4 p.m.</u>, you'll spoil your appetite.
    If <u>you spoil your appetite</u>, you won't eat your dinner.
    You eat your dinner.

56. If you eat a snack at 4 p.m., <u>you'll spoil your appetite</u>.
    If you spoil your appetite, <u>you won't eat your dinner</u>.
    You eat a snack at 4 p.m.

57. Either you write to me, or <u>I'll write to you</u>.
    Either <u>I'll write to you</u>, or I'll call you.
    I don't call you, and you write to me.

58. Either you pass the course, or you won't graduate.
    If you don't graduate, <u>you won't get a job</u>.
    You fail the course.

59. $\angle A \cong \angle B$, and $\angle A \cong \angle C$.
    If $\angle A \cong \angle B$, then <u>$\angle D \cong \angle E$</u>.
    If <u>$\angle E \cong \angle F$</u>, then <u>$\angle A \cong \angle C$</u>.

**60. We have seen that $\sim(p \wedge q) = \sim p \vee \sim q$ and $\sim(p \vee q) = \sim p \wedge \sim q$. Can you determine what statement is equivalent to $\sim(p \rightarrow q)$?

9.4 OTHER FORMS OF THE CONDITIONAL

Given an implication, there are other statements related to it.

> Original statement: $p \rightarrow q$ (p implies q)
> Converse: $q \rightarrow p$ (q implies p)
> Inverse: $\sim p \rightarrow \sim q$ (not p implies not q)
> Contrapositive: $\sim q \rightarrow \sim p$ (not q implies not p)

EXAMPLE 1   Original statement: If I pass the course, then I will go home.

Converse: If I go home, then I pass the course.

Inverse: If I do not pass the course, then I will not go home.

Contrapositive: If I will not go home, then I did not pass the course.

EXAMPLE 2   Original statement: I will study geometry if I do not have a paper to write.

If we let   q: I have a paper to write.
  r: I will study geometry.

The original statement can be symbolized: $\sim q \rightarrow r$

   Converse: $r \rightarrow \sim q$ If I study geometry, I do not have a paper to write.

   Inverse: $q \rightarrow \sim r$ If I have a paper to write, I will not study geometry.

   Contrapositive: $\sim r \rightarrow q$ If I do not study geometry, I have a paper to write.

Not all of these statements are equivalent in meaning. The contrapositive and the original statement are equivalent; they have the same truth table. The converse and the inverse are equivalent, and they have the same entries for their truth tables.

				Original	Converse	Inverse	Contrapositive
p	q	~p	~q	p → q	q → p	~p → ~q	~q → ~p
T	T	F	F	T	T	T	T
T	F	F	T	F	T	T	F
F	T	T	F	T	F	F	T
F	F	T	T	T	T	T	T

> Because a statement and its contrapositive are equivalent, the contrapositive may be used in place of the original statement without changing the truth values of the expression.

EXAMPLE 3   Give the relationship of the following sentences to the sentence "Drink Sanka if you don't want coffee nerves." Determine whether or not the sentence is equivalent or not equivalent to the original sentence.

1. If you drink Sanka, you don't want coffee nerves.
2. If you don't drink Sanka, you want coffee nerves.
3. Don't drink Sanka only if you want coffee nerves.
4. If you want coffee nerves, don't drink Sanka.

Solution   Let   s: Drink Sanka.
                  t: You want coffee nerves.

Original: ~t → s

1.	s → ~t	Converse	Not equivalent
2.	~s → t	Contrapositive	Equivalent
3.	~s → t	Contrapositive	Equivalent
4.	t → ~s	Inverse	Not equivalent

In the special case where a statement $r \to s$ and its converse $s \to r$ ($r \leftarrow s$) are both true or are both false, then r and s are <u>equivalent</u>. In words, we can say "r only if s" and "r if s," which we then combine into one sentence "r if and only if s," which can be written symbolically $r \leftrightarrow s$. This form is particularly useful when defining terms. For instance, "a figure is a triangle if and only if the figure is composed of exactly three noncollinear points connected by three line segments" is a definition of a triangle.

> When $p \to q$ and $q \to p$ are either both true or are both false, we say p and q are <u>equivalent</u>, which we symbolize by $p \leftrightarrow q$. The symbols $p \leftrightarrow q$ are read "p if and only if q."

EXERCISE 9.4

Write Sentences 1 through 10 in symbolic form. Give (in sentence form) the contrapositive, inverse, and converse for each.

1. You are not very wise if you are 10 years old.

2. When spring comes, studying stops.

3. People drink beer only if they are thirsty.

4. If you stop doing homework, your grades suffer.

5. We will go home for the holidays, only if it does not snow.

6. It is hard to stay alert if you do not get enough sleep.

7. Whenever I go home I do not cook.

8. Breathing implies living.

9. If you think these are hard, you should see the ones I omitted.

10. Zallers have stripes only if zillers are not blue.

For each of Statements 11 to 14 (in symbolic form), give the converse, inverse, and contrapositive.

11. $q \to p$ 12. $\sim r \to s$ 13. $p \to \sim q$ 14. $\sim t \to \sim s$

Each of Problems 15 through 18 gives either the converse, the inverse, or the contrapositive of a statement. Write the original statement and the two missing forms.

    Example: $q \to p$ is the converse.

    Answer: Statement: $p \to q$

                  Inverse: $\sim p \to \sim q$

                  Contrapositive: $\sim q \to \sim p$

15. $p \to q$ is the inverse.

16. $r \to s$ is the contrapositive.

17. $q \to \sim r$ is the contrapositive.

18. $\sim p \to q$ is the converse.

For problems 19 through 25 decide whether or not the two statements are equivalent, and give a reason for your answer. For example, $p \to \sim r$ and $\sim r \to p$ are not equivalent because the second is the converse of the first.

19. $p \to q$; $\sim q \to \sim p$
20. $r \to s$; $\sim\sim r \to s$
21. $r \to s$; $s \to r$
22. $p \to \sim q$; $\sim p \to q$
23. $p \to q$; $\sim p \to \sim q$
24. $(p \vee q) \to r$; $\sim r \to \sim(p \vee q)$
25. $(p \wedge q) \to (r \wedge s)$;
    $(r \wedge s) \to (p \wedge q)$

In Problems 26 to 29, a statement is followed by lettered statements. State the relation (converse, inverse, contrapositive, same, or none) of the lettered statements to the original statement. State whether the lettered statements are equivalent or not equivalent to the original statement.

26. If you plan to win the big game, you'll have to work harder.

    a) If you work harder, you'll win the big game.

    b) If you don't plan to win the big game, you won't have to work harder.

    c) If it's false that you don't plan to win the big game, you'll have to work harder.

    d) If you don't work harder, you don't plan to win the big game.

27. If you have the day off, it must be Sunday.

    a) If it isn't Sunday, you don't have the day off.

    b) If you don't have the day off, it isn't Sunday.

    c) If it's Sunday, you have the day off.

28. You're not an insect, if you do not have six legs.

    a) If you don't have six legs, you're not an insect.

    b) If you have six legs, then you are an insect.

    c) If it's false that you're an insect, then it's false that you have six legs.

    d) If you're an insect, then you do have six legs.

29. If the figure is a rectangle, then it is a quadrilateral.

    a) The figure is a rectangle only if it is a quadrilateral.

    b) If the figure is not a rectangle, then it is not a quadrilateral.

    c) If the figure is not a quadrilateral, then it is not a rectangle.

d) If the figure is not a rectangle, then it is a quadrilateral.

9.5 DRAWING CONCLUSIONS

In mathematics we wish to draw conclusions by combining basic definitions and rules. The conclusions we draw must be obtained by using the rules of logic to justify each step. There are several common patterns used in the justification of a conclusion, some of which are valid and some of which are not.

<u>Valid Arguments</u>

The following example illustrates a direct reasoning argument.

EXAMPLE 1 If it is a square, then its sides are all the same length.
<u>It is a square.</u>
Its sides are all the same length.

Let p: It is a square.

q: Its sides are all the same length.

The argument can be symbolized as:

$$p \to q$$
$$\underline{p}$$
$$q$$

In order for this to be a valid argument, we need $[(p \to q) \land p] \to q$ to be true for all possible truth values for p and q. The truth table gives:

p	q	$p \to q$	$(p \to q) \land p$	$[(p \to q) \land p] \to q$
T	T	T	T	T
T	F	F	F	T
F	T	T	F	T
F	F	T	F	T

Therefore this is a valid argument.

$$p \to q$$
$$\underline{p}$$
$$q$$

is a valid argument pattern.

The next example illustrates indirect reasoning.

EXAMPLE 2   If it is a square, then its sides are all the same length.
<u>Its sides are not the same length.</u>
It is not a square.

If we let p and q represent the same statements as in the previous example, this argument can be symbolized:

$$p \rightarrow q$$
$$\underline{\sim q}$$
$$\sim p$$

Once again we need the conjunction of the premises to imply the conclusion. That is, $[(p \rightarrow q) \wedge (\sim q)] \rightarrow (\sim p)$ must be true for all possible truth values of p and q. The truth table is:

p	~p	q	~q	p → q	(p → q) ∧ (~q)	[(p → q) ∧ (~q)] → (~p)
T	F	T	F	T	F	T
T	F	F	T	F	F	T
F	T	T	F	T	F	T
F	T	F	T	T	T	T

Therefore this is a valid argument.

$$p \rightarrow q$$
$$\underline{\sim q}$$
$$\sim p$$

is a valid argument pattern.

EXAMPLE 3   The figure is a square or a rhombus.
<u>The figure is not a square.</u>
The figure is a rhombus.

Let   p: The figure is a square.
       r: The figure is a rhombus.

The argument can by symbolized by:

$$p \vee r$$
$$\underline{\sim p}$$
$$r$$

Let us examine the truth table for $[(p \lor r) \land (\sim p)] \to r$ to see if it is true for all possible truth values of p and r.

p	r	$\sim p$	$p \lor r$	$(p \lor r) \land (\sim p)$	$[(p \lor r) \land (\sim p)] \to r$
T	T	F	T	F	T
T	F	F	T	F	T
F	T	T	T	T	T
F	F	T	F	F	T

Therefore this is a valid argument.

$p \lor r$
$\underline{\sim r}$
p

is a valid argument pattern.

The transitive is another frequently used argument pattern.

EXAMPLE 4   If it is a square, then its sides are all the same length.
<u>If its sides are all the same length, then it is a regular polygon.</u>
If it is a square, it is a regular polygon.

Let   p: It is a square.
       q: Its sides are all the same length.
       s: It is a regular polygon.

Then the argument can be symbolized by:

$p \to q$
$q \to s$
$p \to s$

The truth table for this argument is:

p	q	s	$p \to q$	$q \to s$	$(p \to q) \land (q \to s)$	$p \to s$	$[(p \to q) \land (q \to s)] \to (p \to s)$
T	T	T	T	T	T	T	T
T	T	F	T	F	F	F	T
T	F	T	F	T	F	T	T
T	F	F	F	T	F	F	T
F	T	T	T	T	T	T	T

460

p	q	s	p → q	q → s	(p → q) ∧ (q → s)	p → s	[(p → q) ∧ (q → s)] → (p → s)
F	T	F	T	F	F	T	T
F	F	T	T	T	T	T	T
F	F	F	T	T	T	T	T

Therefore this is a valid argument.

$$p \rightarrow q$$
$$\underline{q \rightarrow s}$$
$$p \rightarrow s$$

is a valid argument pattern.

Using all of the statements given, form a valid conclusion.

EXAMPLE 5 If you survive the week, you will finish logic.
If you finish logic, you will understand it.

Let r: You survive the week.
s: You (will) finish logic.
t: You will understand it (logic).

The sentences can be symbolized:

$$r \rightarrow s$$
$$s \rightarrow t$$

Because the statements fit the transitive pattern, we may conclude:

$$r \rightarrow t, \text{ or}$$

If you survive the week, you will understand logic.

EXERCISE 9.5

Determine the logical conclusions for each of Arguments 1 through 6.

1. ~p → q
    $\underline{\sim p}$

2. r → ~s
    $\underline{s}$

3. t ∨ ~s
    $\underline{s}$

4. r → ~p
    $\underline{\sim p \rightarrow \sim s}$

5. ~s → t
    $\underline{t \rightarrow q}$

6. r ∨ ~t
    $\underline{\sim r}$

For each of Problems 7 through 14, form a valid conclusion using all of the statements given. Express your answer in an English sentence.

7. If I study this exercise, I will learn about logic. I study this exercise.

8. Either the figure is a square or a rectangle. The figure is not a square.

9. If the figures are congruent, the corresponding sides are of equal length. If the corresponding sides are of equal length, then $\overline{AB} \cong \overline{GH}$.

10. The figure is a regular polygon, or the sides are not all the same length. The sides are all the same length.

11. If the figure is an oblique circular cone, the altitude does not pass through the center of the base. The figure is an oblique circular cone.

12. If the slope of a line is 0, then the graph of the line is horizontal. The graph of the line is not horizontal.

13. If $\angle ABC \not\cong \angle BCD$ then $\angle ABC \not\cong \angle CDE$. $\angle ABC \cong \angle CDE$. If $\angle ABC \cong \angle BCD$, then the lines are parallel.

14. If $\angle 1 \not\cong \angle 2$, then the triangles are not congruent. The triangles are congruent.

Consider the premises given in each of the following problems and determine what, if any, conclusion is appropriate. Explain why you believe your answer is correct.

*15. If this animal is a dog, then it eats meat. This animal eats meat.

*16. If you study this book, you will understand logic. You do not study this book.

*17. If the thermostat is set too high, the room is hot. If the thermostat is set too high, too much energy is used.

9.6 INVALID ARGUMENTS

Sometimes argument patterns are used that are not valid.

EXAMPLE 1 If it is a square, then its sides are all the same length.
<u>Its sides are all the same length.</u>
It is a square.

Let p: It is a square.

q: Its sides are all the same length.

Then the argument can be symbolized:

$$p \rightarrow q$$
$$\underline{q}$$
$$p$$

By examining the truth table for $[(p \rightarrow q) \land q] \rightarrow p$, we see that this argument is NOT valid for all possible truth values of p and q.

p	q	$p \rightarrow q$	$(p \rightarrow q) \land q$	$[(p \rightarrow q) \land q] \rightarrow p$
T	T	T	T	T
T	F	F	F	T
F	T	T	T	F
F	F	T	F	T

In particular, it is NOT a valid argument when p is false and q is true. For instance, the figure could be a triangle with sides of equal length. Therefore this is not a valid argument.

$$p \rightarrow q$$
$$\underline{q}$$
$$p$$

is an INVALID argument pattern.

EXAMPLE 2   If it is a square, then its sides are all the same length.
<u>It is not a square.</u>
Its sides are not the same length.

Letting p and q represent the same statements as in the previous example, this argument can be symbolized:

$$p \rightarrow q$$
$$\underline{\sim p}$$
$$\sim q$$

The truth table for $[(p \rightarrow q) \land (\sim p)] \rightarrow (\sim q)$ indicates that this argument is not valid for all possible truth values of p and q.

p	q	~p	~q	p → q	(p → q) ∧ (~p)	[(p → q) ∧ (~p)] → (~q)
T	T	F	F	T	F	T
T	F	F	T	F	F	T
F	T	T	F	T	T	F
F	F	T	T	T	T	T

The argument is NOT valid when p is false and q is true. Once again, the figure could be a triangle with sides of equal length. Therefore this is not a valid argument.

$$p \rightarrow q$$
$$\sim p$$
$$\sim q$$

is an INVALID argument pattern.

EXAMPLE 3   If a figure is a square, then its sides are all the same length.
If a figure is a square, then all angles of the figure are right angles.
If its sides are all the same length, then the angles are all right angles.

Let p and q represent the same statements as before, and

Let r: The angles are all right angles.

This argument can then be symbolized:

$$p \rightarrow q$$
$$p \rightarrow r$$
$$q \rightarrow r$$

The truth table for $[(p \rightarrow q) \wedge (p \rightarrow r)] \rightarrow (q \rightarrow r)$ is:

p	q	r	p → q	p → r	(p → q) ∧ (p → r)	q → r	[(p → q) ∧ (p → r)] → (q → r)
T	T	T	T	T	T	T	T
T	T	F	T	F	F	F	T
T	F	T	F	T	F	T	T
T	F	F	F	F	F	T	T
F	T	T	T	T	T	T	T
F	T	F	T	T	T	F	F
F	F	T	T	T	T	T	T
F	F	F	T	T	T	T	T

This argument is NOT valid when q is true and both p and r are false. The triangle with sides of equal length again satisfies these conditions.

$$p \to q$$
$$\underline{p \to r}$$
$$q \to r$$

is an INVALID argument pattern.

When given premises that match one of these invalid argument patterns, the correct conclusion is "No conclusion is possible."

For the following examples, form a valid conclusion, if possible, using all of the premises.

EXAMPLE 4   If these problems are hard for me, I must be learning something.
I must be learning something.

Solution   First let us symbolize the premises.

Let p: These problems are hard for me.
    q: I must be learning something.

Then the premises can be symbolized:

$$p \to q$$
$$q$$

This is an invalid pattern, and therefore there is no conclusion.

The answer is "No conclusion is possible."

EXAMPLE 5   Vertical angles are equal.
∠AFC and ∠BFD are vertical angles.

First we must symbolize the statements.

Let r: The angles are vertical angles.
    s: The angles are equal.
    t: The angles are ∠AFC and ∠BFD.

Then the premises may be symbolized:

$$r \to s$$
$$t \to r$$

We may reorder the statements:

$$t \to r$$
$$r \to s$$

They now match the transitive pattern and have a valid conclusion of:

$$t \to s$$

We can write this conclusion in sentence form as: If the angles are ∠AFC and ∠BFD, the angles are equal. Or, in smoother English: ∠AFC and ∠BFD are equal. Thus the answer to this argument is "∠AFC and ∠BFD are equal."

You should be sure to review the four valid argument patterns and the three invalid patterns so that you can recognize them as valid or invalid.

EXERCISE 9.6

For Problems 1 through 8 either determine the appropriate conclusion using all the premises, or state that no conclusion is possible.

1. $r \to s$
   $\underline{\sim r}$

2. $t \to s$
   $\underline{s \to \sim r}$

3. $\sim r \to t$
   $\underline{\sim r}$

4. $r \to \sim t$
   $\underline{s \to \sim t}$

5. $\sim p \to s$
   $\underline{s}$

6. $\sim r \to t$
   $\underline{\sim t}$

7. $\sim t \to \sim v$
   $\underline{\sim t \to s}$

8. $\sim v \to w$
   $\underline{v}$

For Problems 9 through 14 determine whether the conclusion is a logical conclusion for the given premises.

9. If you use Brell, your hair will smell.
   Your hair smells. Therefore you used Brell.

10. If two triangles are congruent, the corresponding angles have equal measure. The corresponding angles do not have equal measure. Therefore the triangles are not congruent.

11. If the number is divisible by 4, then the number is not prime. If the number is divisible by 4, then the number is even. Therefore, if the number is not prime, the number is even.

12. If the figure is a square, then the diagonals bisect each other. The figure is not a square. Therefore the diagonals do not bisect each other.

13. All even numbers are divisible by 2. A number divisible by 2 can be written in the form $2n$ where n is some integer. Therefore, all even numbers can be written in the form $2n$ where n is some integer.

14. An odd number can be written in the form $2n + 1$ where n is some integer. 137 is an odd number. Therefore 137 can be written in the form $2n + 1$ where n is some integer.

Form a valid conclusion, if possible, using all of the premises. Write the conclusion in sentence form. If there is no valid conclusion, state "no conclusion."

15. We will go to a movie if it rains.
    We do not go to a movie.

16. If $\triangle ABC$ and $\triangle DEF$ are congruent, then $\overline{AB} \cong \overline{DE}$.
    The two triangles are not congruent.

17. The number N is divisible by 2 or is an odd number.
    N is not an odd number.

18. If $\overline{AB}$ and $\overline{DE}$ intersect at C, then $\angle ACD \cong \angle BCE$.
    If $\angle ACD \cong \angle BCE$, then $\angle ACE \cong \angle BCD$.

19. Numbers divisible by 6 are divisible by 2.
    This number is divisible by 2.

20. Plants wilt because they need water.
    These plants need water.

21. The dog barks if a stranger approaches.
    A stranger approaches only if the cat hides.

22. You will pass the course only if you take the final.
    You take the final.

23. Squares are quadrilaterals.
    Quadrilaterals do not have five sides.

24. Mary is in an English class. All freshmen in college are enrolled in some English class.

For Problems 25 and 26, indicate which of the conclusions logically follow from the given assumptions.

*25. Assumptions: All college students understand elementary mathematics. Anyone who understands elementary mathematics does not have difficulty balancing his checkbook.

   Conclusions:

   a) Some college students have difficulty balancing their checkbook.

   b) Some people who can balance their checkbook do not understand elementary mathematics.

c) College students do not have difficulty balancing their checkbook.

d) Anyone who does not have difficulty balancing his checkbook is a college student.

*26. Assumptions: You get a scholarship only if you are an outstanding student. All outstanding students get publicity.

Conclusions:
a) All students who get publicity get scholarships.
b) All students who get scholarships get publicity.
c) Only students with publicity get scholarships.
d) Some students who do not get publicity get scholarships.

9.7 PROOFS

In mathematics we are frequently concerned about a logical proof for a certain assertion. We start with some undefined terms and definitions. From these, other assertions can be made that must be proved using the rules of logic. In order to prove an argument (or theorem), one must give a reason for each step. For each step you take in any proof, you must be able to show which of the valid argument patterns you used. Recall that the statements $p \to q$ and $\sim q \to \sim p$ are equivalent and may be substituted for one another in logical expressions. Number each premise for easy reference. If a premise is replaced by an equivalent form, use the prime symbol (′) after the number. For example, 1 changes to 1′.

EXAMPLE 1   Form a valid conclusion using all the statements given.

1. $p \to \sim r$
2. $s \to r$
3. $t \to p$

Solution   You can change the order of the statements to:

3. $t \to p$
1. $p \to \sim r$
2. $s \to r$

You use the contrapositive of 2 so that you may apply the transitive pattern to Statements 1 and 2, giving:

$$3. \quad t \to p$$
$$1. \quad p \to \sim r$$
$$2'. \quad \sim r \to \sim s$$

Because Statements 3 and 1 follow the transitive pattern, you get 4:

$$4. \quad t \to \sim r$$
$$2'. \quad \sim r \to \sim s$$

Now statements 4 and 2' display the transitive pattern, so you can conclude that $t \to \sim s$, which gives you a single statement that is your conclusion. The result is:

$$1. \quad p \to \sim r$$
$$2. \quad s \to r$$
$$\underline{3. \quad t \to p}$$
$$t \to \sim s$$

The techniques used in the preceding example can be applied to logical arguments using sentences if the following procedures are followed.

1. Make a chart tabulating what statement each letter will symbolize.
2. Convert each sentence into symbolic form.
3. Evaluate the argument as in the previous example.
4. After the conclusion is reached (if any exists), convert to an English statement (ordinary language).

EXAMPLE 2   Form a valid conclusion using all the statements:

Wet balls are hard to catch.
Basketballs are easy to catch.
Dry balls are hard to find.

Solution   Convert each sentence to the "if . . . then" form:

If it is a wet ball, it is hard to catch.
If it is a basketball, it is easy (not hard) to catch.
If it is a dry (not wet) ball, it is hard to find.

1. Let p: It is a wet ball.
          q: It is hard to catch.
          r: It is a basketball.
          s: It is hard to find.

2. Write each statement in symbolic form:

   1. $p \rightarrow q$
   2. $r \rightarrow \sim q$
   3. $\sim p \rightarrow s$

3. Evaluate the argument:

   $\left. \begin{array}{ll} 1'. & \sim q \rightarrow \sim p \\ 2. & r \rightarrow \sim q \\ 3. & \sim p \rightarrow s \end{array} \right\} \rightarrow \text{reordering} \rightarrow \left. \begin{array}{ll} 2. & r \rightarrow \sim q \\ 1'. & \sim q \rightarrow \sim p \\ 3. & \sim p \rightarrow s \end{array} \right\} \rightarrow r \rightarrow s$

   The conclusion is: "$r \rightarrow s$."

4. Convert to an English sentence:

   If it is a basketball, it is hard to find.

   OR

   Basketballs are hard to find.

To convert a sentence beginning with "no" to the "if . . . then" form, drop the "no," use the subject as the premise and the predicate as the conclusion, and negate the conclusion.

Sentence: No bananas are purple.

"If . . . then" form: If it is a banana, it is not purple.

or, using the contrapositive: If it is purple, it is not a banana.

EXAMPLE 3   No triangle has four sides.
            Figure ABCD has four sides.

The first statement can be reworded into the "if . . . then" form as:

   If the figure is a triangle, then it does not have four sides.

Let r: The figure is a triangle.
     s: The figure has four sides.
     t: The figure ABCD has four sides.

The statements can be symbolized:

$$r \to {\sim}s$$
$$t$$

Now t is a particular case of statement s, so we may replace s by t, giving:

$$r \to {\sim}t$$
$$t$$

This is a valid pattern, and we may conclude ${\sim}r$ for the particular case, or: The figure ABCD is not a triangle.

We can assume that a statement that is true in general will be true for specific cases and may be replaced by the specific case.

EXAMPLE 4   Two congruent triangles have corresponding <u>sides</u> <u>that</u> are congruent. △ABC and △FGH are congruent if $\angle A \cong \angle F$, $\overline{AC} \cong \overline{FH}$ and $\angle C \cong \angle H$.

Let t: The two triangles are congruent.
     s: The triangles have corresponding sides that are congruent.
     v: △ABC and △FGH are congruent.
     w: $\angle A \cong \angle F$, $\overline{AC} \cong \overline{FH}$, and $\angle C \cong \angle H$.

The statements can be symbolized:

$$t \to s$$
$$w \to v$$

Here, v is a particular situation of t, so we may say:

$$w \to v$$
$$v \to s$$

and conclude that:

$$w \to s$$

or:

If $\angle A \cong \angle F$, $\overline{AC} \cong \overline{FH}$, and $\angle C \cong \angle H$, then △ABC and △FGH have corresponding sides that are congruent.

EXERCISE 9.7

If possible, form a valid conclusion using all the statements given. If a valid conclusion is not possible, so state. If the statements are in words, give your conclusion as a grammatically correct sentence.

1. $j \to b$
   $q \to t$
   $b \to q$

2. $r \to t$
   $\sim t$

3. $p \to \sim q$
   $r \to t$
   $t \to p$

4. $p \to q$
   p

5. $r \to \sim p$
   $s \to p$
   $\sim s \to q$

6. $p \to s$
   s

7. $\sim q \to s$
   $s \to p$
   $p \to \sim r$

8. $p \vee q$
   $\sim p$

9. $p \to q$
   $q \to r$
   $s \to r$

10. $\sim t \to r$
    $t \to \sim s$
    $v \to \sim r$

11. $t \to r$
    $\sim t$

12. $p \to \sim s$
    s
    $q \to p$

13. All squares are rectangles. All rectangles are quadrilaterals.

14. No triangles are squares. Squares have four right angles.

15. All ants have six legs. All insects are six-legged.

16. All misers are stingy. No stingy people are generous.

17. Dr. Gauss is a mathematician. No infants are mathematicians.

18. Students who enjoy geometry enjoy mathematics. Students who don't enjoy geometry do not like puzzles.

19. If Molly gets a good job, she will travel.
    If Molly graduates, she will get a good job.
    If Molly passes the course, then she will graduate.
    If Molly does her homework, she will pass the course.
    If Molly travels, she will be appointed to an important position.

20. If you like hamburgers, you will like Big Macs.
    If you don't like hamburgers, eat fried chicken.
    If you are over 20, you won't like Big Macs.

21. People who have a driver's license spend a lot of money on gas.
    People who don't live on the third floor have to carry out their garbage.
    The people who live on the third floor spend little money on gas.

22. People who party don't study much.
    People in this class study a lot.
    People who don't party save money.

23. If the fruit is not sweet, it is on my desk.
    Apples are tart.
    Green fruit is not sweet.
    Bananas are not tart.
    The fruit on my desk is a banana.

24. If two angles are vertical angles, then the angles are congruent. $\angle EOD$ and $\angle COB$ are vertical angles if $\overleftrightarrow{EB}$ and $\overleftrightarrow{CD}$ intersect at O. If $\angle EOD$ is congruent to $\angle COB$, then $\triangle EOD$ is congruent to $\triangle COB$. $\overleftrightarrow{EB}$ and $\overleftrightarrow{CD}$ intersect at O.

CHAPTER 9 REVIEW

1. Mark "S" by the sentences that are statements and "NS" by the sentences that are not statements.

    ____ a)   Stop writing.

    ____ b)   Jack and Henry are eating pears.

    ____ c)   5 + 4 = 8

    ____ d)   How many persons are in class?

    ____ e)   If squares are round, then the area is measured in gallons.

2. Using the statements: i) $x^2 + 5x + 4 = (x + 1)(x + 4)$.

    ii) 81 is a prime number.

    a) Form the conjunction.

    b) Form the disjunction.

    c) Determine the truth value of sentence (a).

    d) Determine the truth value of sentence (b).

3. Write the following sentences in symbolic form. Be sure to indicate what each letter stands for.

    a) Geometry and algebra are hard courses.

    b) Most freshmen take English 10, English 20, Math 4, or Math 5 during their second semester.

4. Given   p: Logic is fun.

    q: Math is exciting.

    Write the following sentences. Be sure your answer is in the simplest form and in correct English.

a) $p \wedge \sim q$

b) $\sim(\sim p \wedge q)$

5. Negate the following sentences. Incorporate the negative into the sentence.

   a) The geometry book is green.

   b) The door is not closed.

   c) Jack will sing, or I will not play the piano.

   d) Mathematics courses are hard.

   e) Some days the sun does not shine.

   f) All doors and windows should be closed.

6. Write the following sentences in symbolic form. Be sure to indicate what each letter represents.

   a) If we go tomorrow, we will call you.

   b) The streets are slick when it rains.

   c) A rectangle is a square only if all sides are the same length.

7. Give the inverse, converse, and contrapositive for the following:

   a) $(\sim r) \rightarrow t$

   b) Two triangles are congruent if corresponding sides are the same length.

8. Accept each sentence as true, and give the truth value (T, F, or cannot be determined) for the underlined part.

   a) If <u>the corresponding angles are equal</u>, then the two triangles are similar. The two triangles are not similar.

   b) Triangles and squares are polygons. If triangles are polygons, then <u>some polygons have only three sides</u>.

   c) If <u>the two lines are parallel</u>, the two triangles are not congruent. The two triangles are not congruent.

9. Using facts that you know, determine if the following sentences are true or false.

   a) $(3x + 2)(5x - 1) = 15x^2 - 7x - 2$, or a square is a circle.

   b) If 15 is a prime number, then 12 is an odd number.

   c) If $x^2 - 4 = 0$ has two solutions, then 18 is divisible by 7.

10. Given the statement: If you want a light beer, you'll drink Miller's.

   a) Determine the relation (converse, inverse, contrapositive, same, or none) of each of the following statements to the statement above.

   1. You want a light beer only if you drink Miller's.

   2. You don't drink Miller's if you don't want a light beer.

   3. If you drink Miller's you want a light beer.

   4. If you don't want a light beer, you drink Miller's.

   5. If you don't drink Miller's, you don't want a light beer.

   b) List the number or numbers OF THE STATEMENTS ABOVE THAT ARE EQUIVALENT to the original statement.

For Problems 11 and 12, form a valid conclusion using all statements. If no conclusion is possible, write "no conclusion."

11. a) $\sim p \to r$
    $\underline{\sim r}$

    b) $s \to t$
    $\underline{\sim s}$

    c) $t \to \sim v$
    $\underline{\sim v}$

    d) $t \vee \sim q$
    $\underline{q}$

    e) $j \to \sim v$
    $\sim j \to \sim t$
    $\underline{\sim s \to t}$

12. a) If this test is not hard, you understand logic.
    This test is not hard.

    b) If you enjoy good weather, you'll love Quitzells.
    You love Quitzells.

    c) If you don't study, your grades will suffer.
    Your grades are excellent.

    d) Squares have opposite sides that are parallel.
    Squares have diagonals that bisect each other.

    e) You won't earn extra money if you don't talk to Mrs. Moneybags.
    You will have a good weekend only if you earn some extra money.

13. Find the best conclusion using all the hypotheses. Use symbolic logic. Express your answer in sentence form.

    People who don't carry book bags have weak backs.
    Students at State University do not like dormitory food.
    People who like rain don't carry book bags.
    People who hate dormitory food have strong backs.

CUMULATIVE REVIEW

Give exact answers unless there is a "C" by the problem. For those problems, give the answer correct to hundredths.

1. Can the intersection of a ray and ray be:
   (a) a point?   (b) a segment?   (c) a line?

2. Find the complement of an angle of $\frac{\pi}{5}$ rad.

3. Construct an angle containing 82.5°.

4. Convert 73.45° to degrees and minutes.

5. The angles of a triangle are in the ratio of 3:4:5; find the number of degrees in each angle.

6. Two ships leave an island in the Pacific at the same time at a 90° angle to each other. The first ship travels at a speed of 20 knots (n mi/hr) and the second travels at 15 knots; after 3 hours how far apart are the ships?

**7. After t hours, how far apart are the ships in Problem 6?

8. Construct a 45°-30°-105° triangle with segment AB as the common side between the 45° and the 30° angles.

$$\overline{}$$
A   B

9. Find the value for "x" (y, or z) in each figure below.

a)
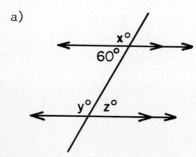

b)
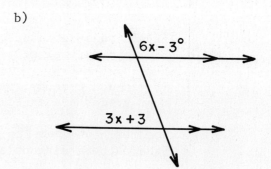

c)
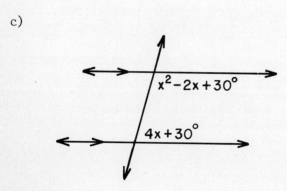

476

10. In the figure below find x, y, z, and w.

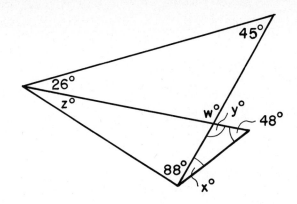

11. Given: △ABC, ∠A = 45°, ∠B = 30°,
    $\overline{CD} \perp \overline{AB}$, and CD = 6 cm

    Find the length of
    (a) AC   (c) DB
    (b) BC   (d) AB

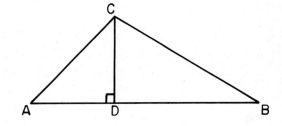

12. Given isosceles right triangle, △DEF where
    ∠F = 90°, find the remaining two sides in
    each of the following cases.

    (a) x = 2√2 cm   y = _____   z = _____
    (b) x = _____   y = 4 ft   z = _____
    (c) x = 3 in.   y = _____   z = _____
    (d) x = _____   y = _____   z = 2√2 m

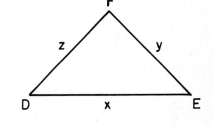

13. Given △ABC, where ∠ACB = 90° and $\overline{CD}$ is
    the altitude of the triangle, find the
    length of

    (a) $\overline{AB}$ if CD = 6", DB = 3"
    (b) $\overline{BD}$ if AD = 4', CD = 8'
    (c) $\overline{CD}$ if AB = 20', DB = 4'
    (d) $\overline{AC}$ if BD = 5", AD = 3"
    (e) $\overline{BD}$ if CB = 6', AD = 5'

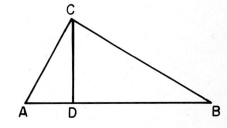

14. Find the number of degrees in ∠4 below, if ℓ∥m

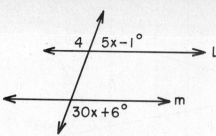

15. If $AE = 2x + 2'$, $EC = 3'$, $\overline{DE} \parallel \overline{BC}$, $AD = 8'$, and $DB = x - 3'$, find the length of AB.

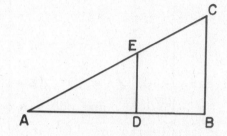

16. A woman 5 ft tall walks away from the base of a streetlight mounted 8 ft above the sidewalk. How long is her shadow after she has walked for 3 seconds if she walks at a rate of 5 feet per second?

**17. How long is the woman's shadow in Problem 16 after she has walked t seconds?

18. Divide AB into parts with the ratio of 1:3:2.

   ———————————————————
   A   B

19. Given: $\overline{CE}$ and $\overline{DG}$ intersect at F
    $\overline{CG} \parallel \overline{DE}$

    Prove: $\triangle CFG \sim \triangle DEF$

20. Given: $\overline{AS} \cong \overline{BR}$, $\overline{AS} \perp \overline{BC}$, $\overline{BR} \perp \overline{AC}$

    Prove: $\triangle ABC$ is isosceles

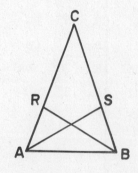

21. Given: W is the midpoint of $\overline{AT}$ and $\overline{SR}$

    Prove: $\overline{AR} \parallel \overline{ST}$

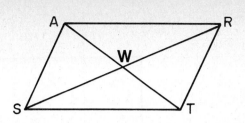

22. Given: $\overline{CE} \cong \overline{BE}$, $\overline{DE} \cong \overline{FE}$

    Prove: $\overline{DC} \cong \overline{FB}$

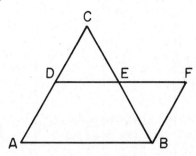

23. Find the number of degrees in each angle of the polygon below.

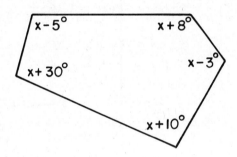

24. Given: parallelogram ABCD

    (a) If AE = 7x - 1' and EC = 5x + 5', find the length of $\overline{AC}$.

    (b) If $\angle DCB = 4x - 60°$ and $\angle DAB = 30 - x$, find the number of degrees in $\angle ABC$.

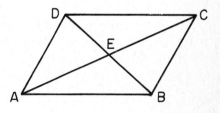

25. If the length of a side of a rhombus is 15' and the length of one diagonal is 20', find:

    (a) length of the other diagonal

    (b) the area of the rhombus

479

26. Find the area and perimeter of the figure below.

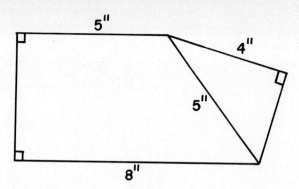

27. Given isosceles trapezoid ABCD with bases $\overline{BC}$ and $\overline{AD}$, $\angle A = 45°$, BC = 7", and AD = 17",

    (a) Find the height of the trapezoid.
    (b) Find the area of the trapezoid.
    (c) Find the length of $\overline{BD}$.

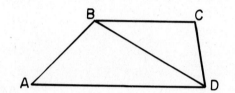

28. Given rhombus ABCD, find the length of $\overline{AD}$.

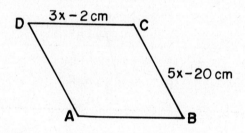

29. Find the length of a diagonal of a square whose side is (a) 6" (b) 8 cm (c) $\sqrt{3}$ m

30. Find the length of a diagonal of a rectangle whose dimensions are
    (a) 3 by 8 m (b) 7 by 4 cm (c) 6 by $\sqrt{5}$ in.

31. Find the area of the following sectors with the given central angle in a circle with the given radius or diameter. Also find the lengths of their arcs.

    (a) 30°, r = 7'   (d) 0.942, r = 14 m
    (b) $\frac{\pi}{4}$, r = 4"   (e) 127°, d = 12'
    (c) 28°, r = 12 cm   (f) 235°, d = 16 yd

32. In the given circle, ∠BAC = 30°
    and BC = 14'. Find:

    (a) $\stackrel{\frown}{BC}$

    (b) $\stackrel{\frown}{AB}$

    (c) radius of the circle.

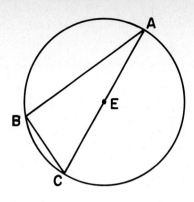

33. Given: Line segment $\overline{EB}$

    $\overline{AD} \cong \overline{DC}$

    $\angle ADE \cong \angle CDE$

    Prove: $\overline{AB} \cong \overline{BC}$

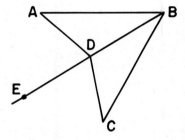

34. Given: KOFA is a rectangle

    OUFA is a parallelogram

    Prove: △KUF is isosceles

    (Show $\overline{KF} \cong \overline{FU}$)

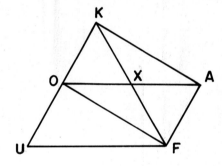

35. Find the measure of each angle in a regular 10-sided polygon (in radians).

36. Given ABCD is a parallelogram,
    find the measure of each angle
    (in degrees).

    ∠A = 2x + y   ∠C = 3y - x

    ∠B = 13x + y   ∠D = 7y + 4x

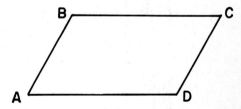

37. Find the area and perimeter:

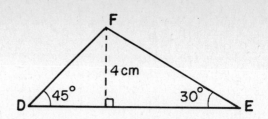

38. Find the cost of carpeting a room 12 feet by 8 feet 3 inches if the cost of carpeting is $4.25 per square yard.

39. Given: $\overline{AC} \cong \overline{CE}$
   $\overline{BC} \cong \overline{CD}$

   Prove: $\angle ABE \cong \angle EDA$

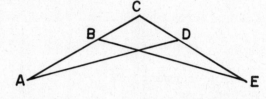

40. Given: $\triangle ABC$ is isosceles with $\overline{AC} \cong \overline{CB}$
   $\angle A \cong \angle CDE$

   Prove: $\overline{AB} \parallel \overline{ED}$

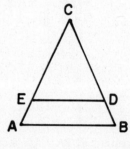

41. $\angle 2$ is supplementary to $\angle 3$
   Prove: $\ell_1 \parallel \ell_2$

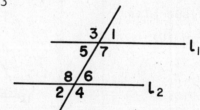

*42. Given: ABCD is a parallelogram
   $\overline{AM} \cong \overline{NC}$
   AC a diagonal

   Prove: MBND is a parallelogram

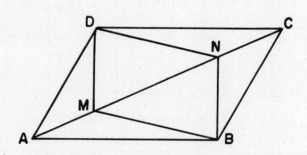

43. Given: ABCD is a parallelogram
    DC = 24″

    Show: △BPC ~ △BQA

    Find: x

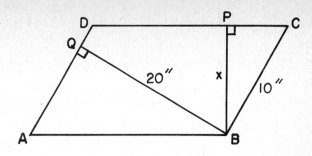

44. I have decided to build a projection television. The plans say that the projected picture will be seven feet diagonally. I need to know the length and the width of the projected picture so that I can build a screen. I asked a friend (who had built one) what the dimensions I needed were, and he said that the length is 0.8 of the diagonal. Find the height of the projected picture.

45. ABCD is a rectangle with $\overline{AC}$ intersecting $\overline{CD}$ at E. Find x and y if $\overline{CE} = 4x + 2y$, $\overline{ED} = 3x + 4y$, and $\overline{EB} = 50$ mm.

46. Given the figure at the right, find the length of $\overline{AC}$.

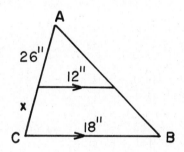

47. A swimming pool length is one more foot than two times its width. How many feet would you have to swim if you went diagonally and the perimeter is 188 feet?

48. A painter painted a picture of a diamond which had the shape of a rhombus with the angle of 60° at one vertex. The perimeter of the diamond in the painting was 20 inches. What is the length of the diamond's diagonals?

49. Given the figure to the right and that the figure is a regular octagon
    AF = 3y + 2x
    BE = x + 5y + 1 cm
    HC = 6x − 8y + 2 cm
    GD = 16 cm
    AM = r

    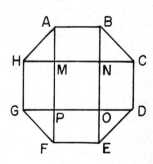

    Find:
    The values of x and y.
    Define the area of the quadrilateral MNOP in terms of r.

50. ABCD is a rectangle with diagonals $\overline{AC}$ and $\overline{BD}$ intersecting at E. BE = 5x + 2y, DE = 11x − 4y, and AE = 70 ft. Find x and y.

51. The largest side of a scalene triangle is 1 in. less than 4 times the smallest side. The other side is 1 in. greater than 2 times the smallest side. The perimeter is equal to 142 in. Find the length of the sides.

52. Given: $\overline{AB} \perp \overline{AE}$

    $\overline{AE} \perp \overline{DE}$

    $\angle CDE \cong \angle BCA$

    BC = 12 m

    CD = 8 m,

    find x.

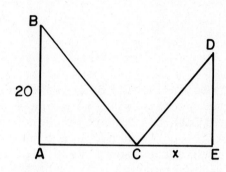

53. A parallelogram has opposite sides given by 4 + 12x and 4x + 7y each with length 145 cm. Find x and y.

*54. In a parallelogram $\underline{ABCD}$, the area is 60 cm^2, AB = 15 cm, and $\angle A = 60°$, find the length of $\overline{AD}$.

55. A rhombus with area 50 cm^2 has one diagonal equal to 20 cm. Find the length of the other diagonal.

56. A rhombus with area 24 ft^2 has one diagonal equal to 12 ft. Find the length of the other diagonal.

57. In a trapezoid A = 360 m^2, b_1 = 20 m, h = 10 m; find b_2.

58. The area of a trapezoid is 120 in^2. The length of the bases are 18 in. and 20 in. Find the length of the altitude.

59. Given the figure to the right, find the area of $\triangle DBC$.

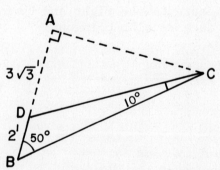

60. Find the value for x and y for the figure to the right.

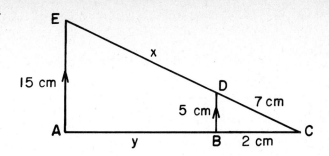

61. Two similar triangles have sides in a ratio of 1:2. Both are right triangles. The smaller triangle has legs given by 5x - 1' and 3x + 3'. The hypotenuse is given by 6x. Find the perimeter of the larger triangle.

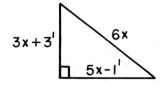

62. The diagonal measure of a box is 30 feet. Find the sides if they are x and x + 6.

63. a) For the figure at the right, find the value of y.

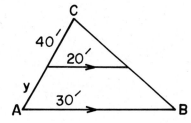

b) If O is the center of the circle, $\overparen{BC} = 3x + 6°$ and $\angle A = 2x - 6°$ find the number of degrees in $\angle BOC$

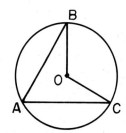

64. If $\angle ABO = 35°$, find the number of degrees in $\angle BCA$.

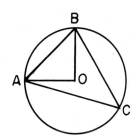

65. P is center of the circle and $\overline{PM} \perp \overline{AB}$.

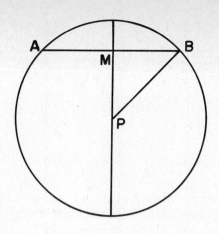

a) If PM = 6' and r = 10' find the length of $\overline{AB}$.

b) If AB = 16" and r = 12", find the length of $\overline{PM}$.

c) If ∠PBM = 30° and r = 20", find the length of $\overline{PM}$ and $\overline{AB}$.

d) If ∠MPB = 60° and AB = 22', find the length of r, and $\overline{PM}$.

66. If ℓ is tangent to the circle (with center O) at K, $\overline{AB} \| \ell$, AB = 12 cm, and C is midpoint of $\overline{OK}$, find the radius of circle.

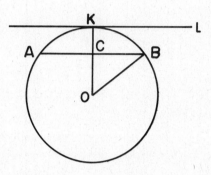

67. In the circle O, ∠AFC = 90° and $\overleftrightarrow{ED}$ is tangent to the circle at D.

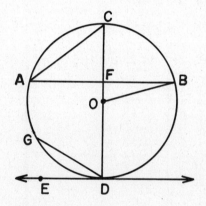

(a) If AF = 5', find the length of $\overline{AB}$.

(b) If OB = 7', find the length of $\overline{CD}$.

(c) If OB = 13' and OF = 5', find the length of $\overline{AB}$.

(d) If $\widehat{GAC}$ = 88°, find the number of degrees in ∠EDG.

(e) If CF = 3', and the radius is 12', find the length of $\overline{FB}$.

68. In quadrilateral ABCD inscribed in Circle O,

$\widehat{AB} = 4x - 25°$ $\widehat{CD} = 2x + 20°$

$\widehat{BC} = x°$ $\widehat{AD} = 3x - 55°$

Find the number of degrees in:

(a) $\widehat{AB}$ (c) ∠ACB
(b) $\widehat{BC}$ (d) ∠ADC
              (e) ∠DAB

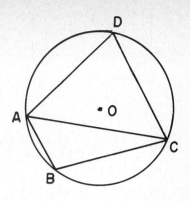

69. In circle P, $\overrightarrow{EC}$ and $\overrightarrow{ED}$ are tangent to the circle at C and D, respectively, $\widehat{AB} = 60°$, and $\overline{PD}$ bisects ∠BPE.

Find:

(a) ∠1   (d) ∠4
(b) ∠2   (e) ∠APC
(c) ∠3   (f) ∠5

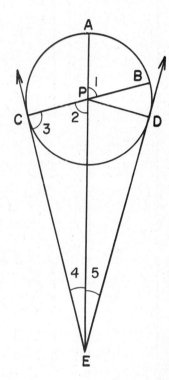

70. Given △ABC circumscribed about circle P, AB = 20 m, BC = 48 m, AC = 44 m, D, E, and F the points of tangency, find the length of $\overline{AF}$, $\overline{CE}$, and $\overline{BD}$.

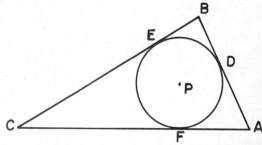

487

71. The vertices of a triangle, inscribed in a circle, divide the circle into three arcs represented by x + 15°, 6x + 10°, and 8x - 40°. Show that the triangle is isosceles. (Draw a picture!).

72. Find the area of the figure.

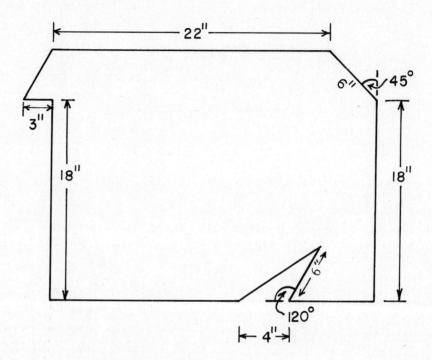

73. Find the area of the figure.

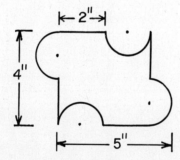

74. Circumscribe an equilateral triangle about the circle.

75. Inscribe a regular hexagon in the given circle.

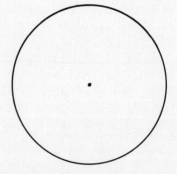

76. Inscribe a circle in △ABC.

77. Find the circumcenter of △ABC.

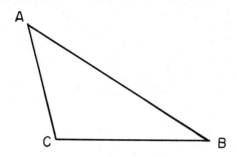

C,T 78. What is the angle of elevation of the sun when a 72 foot building casts a 40 foot shadow?

C,T 79. From the top of a building 180 feet high, the angle of depression of a car on the street is 35°. Find the distance from the car to the base of the building.

C,T 80. Find the area of a regular 15-sided figure whose side is 10 inches.

C,T 81. A 6 ft tall man stands on the roof of a 75 ft tall building and looks at a flagpole on top of a building 58 ft tall. If he is 90 ft away from the flagpole and the angle of elevation to the top of the flagpole is 10°, determine the length of the flagpole.

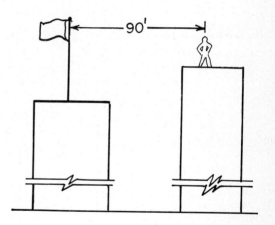

82. Find the measure of each interior angle of a regular 20-sided polygon.

83. If the area of a regular polygon is 420 cm^2 and its apothem is 12 cm, find its perimeter.

84. Given: a regular hexagon inscribed in Circle A.
Find the area of the shaded region if:
(a) a side of the hexagon is 4".
(b) the radius of the circle is 5".

85. Find the volume and surface area of Figure A below, given r = 5 cm and h = 6 cm.

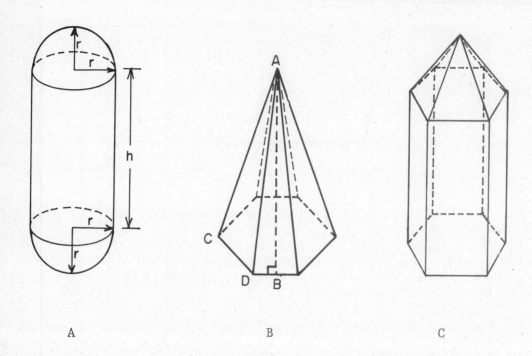

A B C

86. Find the surface area and volume of the regular hexagonal pyramid shown in Figure B if AB = 10' and DC = 6'.

87. Find the surface area and volume of Figure C. The base is a regular hexagon and the tip is a right hexagonal pyramid, the apothem is 3", height of the pyramid is 4", and height of the prism is 5".

88. Which of the following are statements?
    (a) Shut the door.
    (b) If Otto does not love to do long division, then he is perfectly normal.
    (c) Where do you live?
    (d) Snow falls up.
    (e) If I were you, then I wouldn't be me.
    (f) How much money do you have?
    (g) Five plus seven equals nine.

89. Write the converse, inverse, and contrapositive of each of the following.
    (a) If you break the law, then you go to jail.
    (b) The sun shines only if you are happy.
    (c) I will go on Saturday, if I get paid.

90. Negate the following.
    (a) Some greedy creatures cannot fly.
    (b) No antelopes fail to delight the eye.
    (c) All black rabbits are young.
    (d) Englishmen are not Frenchmen.
    (e) Not all oysters are amusing.
    (f) Every line is a set of points.
    (g) John or Bob will buy a calculator.

(h) John and Bob will not buy a calculator.
(i) Every day is Christmas and some cats are fickle.
(j) Nice guys finish last or some students will not pass this course.

91. True or false?
  (a) If $7 + 9 = 16$, then $16 - 9 = 7$.
  (b) If $(-2) \cdot (-3) = -6$, then $2 \cdot 3 = -6$
  (c) $15 + 3 = 18$ if $72 - (-10) = 82$.

92. Let p: You like mathematics.
    q: All students like this text.

   Write out each of the following into sentences.
   (a) $p \to q$
   (b) $\sim p \vee q$
   (c) $\sim q \to \sim p$
   (d) $\sim p \to q$
   (e) $\sim(\sim p \vee q) \to q$

93. State a valid conclusion, if possible, for each of the following groups of statements. If no conclusion is possible write "no conclusion".
   (a) If you brush your teeth with Brylcreem, you misunderstood the commercial.
       You didn't misunderstand the commercial.

   (b) If you see spots in front of your eyes, you are looking at a leopard.
       You are looking at a leopard.

   (c) If your dog has fleas, fleas get in your bed.
       You do not have time to eat breakfast if you brush your dog.
       Fleas are not on your bed if they do not crawl on you.
       If your dog doesn't have fleas, you brush him.

   (d) If you smoke pot, you will get arrested.
       When you get arrested, it is written on your record.
       You will never get a job if there is an arrest on your record.
       If you don't get a job, you will have to move to Bellefonte.

   (e) If Harry throws a party, Carol won't attend.
       If Carol does not attend, her husband, Bob, won't come either.
       John will come only if Bob comes.
       John's wife, Mary, will go only if John goes.
       If Mary doesn't come to the party, Harry will have a miserable time.

   (f) If John goes to the store, then Bob won't go.
       If Jane comes home, then Bob will go to the store.
       If Janes does not come home, Harry will have to pick her up.
       If Sam comes with Jane, then Harry won't have to pick her up.

APPENDIX A
TRIGONOMETRIC FUNCTIONS

DEGREES

θ	$\sin \theta$	$\cos \theta$	$\tan \theta$	θ	$\sin \theta$	$\cos \theta$	$\tan \theta$
0°	.0000	1.0000	.0000	**46°**	.7193	.6947	1.0355
1°	.0175	.9998	.0175	**47°**	.7314	.6820	1.0724
2°	.0349	.9994	.0349	**48°**	.7431	.6691	1.1106
3°	.0523	.9986	.0524	**49°**	.7547	.6561	1.1504
4°	.0698	.9976	.0699	**50°**	.7660	.6428	1.1918
5°	.0872	.9962	.0875	**51°**	.7771	.6293	1.2349
6°	.1045	.9945	.1051	**52°**	.7880	.6157	1.2799
7°	.1219	.9925	.1228	**53°**	.7986	.6018	1.3270
8°	.1392	.9903	.1405	**54°**	.8090	.5878	1.3764
9°	.1564	.9877	.1584	**55°**	.8192	.5736	1.4281
10°	.1736	.9848	.1763	**56°**	.8290	.5592	1.4826
11°	.1908	.9816	.1944	**57°**	.8387	.5446	1.5399
12°	.2079	.9781	.2126	**58°**	.8480	.5299	1.6003
13°	.2250	.9744	.2309	**59°**	.8572	.5150	1.6643
14°	.2419	.9703	.2493	**60°**	.8660	.5000	1.7321
15°	.2588	.9659	.2679	**61°**	.8746	.4848	1.8040
16°	.2756	.9613	.2867	**62°**	.8829	.4695	1.8807
17°	.2924	.9563	.3057	**63°**	.8910	.4540	1.9626
18°	.3090	.9511	.3249	**64°**	.8988	.4384	2.0503
19°	.3256	.9455	.3443	**65°**	.9063	.4226	2.1445
20°	.3420	.9397	.3640	**66°**	.9135	.4067	2.2460
21°	.3584	.9336	.3839	**67°**	.9205	.3907	2.3559
22°	.3746	.9272	.4040	**68°**	.9272	.3746	2.4751
23°	.3907	.9205	.4245	**69°**	.9336	.3584	2.6051
24°	.4067	.9135	.4452	**70°**	.9397	.3420	2.7475
25°	.4226	.9063	.4663	**71°**	.9455	.3256	2.9042
26°	.4384	.8988	.4877	**72°**	.9511	.3090	3.0777
27°	.4540	.8910	.5095	**73°**	.9563	.2924	3.2709
28°	.4695	.8829	.5317	**74°**	.9613	.2756	3.4874
29°	.4848	.8746	.5543	**75°**	.9659	.2588	3.7321
30°	.5000	.8660	.5774	**76°**	.9703	.2419	4.0108
31°	.5150	.8572	.6009	**77°**	.9744	.2250	4.3315
32°	.5299	.8480	.6249	**78°**	.9781	.2079	4.7046
33°	.5446	.8387	.6494	**79°**	.9816	.1908	5.1446
34°	.5592	.8290	.6745	**80°**	.9848	.1736	5.6713
35°	.5736	.8192	.7002	**81°**	.9877	.1564	6.3138
36°	.5878	.8090	.7265	**82°**	.9903	.1392	7.1154
37°	.6018	.7986	.7536	**83°**	.9925	.1219	8.1443
38°	.6157	.7880	.7813	**84°**	.9945	.1045	9.5144
39°	.6293	.7771	.8098	**85°**	.9962	.0872	11.4301
40°	.6428	.7660	.8391	**86°**	.9976	.0698	14.3007
41°	.6561	.7547	.8693	**87°**	.9986	.0523	19.0811
42°	.6691	.7431	.9004	**88°**	.9994	.0349	28.6363
43°	.6820	.7314	.9325	**89°**	.9998	.0175	57.2900
44°	.6947	.7193	.9657	**90°**	1.0000	.0000	—
45°	.7071	.7071	1.0000				

RADIANS

t	$\sin t$	$\cos t$	$\tan t$
.00	.0000	1.0000	.0000
.01	.0100	1.0000	.0100
.02	.0200	.9998	.0200
.03	.0300	.9996	.0300
.04	.0400	.9992	.0400
.05	.0500	.9988	.0500
.06	.0600	.9982	.0601
.07	.0699	.9976	.0701
.08	.0799	.9968	.0802
.09	.0899	.9960	.0902
.10	.0998	.9950	.1003
.11	.1098	.9940	.1104
.12	.1197	.9928	.1206
.13	.1296	.9916	.1307
.14	.1395	.9902	.1409
.15	.1494	.9888	.1511
.16	.1593	.9872	.1614
.17	.1692	.9856	.1717
.18	.1790	.9838	.1820
.19	.1889	.9820	.1923
.20	.1987	.9801	.2027
.21	.2085	.9780	.2131
.22	.2182	.9759	.2236
.23	.2280	.9737	.2341
.24	.2377	.9713	.2447
.25	.2474	.9689	.2553
.26	.2571	.9664	.2660
.27	.2667	.9638	.2768
.28	.2764	.9611	.2876
.29	.2860	.9582	.2984
.30	.2955	.9553	.3093
.31	.3051	.9523	.3203
.32	.3146	.9492	.3314
.33	.3240	.9460	.3425
.34	.3335	.9428	.3537
.35	.3429	.9394	.3650
.36	.3523	.9359	.3764
.37	.3616	.9323	.3879
.38	.3709	.9287	.3994
.39	.3802	.9249	.4111
.40	.3894	.9211	.4228
.41	.3986	.9171	.4346
.42	.4078	.9131	.4466
.43	.4169	.9090	.4586
.44	.4259	.9048	.4708
.45	.4350	.9004	.4831
.46	.4439	.8961	.4954
.47	.4529	.8916	.5080
.48	.4618	.8870	.5206
.49	.4706	.8823	.5334

t	$\sin t$	$\cos t$	$\tan t$
.50	.4794	.8776	.5463
.51	.4882	.8727	.5594
.52	.4969	.8678	.5726
$\dfrac{\pi}{6}$	.5000	.8660	.5774
.53	.5055	.8628	.5859
.54	.5141	.8577	.5994
.55	.5227	.8525	.6131
.56	.5312	.8473	.6269
.57	.5396	.8419	.6410
.58	.5480	.8365	.6552
.59	.5564	.8309	.6696
.60	.5646	.8253	.6841
.61	.5729	.8196	.6989
.62	.5810	.8139	.7139
.63	.5891	.8080	.7291
.64	.5972	.8021	.7445
.65	.6052	.7961	.7602
.66	.6131	.7900	.7761
.67	.6210	.7838	.7923
.68	.6288	.7776	.8087
.69	.6365	.7712	.8253
.70	.6442	.7648	.8423
.71	.6518	.7584	.8595
.72	.6594	.7518	.8771
.73	.6669	.7452	.8949
.74	.6743	.7385	.9131
.75	.6816	.7317	.9316
.76	.6889	.7248	.9505
.77	.6961	.7179	.9697
.78	.7033	.7109	.9893
$\dfrac{\pi}{4}$	.7071	.7071	1.000
.79	.7104	.7038	1.009
.80	.7174	.6967	1.030
.81	.7243	.6895	1.050
.82	.7311	.6822	1.072
.83	.7379	.6749	1.093
.84	.7446	.6675	1.116
.85	.7513	.6600	1.138
.86	.7578	.6524	1.162
.87	.7643	.6448	1.185
.88	.7707	.6372	1.210
.89	.7771	.6294	1.235
.90	.7833	.6216	1.260
.91	.7895	.6137	1.286
.92	.7956	.6058	1.313
.93	.8016	.5978	1.341
.94	.8076	.5898	1.369

t	$\sin t$	$\cos t$	$\tan t$
.95	.8134	.5817	1.398
.96	.8192	.5735	1.428
.97	.8249	.5653	1.459
.98	.8305	.5570	1.491
.99	.8360	.5487	1.524
1.00	.8415	.5403	1.557
1.01	.8468	.5319	1.592
1.02	.8521	.5234	1.628
1.03	.8573	.5148	1.665
1.04	.8624	.5062	1.704
$\frac{\pi}{3}$	.8660	.5000	1.732
1.05	.8674	.4976	1.743
1.06	.8724	.4889	1.784
1.07	.8772	.4801	1.827
1.08	.8820	.4713	1.871
1.09	.8866	.4625	1.917
1.10	.8912	.4536	1.965
1.11	.8957	.4447	2.014
1.12	.9001	.4357	2.066
1.13	.9044	.4267	2.120
1.14	.9086	.4176	2.176
1.15	.9128	.4085	2.234
1.16	.9168	.3993	2.296
1.17	.9208	.3902	2.360
1.18	.9246	.3809	2.427
1.19	.9284	.3717	2.498
1.20	.9320	.3624	2.572
1.21	.9356	.3530	2.650
1.22	.9391	.3436	2.733
1.23	.9425	.3342	2.820
1.24	.9458	.3248	2.912
1.25	.9490	.3153	3.010
1.26	.9521	.3058	3.113
1.27	.9551	.2963	3.224
1.28	.9580	.2867	3.341
1.29	.9608	.2771	3.467
1.30	.9636	.2675	3.602
1.31	.9662	.2579	3.747
1.32	.9687	.2482	3.903
1.33	.9711	.2385	4.072
1.34	.9735	.2288	4.256
1.35	.9757	.2190	4.455
1.36	.9779	.2092	4.673
1.37	.9799	.1994	4.913
1.38	.9819	.1896	5.177
1.39	.9837	.1798	5.471

t	$\sin t$	$\cos t$	$\tan t$
1.40	.9854	.1700	5.798
1.41	.9871	.1601	6.165
1.42	.9887	.1502	6.581
1.43	.9901	.1403	7.055
1.44	.9915	.1304	7.602
1.45	.9927	.1205	8.238
1.46	.9939	.1106	8.989
1.47	.9949	.1006	9.887
1.48	.9959	.0907	10.983
1.49	.9967	.0807	12.350
1.50	.9975	.0707	14.101
1.51	.9982	.0608	16.428
1.52	.9987	.0508	19.670
1.53	.9992	.0408	24.498
1.54	.9995	.0308	32.461
1.55	.9998	.0208	48.078
1.56	.9999	.0108	92.620
1.57	1.0000	.0008	1255.8
$\frac{\pi}{2}$	1.0000	.0000	—

APPENDIX B
GREEK ALPHABET

A	α	alpha		N	ν	nu
B	β	beta		Ξ	ξ	xi
Γ	γ	gamma		O	o	omicron
Δ	δ	delta		Π	π	pi
E	ε	epsilon		P	ρ	rho
Z	ζ	zeta		Σ	σ	sigma
H	η	eta		T	τ	tau
Θ	θ	theta		Υ	υ	upsilon
I	ι	iota		Φ	φ	phi
K	κ	kappa		X	χ	chi
Λ	λ	lambda		Ψ	ψ	psi
M	μ	mu		Ω	ω	omega

GLOSSARY

ACUTE ANGLE   an angle that contains less than 90° or $\frac{\pi}{2}$ rad (p. 30).

ALTERNATE EXTERIOR ANGLES   angles on opposite sides of the transversal that are also outside the two lines the transversal intersects (p. 139).

ALTERNATE INTERIOR ANGLES   angles on opposite sides of the transversal that are also inside the two lines the transversal intersects (p. 139).

ALTITUDE   the perpendicular distance between a base and a vertex, or the other base (p. 235, 238, 376).

ANGLE   formed by two rays having a common endpoint (p. 27).

ANGLE OF DEPRESSION   angle measured from the horizontal down to the given object (p. 354).

ANGLE OF ELEVATION   the angle measured from the horizontal up to the given object (p. 354).

ANNULUS   the area between two concentric circles (p. 302).

APOTHEM   the perpendicular distance from a side of a regular polygon to its center (p. 255).

ARC   a segment of a circle (a part of its circumference) (p. 267).

AREA   the number of square units needed to cover a two-dimensional figure exactly (p. 225).

BISECT   divide into two congruent (equal) parts (p. 15).

CENTRAL ANGLE   an angle formed by two radii of a circle (p. 265).

CENTROID   the point of intersection of the medians of a triangle (p. 317).

CHORD   the portion of a secant that lies within a circle (p. 268).

CIRCLE   the set of points in a plane that are the same distance from a fixed point (pp. 2, 265).

CIRCUMCENTER   the point of intersection of the perpendicular bisectors of the sides of a triangle (p. 313).

CIRCUMFERENCE   the distance around a circle; its perimeter (p. 267).

CLOSED CURVE   a curve that has no unique beginning and end (p. 1).

COLLINEAR   lying on the same line (p. 3).

COMPLEMENTARY ANGLES   two angles containing a sum of 90° or $\frac{\pi}{2}$ rad (p. 35).

CONCENTRIC   having the same center (p. 265).

CONDITIONAL   a statement that can be written in the form of "if p then q" (p. 444).

CONE   a simple closed curve and the line segments that join the closed curve to a point called the vertex (p. 379).

CONGRUENT FIGURES   two figures that are the same size and shape (p. 102).

CONJUNCTION   a compound statement where the simple statements are joined by "and" (p. 429).

CONVEX   a curve where all possible line segments with endpoints in the interior lie entirely within or on the simple closed curve (p. 2).

COPLANAR   lying in the same plane (p. 7).

COSINE θ   the cosine of θ, an acute angle in a right triangle, is the ratio of the leg adjacent to θ to the hypotenuse (p. 341).

CUBE   a right rectangular prism with faces that are squares (p. 377).

CORRESPONDING ANGLES   angles in the same position in two figures, usually in two triangles or when two lines are cut by a transversal (pp. 103, 139).

CURVE   a connected pathway (p. 1).

CYLINDER   a simple closed surface with parallel and congruent bases with a curved surface connecting the two bases (p. 378).

DISJUNCTION   a compound statement where the simple statements are joined by "or" (p. 430).

DIAMETER   a line segment that passes through the center of a circle and has endpoints on the circle (p. 266).

DIAGONALS   line segments that join nonadjacent vertices of a polygon (p. 185).

DEGREE   a unit used to measure angles; 1° is $\frac{1}{360}$ of a full rotation (p. 30).

EDGE   the intersection of two faces of a polyhedron (p. 375).

FACE   each polygonal region of a polyhedron (p. 375).

EQUIANGULAR   having all angles the same size (p. 61).

EQUILATERAL   having all sides the same length (p. 61).

EXTERIOR ANGLE   an angle formed by one side of a polygon and the adjacent side extended (p. 217).

HEXAGON   a six-sided polygon (p. 215).

HYPOTENUSE   the side opposite the right angle in a right triangle (p. 70).

INCENTER   the point of intersection of the angle bisectors of a triangle (p. 316).

INSCRIBED ANGLE   two chords that have one common endpoint (p. 269).

ISOSCELES TRAPEZOID   a trapezoid with two nonparallel sides that are the same length (p. 191).

ISOSCELES TRIANGLE   a triangle with at least two sides congruent (p. 63).

INTERSECTING   having one or more points in common (pp. 5, 7).

LEGS   the sides that form the right angle in a right triangle (p. 70).

LINE   a straight line (connected pathway) that extends indefinitely in opposite directions (p. 3).

LINE SEGMENT   a portion of a line that lies between and contains two distinct points on the line (p. 4).

MEDIAN   a line segment connecting the midpoint of a side of a triangle with the opposite vertex (p. 317).

MIDPOINT   the point on a segment that bisects the segment (p. 15).

MAJOR ARC   an arc measuring more than a semicircle (p. 267).

MINOR ARC   an arc measuring less than a semicircle (p. 267).

MINUTE   a unit used to measure angles; 1' is $\frac{1}{60}$ of a degree (p. 32).

NEGATION   a statement with opposite truth value (p. 435).

NONCOLLINEAR   not lying on the same line (p. 4).

OBTUSE ANGLE   an angle greater than 90° (or $\frac{\pi}{2}$ rad) and less than 180° (or π rad) (p. 30).

OCTAGON   an eight-sided polygon (p. 215).

OPPOSITE RAYS   two rays that lie on the same line, having a common endpoint and point in opposite directions (p. 5).

OPPOSITE SIDES   two sides of a quadrilateral that do not share a common vertex (p. 185).

ORTHOCENTER   the point of intersection of the altitudes of a triangle (p. 317).

PARALLEL LINES   two lines that lie in the same plane and do not intersect (p. 6).

PARALLEL PLANES   two planes that do not intersect (p. 7).

PARALLELOGRAM   a quadrilateral with each pair of sides parallel (p. 183).

PENTAGON   a five-sided polygon (p. 215).

PERIMETER   the sum of the length of the sides of a polygon (p. 225).

PERPENDICULAR   two lines or segments that form a right angle (p. 30).

PLANE   a surface such that any two points in it can be joined by a line and that line will be contained in the surface (p. 5).

POLYGON   a simple closed curve that consists entirely of line segments (pp. 6, 61).

POLYHEDRON   a union of a finite number of polygonal regions so that (1) the interiors of any two regions do not intersect, and (2) every side of any of the polygons is the side of exactly one of the other polygons (pp. 374-75).

PRISM   a polygon having two parallel and congruent bases and lateral faces that are parallelograms (p. 376).

PROTRACTOR   the instrument used to measure the number of degrees or radians in an angle (p. 30).

PYRAMID   a polyhedron with one base and lateral faces that are triangles that meet in a common point (p. 377).

QUADRILATERAL   a polygon with four sides (p. 61).

RADIAN   the measure of a central angle subtended by an arc equal in length to the radius of the circle (p. 274).

RADIUS (plural, RADII)   a line segment from the center of a circle (or a sphere); the distance from the center of the circle (or the sphere) to the circle (or sphere) itself (pp. 265, 374).

REGULAR POLYGON   a polygon with equal sides and equal angles (p. 216).

RHOMBUS   a parallelogram with equal sides (p. 183).

RECTANGLE   a parallelogram with right angles (p. 183).

RIGHT TRIANGLE   a triangle with a right (90° or $\frac{\pi}{2}$ rad) angle (p. 64).

REFLEX ANGLE   an angle containing more than 180° or π rad (p. 30).

RIGHT ANGLE   an angle containing 90° or $\frac{\pi}{2}$ rad (p. 30).

RAY   that portion of a line that lies on one side of a point P (on the line) and includes P (p. 4).

SCALENE TRIANGLE   a triangle with no sides congruent (p. 63).

SECANT   a line that intersects a circle at two points (p. 268).

SECTOR   a region of a circle formed by two radii and the intercepted arc (p. 299).

SEGMENT   the region of a circle between a chord and its arc (p. 299).

SEMICIRCLES   two equal arcs formed by a diameter cutting a circle (p. 267).

SIMILAR FIGURES   figures that are the same shape, but may not be the same size (p. 155).

SIMPLE CLOSED CURVE   a closed curve that does not intersect itself (p. 61).

SINE θ   the sine of θ, an acute angle in a right triangle, is the ratio of the leg opposite θ to the hypotenuse (p. 341).

SKEW LINES   two unique lines that do not intersect, do not lie in the same plane, and are not parallel (p. 6).

SLANT HEIGHT   the altitude of a lateral face of a pyramid (p. 377), or the length of an element of a cone (p. 379).

SPHERE   the set of all points in space equidistant from a fixed point (p. 374).

SQUARE   a rectangle with equal sides (p. 183).

STATEMENT   a declarative sentence that is true or false but not both true and false (p. 428).

STRAIGHT ANGLE   an angle containing 180° or π rad (p. 30).

SUPPLEMENTARY ANGLES   two angles with measures that total 180° or π rad (p. 34).

SURFACE AREA   the sum of the measure of the areas of a three-dimensional figure (p. 384).

TANGENT   a line that intersects a circle at one and only one point (p. 268).

TANGENT θ   the tangent of θ, an acute angle in a right triangle, is the ratio of the side opposite θ to the side adjacent to θ (p. 341).

TETRAHEDRON   a polygon with four faces (p. 376).

TRANSVERSAL   a line that crosses two lines (p. 138).

TRAPEZOID   a quadrilateral with at least one pair of sides parallel (p. 183).

TRIANGLE   a polygon with three sides (p. 61).

VERTEX   the common endpoint of two rays forming an angle (p. 27), the intersection of two edges of a polyhedron (p. 375).

VERTICAL ANGLES   non-adjacent angles formed by the intersection of two lines (p. 35).

VOLUME   the measure of the number of cubic units a figure contains (p. 409).

ANSWERS TO SELECTED PROBLEMS

CHAPTER 1

Exercise 1.1 (page 10)

The drawings given represent one way of illustrating the problems. For some problems, other drawings may also be correct.

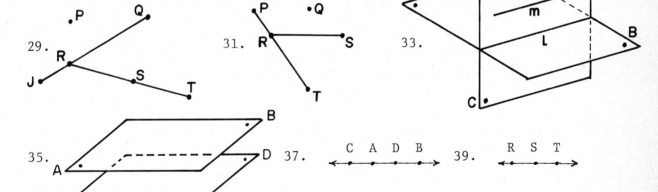

1. A-I, K-Q
3. A, B, E, F, G, I, M, N, O, Q
5. H
7. a) P Q   b) P Q
9. A B S
11. a) [figure: quadrilateral ABCD with diagonals] 6 segments: AB, AC, AD, BC, CD, BD
   b) A B C D — minimum number of non-coincident segments. Maximum is 6, illustrated in part (a).
13. No
15. No
17. Q R P / S
19. [figure with points C, E, A, D, B]
21. [figure with points R, Q, P, U, S, T]
23. False
25. True
27. False
29. [figure with points P, Q, R, S, T, J]
31. [figure with points P, Q, R, S, T]
33. [figure showing two intersecting planes with line L and line m, points A, B, C, D]
35. [figure showing two parallel planes with points A, B, C, D]
37. C A D B
39. R S T

41.

These two sets of points in the same plane	Point?	Segment?	Ray?	Line?	Plane?
Point and segment	yes	no	no	no	no
Point and ray	yes	no	no	no	no
Point and line	yes	no	no	no	no

Point and plane	▱• yes	no	no	no	no
Segment and segment	✕ yes	— yes	no	no	no
Segment and ray	↘• yes	▬→ yes	no	no	no
Segment and line	✕ yes	←▬→ yes	no	no	no
Segment and plane	no	▱∕ yes	no	no	no
Ray and ray	✕ yes	←▬→ yes	▬→ yes	no	no
Ray and line	✕ yes	no	←▬→ yes	no	no
Ray and plane	no	no	▱→ yes	no	no
Line and line	✕ yes	no	no	←→ yes	no
Line and plane	no	no	no	▱↔ yes	no

Exercise 1.2 (p. 24)

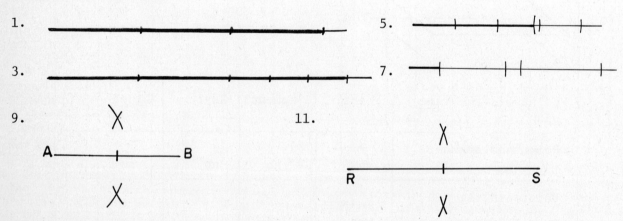

13.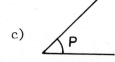

15. 4 cm

17. a) B A C

    b) A is midpoint of $\overline{BC}$

19. $\overline{CB}$, $\overline{CA}$, $\overline{AB}$, $\overline{AF}$, $\overline{EF}$, $\overline{CD}$, $\overline{DE}$, $\overline{DG}$, $\overline{CG}$, $\overline{CE}$

21. $\overline{CE}$, $\overline{CD}$, $\overline{CG}$, $\overline{DE}$, $\overline{DG}$ bisect $\overline{IH}$ at D
$\overline{EF}$, $\overline{AE}$, $\overline{AF}$ bisect $\overline{DG}$ at E
$\overline{HD}$, $\overline{DF}$, $\overline{ID}$, $\overline{BD}$ bisect $\overline{CE}$ at D

23. 15 in.   25. 29.5 ft   27. 76 in. or $12\frac{3}{4}$ in.   29. 430 cm or 534 cm

31. 6 ft, 9 ft   33. 5 in.   35. $L = \dfrac{10x + 2}{3}$ 37. $\dfrac{L-2x}{2}$

Exercise 1.3 (p. 38)

1. a) 35°, b) 120°

3. a) 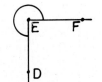   b) (angle ABC figure)

c) (angle P figure)   d) (reflex angle DEF figure)

5. a) 127.30°
   b) 24.35°
   c) 87.89°

7. a) 86°26'
   b) 183°48'
   c) 34°13'

9.

Degrees	0°	30°	45°	60°	90°	120°	135°	150°	180°
Radians	0	$\dfrac{\pi}{6}$	$\dfrac{\pi}{4}$	$\dfrac{\pi}{3}$	$\dfrac{\pi}{2}$	$\dfrac{2\pi}{3}$	$\dfrac{3\pi}{4}$	$\dfrac{5\pi}{6}$	π

11. a) 0.7156 rad
    b) 5.3407 rad
    c) 2.9496 rad
    d) 1.3963 rad

13. a) 25.71°
    b) 78.50°
    c) 56.25°
    d) 17.19°

15. 1a, 3a, 3c

17. a) 76°
    b) $\dfrac{\pi}{14}$ rad
    c) 58°
    d) 90 − y°
    e) 90 − 2x°

19. ∠1 = 22°, ∠2 = 158°, ∠4 = 158°   21. ∠2 = 95°, ∠1 = 85°

23. 71°, 109°   25. 27°, 63°   27. 12°, 78°

Exercise 1.4 (p. 54)

1.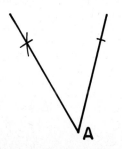

3.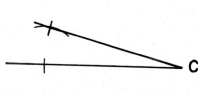

5.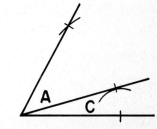

7.
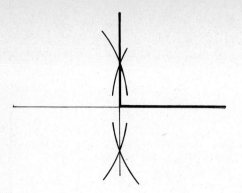

9.
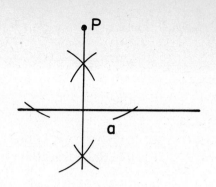

11.
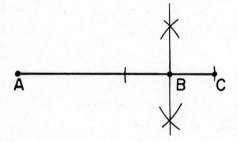

13.

15.
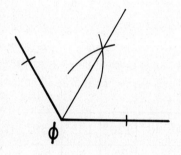

17.
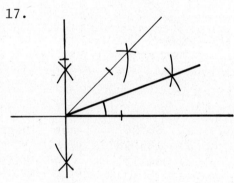

19.
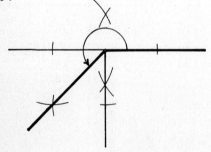

21.
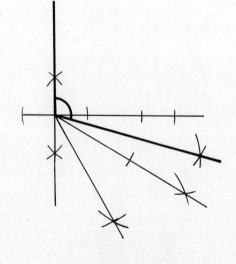

23.
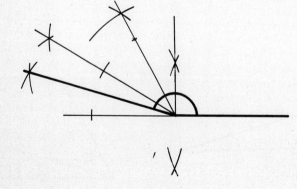

504

Chapter 1 Review (p. 56)

1. a) curve, closed curve, simple closed curve, convex simple closed curve b) curve, closed curve c) curve d) curve, line e) curve f) curve, closed curve, simple closed curve g) curve, closed curve, simple closed curve, convex simple closed curve, circle h) none

2. 1-I, 2-K, 3-I, 4-F, 5-B, 6-J, 7-E, 8-C   3. collinear   4. skew lines   5. coincident

6. Two   7. Three   8. R P Q   9. T, Q S R   10. B S C / A T

11. A C F B E D

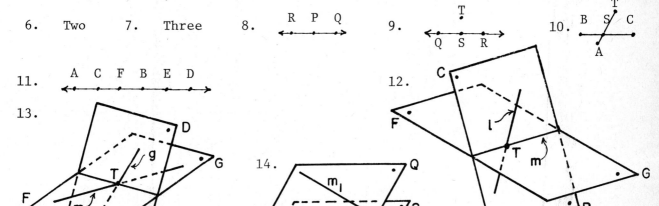

12., 13., 14. , NO

15. No   16. Yes

17. No   18. No   19. No   20. $\overline{HB}, \overline{HF}, \overline{HJ}, \overline{BF}, \overline{BJ}, \overline{GC}, \overline{GF}, \overline{GE}, \overline{CF}, \overline{CE}$

21. $\overline{HF}, \overline{EC}, \overline{HJ}$ 22. 28 ft   23. 2" or 37"   24. $L = \frac{3x-1}{2}$

25. (side views) parallel — intersecting in one line — intersecting in three lines — intersect in two lines (two are parallel)

26. 121°21'   27. 74.65°   28. $\frac{10\pi}{9}$ rad

29. 80°   30. 65.4891°   31. obtuse, acute, reflex, acute, acute

32. 71°   33. $\angle 2 = \angle 4 = \frac{3\pi}{7}$ rad   34. $\angle 1 = 135.50°, \angle 2 = 44.50°$

35. 27°, 153°   36. $\angle A = 49°$ 37. $\angle 1 = \angle 3 = 37°, \angle 2 = \angle 4 = 143°$

38. $3a + c - 2b$

39.
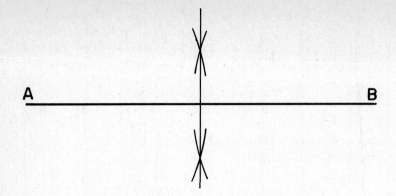

40.
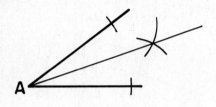

41.

42.
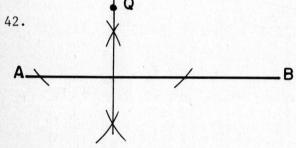

43.
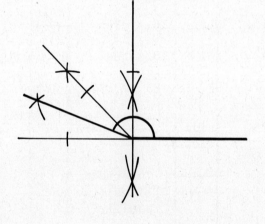

44.
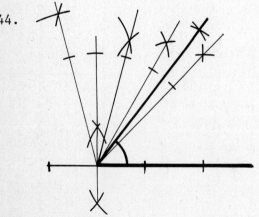

CHAPTER 2

Exercise 2.1 (p. 62)

1. All   3. A, C, E-K, M-R   5. C, G, M   7. H, J, Q

Exercise 2.2 (p. 67)

1. Vertex   : $\underline{P}$ $\underline{Q}$ $\underline{R}$ 3. No   5. a) obtuse b) acute c) right
   Opposite side: $\overline{QR}$ $\overline{PR}$ $\overline{PQ}$ d) right e) acute f) acute

7. △ACD - obtuse; △ABD-obtuse   9. 16 (triangle ABG, BCG, CDG, DEG, EFG,
   △ACD - right; △EDF - obtuse   AFG, ACF, BEC, CFE, DEA, EAB, CGE, AGC,
                                      AGE, CDA, ACE)

11. three   13. a) 50°, b) 59.96°   c) 71°55'   15. a) $\frac{9\pi}{16}$ rad

b) $\frac{5\pi}{18}$ rad   c) 1.148 rad   17. 23°29'   19. 51.5°

21. two ≅ ∠s = $\frac{\pi}{4}$ rad, third∠ = $\frac{\pi}{2}$ rad; right isosceles triangle

23. 30°, 72°, 78°   25. 60° or $\frac{\pi}{3}$ rad   27. 75°, 25°, 80°

29. ∠B = 54°, ∠A = 111°, ∠C = 15°

Exercise 2.3 (p. 83)

1. $\frac{\pi}{4} - \frac{\pi}{4} - \frac{\pi}{2}$ rad   3.   legs   hypotenuse   5. first column: 1, 2, 1, 3,
                                   a) $\overline{AC}, \overline{CB}$ $\overline{AB}$ 4, 1, 5, 2, 6, 1, 7, 8, 3,
                                   b) $\overline{TR}, \overline{TS}$ $\overline{RS}$ 1, 2, 4, 9, 1, 5, 10, 2, 1
                                   c) $\overline{KN}, \overline{NL}$ $\overline{KL}$ second column: 3, 1, 4, 7,
                                      $\overline{NL}, \overline{NM}$ $\overline{ML}$ 2, 1, 11, 3, 1, 2, 4, 1, 12,
                                      $\overline{KL}, \overline{LM}$ $\overline{MK}$ 2, 1, 13, 1, 14, 10, 4, 1, 1
                                                                                               third column: 15, 1, 1, 1,
                                                                                               1, 10, 1, 16

7. a) 0.9 mm   b) $\sqrt{7}$ in.   c) $\sqrt{118}$ cm   9. 11.0905 in.   11. yes

13. a) $3\sqrt{3}$ cm, 3 cm   15. a) 7 cm, $7\sqrt{2}$ cm   17. a) 8.7988 m, 10.16 m
    b) $8\sqrt{3}$ in., $4\sqrt{3}$ in.   b) 11 cm, 11 cm   b) 6.5241 in.,
                                                                        13.0481 in.

19. a) 14.09 ft, 19.9263 ft   21. AB = DC = AD = BC = $7\sqrt{2}$ mm
    b) 4.9710 mm, 4.9710 mm   DE = EC = 7 mm

23. EF = $4x\sqrt{3}$ 25. WY = $5x\sqrt{2}$ 27. leg = 9 ft   29. 20 ft
    DF = 8x   XY = 5x   hyp. = 15 ft

31. $50\sqrt{2}$ = 70.71 ft   33. $240\sqrt{2}$ = 339.41 mi   35. $2\sqrt{14}$ = 7.48 in.

37. 1 in. and 4 in.   39. 7 in. and 24 in.   41. AB = 6 ft, AD = 30 ft

43. $\sqrt{320} = 17.89$ ft   45. $\sqrt{560-32t-16t^2}$ ft   47.   7.42 sec

49. $9-\left(25-\sqrt{(2t)^2+6^2}\right) = -16 + 2\sqrt{t^2+9}$ ft

Exercise 2.4 (p. 99)

1.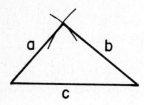

3.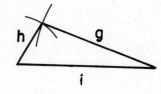

5. not possible

7.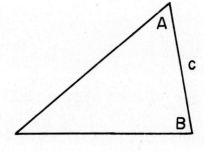

9.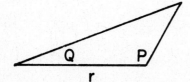

11. not possible

13.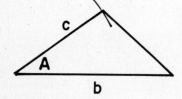

15.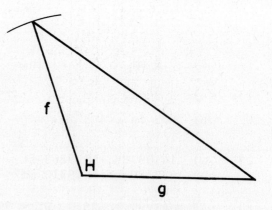

508

17. two triangles are possible

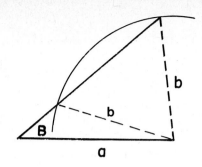

19.

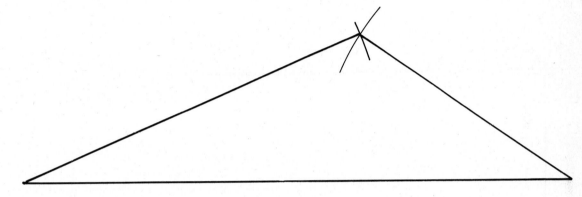

21.

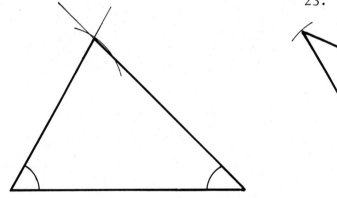

23.

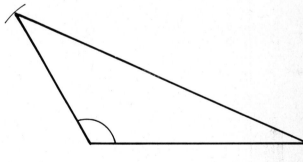

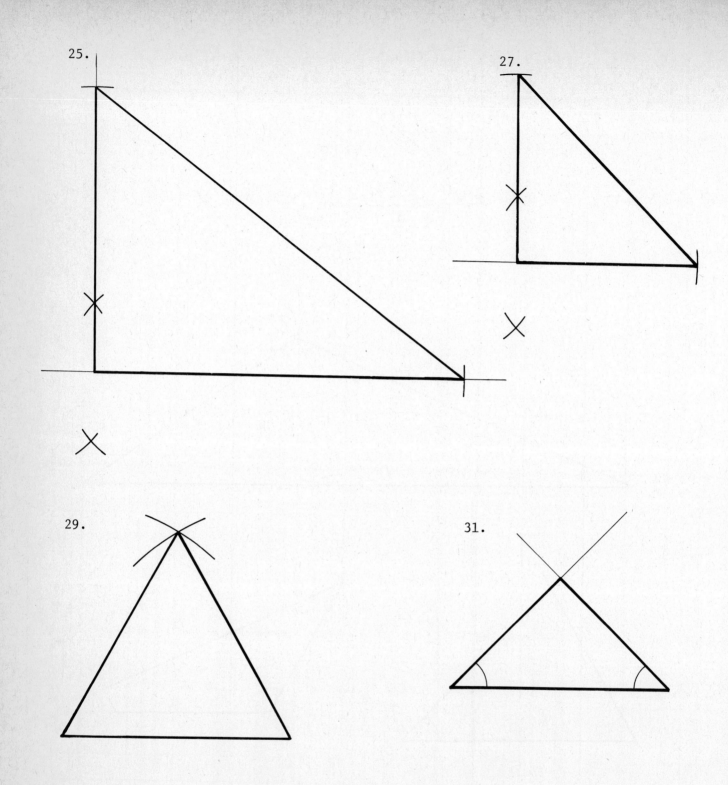

33. Construct a triangle given one leg a and hypotenuse c.

———— a ————   ———————— c ————————

Step 1:   Construct a segment congruent to a. Call its endpoints B and C.
Step 2:   Construct a right angle at c.

510

Step 3: Set the compass to the length of segment c, and mark off that length from point B to the side of ∠C. Call the point of intersection A.
Step 4: Draw the segment from A to B.
Step 5: △ABC just constructed has one leg a and hypotenuse c.

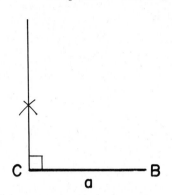 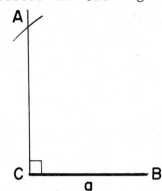 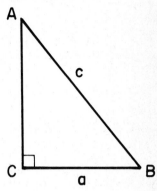

35. Construct a triangle containing ∠A and ∠B where a is the side opposite ∠A.

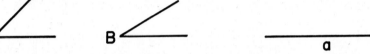

Step 1: Construct an angle congruent to ∠A + ∠B.

When ∠A + ∠B is subtracted from a straight angle, then ∠C remains. This ∠C is also the third angle in the triangle because the sum of the angles in a triangle is 180°. The problem is now reduced to constructing a triangle given ∠B and ∠C and segment a, the side included between the two angles. This procedure is given on page 93.

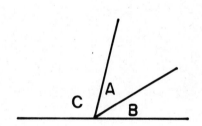

Exercise 2.5 (p.110)

1. a) ∠A ≅ ∠F $\overline{BC} \cong \overline{DE}$
     ∠B ≅ ∠D $\overline{AC} \cong \overline{EF}$
     ∠C ≅ ∠E $\overline{AB} \cong \overline{DF}$

   b) ∠P ≅ ∠R $\overline{QS} \cong \overline{QS}$
     ∠PQS ≅ ∠SQR $\overline{PS} \cong \overline{SR}$
     ∠PSQ ≅ ∠RSQ $\overline{PQ} \cong \overline{QR}$

   c) ∠L ≅ ∠Q $\overline{MN} \cong \overline{OP}$
     ∠M ≅ ∠O $\overline{LN} \cong \overline{PQ}$
     ∠N ≅ ∠P $\overline{LM} \cong \overline{OQ}$

   d) ∠J ≅ ∠X $\overline{KL} \cong \overline{ZY}$
     ∠K ≅ ∠Z $\overline{JL} \cong \overline{XY}$
     ∠L ≅ ∠Y $\overline{JK} \cong \overline{XZ}$

3. a) congruent - SAS
   b) congruent - ASA
   c) congruent - SSS
   d) cannot show congruence

5. SAS: $\overline{BC} \cong \overline{CD}$
   ASA: ∠A ≅ ∠E

7. SAS: $\overline{EB} \cong \overline{AD}$
    ASA: $\angle BAE \cong \angle DBA$

9. ASA: $\angle B \cong \angle D$
    SAS: $\overline{AC} \cong \overline{EF}$

11. a, b & c: $\triangle ABR \cong \triangle DCR$ or $\triangle ABC \cong \triangle BRD$

13. No, sides could be different lengths

15. $\angle DCE = 29°$
    $\angle AEC = 64°$

17. $\angle BFC = 72°$
    $\angle FBC = 54°$

19. $\angle ABE = 69°$
    $\angle AED = 112°$

21. $\angle DCB = 1.43$ rad
    $\angle EBA = 0.69$ rad

Exercise 2.6 (p.123)

1.

		Statement	Reason
	1.	$\triangle ARS$ with W midpoint of $\overline{AR}$	Given
A	2.	$\angle AWS \cong \angle RWS$	Given
S	3.	$\overline{AW} \cong \overline{WR}$	1 & Def of midpoint
S	4.	$\overline{SW} \cong \overline{SW}$	Common side
	5.	$\triangle AWS \cong \triangle RWS$	2,3,4 & SAS

3.

		Statement	Reason
A	1.	$\angle BDA \cong \angle CDA$	Given
S	2.	$\overline{BD} \cong \overline{CD}$	Given
S	3.	$\overline{AD} \cong \overline{AD}$	Common side
	4.	$\triangle ABD \cong \triangle ADC$	1,2,3 & SAS
	5.	$\angle ABD \cong \angle ACD$	4 & CPCTC

5.

		Statement	Reason
	1.	$\overline{CD} \perp \overline{AB}$	Given
S	2.	$\overline{AD} \cong \overline{BD}$	Given
	3.	$\angle ADC = 90°$	1 & Def of $\perp$
	4.	$\angle BDC = 90°$	1 & Def of $\perp$
A	5.	$\angle ADC \cong \angle BDC$	3,4 & Def of $\cong$
S	6.	$\overline{CD} \cong \overline{CD}$	Common side
	7.	$\triangle ADC \cong \triangle BDC$	2,5,6 & SAS

7.

		Statement	Reason
	1.	$\overline{BD}$ and $\overline{AE}$ intersect at C	Given
S	2.	$\overline{CE} \cong \overline{CB}$	Given
A	3.	$\angle E \cong \angle B$	Given
A	4.	$\angle DCE \cong \angle ACB$	Vertical $\angle$s
	5.	$\triangle DCE \cong \triangle ACB$	2,3,4 & ASA
	6.	$\angle D \cong \angle A$	5 & CPCTC

9.

		Statement	Reason
	1.	$\overline{KG} \perp \overline{GH}$	Given
	2.	$\overline{LH} \perp \overline{GH}$	Given
A	3.	$\angle KHG \cong \angle LGH$	Given
	4.	$\angle KGH = 90°$	1 & Def of $\perp$
	5.	$\angle LHG = 90°$	2 & Def of $\perp$
A	6.	$\angle KGH \cong \angle LHG$	4,5 & Def of $\cong$
S	7.	$\overline{GH} \cong \overline{GH}$	Common side
	8.	$\triangle KGH \cong \triangle GHL$	3,6,7 & ASA
	9.	$\overline{KH} \cong \overline{LG}$	8 & CPCTC

11.

		Statement	Reason
	1.	$\overline{QS}$ and $\overline{RT}$ bisect each other	Given
	2.	$\overline{AB}, \overline{QR}, \overline{TS}$ are segments	Given
S	3.	$\overline{PQ} \cong \overline{PS}$	1 & Def of bisect
S	4.	$\overline{RP} \cong \overline{PT}$	1 & Def of bisect
A	5.	$\angle QPR \cong \angle TPS$	Vertical $\angle$s
	6.	$\triangle QPR \cong \triangle TPS$	3,4,5 & SAS
A	7.	$\angle RQP \cong \angle TSP$	6 & CPCTC
A	8.	$\angle APQ \cong \angle SPB$	Vertical $\angle$s
	9.	$\triangle APQ \cong \triangle SPB$	3,7,8 & ASA
	10.	$\overline{AP} \cong \overline{BP}$	9 & CPCTC

Exercise 2.7 (p.129)

1.

		Statement	Reason
	1.	$\overline{AE}$ intersects $\overline{BD}$ at C	Given
S	2.	$\overline{AC} \cong \overline{CE}$	Given
S	3.	$\overline{DC} \cong \overline{CB}$	Given
A	4.	$\angle DCE \cong \angle ACB$	Vertical $\angle$s

3.

		Statement	Reason
	1.	Segments $\overline{MK}, \overline{SV}, \overline{KT},$ and $\overline{MN}$	Given
S	2.	$\angle K \cong \angle T$	Given
A	3.	$\overline{KR} \cong \overline{TR}$	Given
S	4.	$\angle SRK \cong \angle TRV$	Vertical $\angle$s

		5. △DCE ≅ △ACB	2,3,4 & SAS			5. △SRK ≅ △TRV	2,3,4 & SAS
		6. ∠B ≅ ∠D	5 & CPCTC			6. SK ≅ TV	5 & CPCTC

5.
		Statement	Reason
S	1.	CD ≅ CB	Given
	2.	AB ⊥ CE	Given
	3.	ED ⊥ AC	Given
	4.	∠CDE = 90°	3 & Def of ⊥
	5.	∠ABC = 90°	2 & Def of ⊥
A	6.	∠CDE ≅ ∠ABC	4,5 & Def of ≅
A	7.	∠C ≅ ∠C	Common angle
	8.	△ABC ≅ △EDC	1,6,7 & ASA

7.
	Statement	Reason
1.	Q,R,S lie in plane EF	Given
S 2.	QS ≅ QR	Given
A 3.	∠PQS ≅ ∠PQR	Given
S 4.	PQ ≅ PQ	Common side
5.	△PQR ≅ △PQS	2,3,4 & SAS

9.
		Statement	Reason
	1.	M is midpoint of AC	Given
	2.	BM ⊥ AC	Given
S	3.	AM ≅ MC	1 & Def of midpoint
	4.	∠AMB = 90°	2 & Def of ⊥
	5.	∠CMB = 90°	2 & Def of ⊥
A	6.	∠AMB ≅ ∠CMB	4,5 & Def of ≅
S	7.	BM ≅ BM	Common side
	8.	△AMB ≅ △CMB	3,6,7 & SAS
	9.	AB ≅ BC	8 & CPCTC
	10.	△ABC is isosceles	9 & Def of isosceles

11.

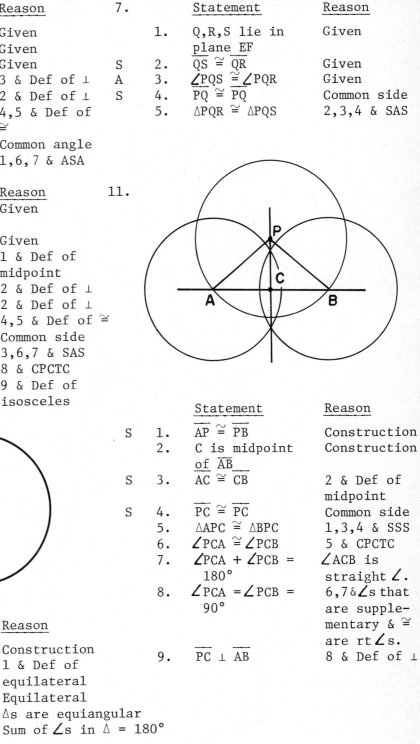

		Statement	Reason
S	1.	AP ≅ PB	Construction
	2.	C is midpoint of AB	Construction
S	3.	AC ≅ CB	2 & Def of midpoint
S	4.	PC ≅ PC	Common side
	5.	△APC ≅ △BPC	1,3,4 & SSS
	6.	∠PCA ≅ ∠PCB	5 & CPCTC
	7.	∠PCA + ∠PCB = 180°	∠ACB is straight ∠.
	8.	∠PCA = ∠PCB = 90°	6,7 & ∠s that are supplementary & ≅ are rt ∠s.
	9.	PC ⊥ AB	8 & Def of ⊥

13.

	Statement	Reason
1.	AD ≅ DC ≅ AC	Construction
2.	△ADC is equilateral	1 & Def of equilateral
3.	∠DAC ≅ ∠ACD ≅ ∠CDA	Equilateral △s are equiangular
4.	∠DAC + ∠ACD + ∠CDA = 180°	Sum of ∠s in △ = 180°
5.	∠DAC + ∠DAC + ∠DAC = 180°	3,4 & Subst.
6.	3·∠DAC = 180°	Restatement of 5
7.	∠DAC = 60°	6 & Div of =s
8.	∠DAC = ∠ACD = ∠CDA = 60°	7,3 & trans. prop.

15.

		Statement	Reason
	1.	$\overline{BD} \perp \overline{AC}$	Given
S	2.	$\overline{AB} \cong \overline{BC}$	Given
	3.	$\overline{BD} \cong \overline{BD}$	Common side
	4.	△ABC is isosceles	2 & Def of isosceles
A	5.	∠A ≅ ∠C	4 & Base ∠s of isosceles △ are ≅
	6.	∠ADB = 90°	1 & Def of ⊥
	7.	∠CDB = 90°	1 & Def of ⊥
	8.	∠ADB ≅ ∠CDB	6,7 & Def of ≅
S	9.	∠ABD ≅ ∠DBC	5,8 & if 2 pairs of ∠s in 2△s are ≅, then 3rd pair are ≅.
	10.	△ABD ≅ △DBC	2,5,9 & SAS

Chapter 2 Review (p. 132)

1. (a) A, B, G, J, K, L (b) G (c) D (d) K

   (e) A, B, F, G, J, K, L (f) E (g) J (h) A, K, L

   (i) G, J

2. △ABD: right, scalene
   △ADC: obtuse, isosceles
   △ABC: obtuse, scalene

3. △SPU, △PUR, △TUR, △QTP
   △SPR, △PTR, △QSR, △PQR

4. 82°54'

5. $\frac{7\pi}{18}$ rad

6. 105.36°

7. 1.03 rad

8. 37°, 52°, 91°

9. 24°, 48°, 108°

10. $\frac{\pi}{6}, \frac{\pi}{3}, \frac{\pi}{2}$ rad

11. (a) $7\sqrt{3}$ ft
    (b) 6 m
    (c) $2\sqrt{7}$ cm

12. (a) 2.4 ft
    (b) 11.27 mm
    (c) 3.10 m

13. (a) x = 2 in.
        y = $2\sqrt{3}$ in.
    (b) x = $\sqrt{c}$ cm
        y = $\sqrt{3}$ cm
    (c) x = $2\sqrt{15}$ mm
        y = $4\sqrt{5}$ mm

14. 35 ft, 37 ft, 12 ft

15. 20 cm
    21 cm
    29 cm

16. 174.93 mi

17. Umbrella will fit. (The diagonal is 21.63 in.)

18. 2125.6 m

19. 421.3 m under surface

20. $300t\sqrt{2} - 3$ m under surface

21. 64.03 ft

22. d = $2500 + 400t^2$ ft

23.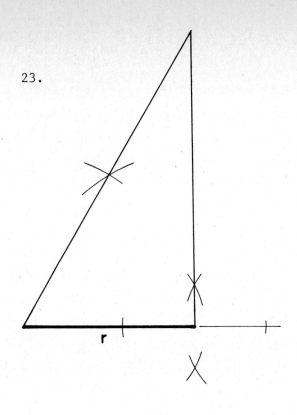

24.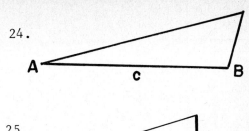

25.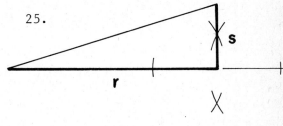

26.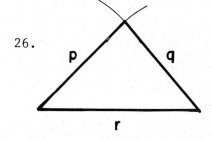

27.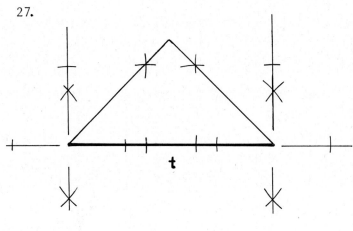

28.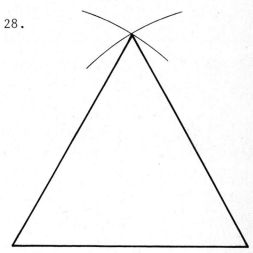

29. SAS: $\overline{AE} \cong \overline{EB}$
    ASA: $\angle ADE \cong \angle BCE$

30. SSS: $\overline{AD} \cong \overline{BE}$
    SAS: $\angle F \cong \angle C$

31.

		Statement	Reason
	1.	$\overline{AE} \perp \overline{AD}$	Given
	2.	$\overline{BF} \perp \overline{AD}$	Given
	3.	$\overline{AB} \cong \overline{CD}$	Given
A	4.	$\angle BDF \cong \angle ACE$	Given
	5.	$\overline{BC} \cong \overline{BC}$	Same segment
	6.	AB + BC = AC CD + BC = BD	Whole = sum of parts
	7.	AB + BC = CD+BC	3,5 & Add. of =
S	8.	$\overline{AC} \cong \overline{BD}$	6,7 & Trans. property
	9.	$\angle EAC = 90°$	1 & Def of $\perp$
	10.	$\angle FBD = 90°$	2 & Def of $\perp$
A	11.	$\angle EAC \cong \angle FBD$	9, 10 & Def of $\cong$
	12.	$\triangle AEC \cong \triangle BFD$	4,8,11 & ASA

32.

		Statement	Reason
S	1.	$\overline{AB} \cong \overline{AD}$	Given
	2.	$\overline{BC} \cong \overline{DE}$	Given
	3.	AB + BC = AC AD + DE = AE	Whole = sum of parts
	4.	AB + BC = AD+AE	1,2 & Add. of =
S	5.	$\overline{AC} \cong \overline{AE}$	3,4 & Def of $\cong$
A	6.	$\angle A \cong \angle A$	Common angle
	7.	$\triangle ADC \cong \triangle ABE$	1,5,6 & SAS
	8.	$\angle C \cong \angle E$	7 & CPCTC

33.

		Statement	Reason
	1.	$\overline{BX} \cong \overline{CY}$	Given
	2.	$\overline{AB} \perp \overline{BC}$	Given
	3.	$\overline{DC} \perp \overline{BC}$	Given
	4.	$\angle XBC = 90°$	2 & Def of $\perp$
	5.	$\angle YCB = 90°$	3 & Def of $\perp$
	6.	$\angle XBC \cong \angle YCB$	4,5 & Def of $\cong$
	7.	$\overline{BC} \cong \overline{BC}$	Common side
	8.	$\triangle XBC \cong \triangle YCB$	1,6,7 & SAS
	9.	$\overline{XC} \cong \overline{YB}$	8 & CPCTC

34.

		Statement	Reason
	1.	$\overline{BD}$ is $\perp$ bisector of $\overline{AC}$	Given
	2.	$\angle DBA = 90°$	1 & Def of $\perp$
	3.	$\angle DBC = 90°$	1 & Def of $\perp$
A	4.	$\angle DBA \cong \angle DBC$	2,3 & Def of $\cong$
S	5.	$\overline{AB} \cong \overline{BC}$	1 & Def of bisect
S	6.	$\overline{BD} \cong \overline{BD}$	Common side
	7.	$\triangle ABD \cong \triangle BDC$	4,5,6 & SAS
	8.	$\angle A \cong \angle C$	7 & CPCTC

35.

		Statement	Reason
	1.	$\overline{AG}$ and $\overline{CE}$ bisect each other	Given
S	2.	$\overline{AD} \cong \overline{DG}$	1 & Def of bisect
S	3.	$\overline{ED} \cong \overline{DC}$	1 & Def of bisect
A	4.	$\angle EDG \cong \angle ADC$	Vertical $\angle$s
	5.	$\triangle ACD \cong \triangle EDG$	2,3,4 & SAS
	6.	$\angle DEG \cong \angle DCA$	5 & CPCTC

36.

		Statement	Reason
	1.	D is the midpoint of $\overline{BF}$	Given
	2.	Segment $\overline{AG}$	Given
	3.	$\overline{AC} \perp \overline{BF} \perp \overline{EG}$	Given
S	4.	$\overline{BD} \cong \overline{DF}$	1 & Def of midpoint
	5.	$\angle ABD = 90°$	3 & Def of $\perp$
	6.	$\angle DFG = 90°$	3 & Def of $\perp$
A	7.	$\angle ABD \cong \angle DFG$	5,6 & Def of $\cong$
A	8.	$\angle ADB \cong \angle FDG$	Vertical $\angle$s
	9.	$\triangle ABD \cong \triangle FDG$	4,7,8 & ASA
	10.	$\overline{DG} \cong \overline{AD}$	9 & CPCTC

37.

		Statement	Reason
	1.	$\triangle ADE$	Given
S	2.	$\overline{AE} \cong \overline{ED}$	Given
S	3.	$\overline{AC} \cong \overline{BD}$	Given
	4.	$\triangle ADE$ is isosceles	2 & Def of isosceles
A	5.	$\angle A \cong \angle D$	4 & base $\angle$s of isosceles $\triangle$ are $\cong$

38.

		Statement	Reason
	1.	$\angle AEB \cong \angle DEC$	Given
S	2.	$\overline{AB} \cong \overline{CD}$	Given
	3.	$\angle ACE \cong \angle DBE$	Given
	4.	$\angle ABE$ suppl. to $\angle DBE$; $\angle ACE$ suppl. to $\angle DCE$	Def of suppl $\angle$s
A	5.	$\angle ABE \cong \angle DCE$	Suppl. to $\cong \angle$s are $\cong$

	6.	△AEC ≅ △DEB	2,3,5 & SAS	A	6.	∠A ≅ ∠D	1,5 & if 2 pair of ∠s in 2△s are ≅, then 3rd pair are ≅.
	7.	$\overline{BE}$ ≅ $\overline{CE}$	6 & CPCTC				
					7.	△AEB ≅ △ECD	2,5,6 & ASA
					8.	$\overline{AE}$ ≅ $\overline{ED}$	7 & CPCTC

CHAPTER 3

Exercise 3.1 (page 147)

1. a) ∠3 and ∠5   b) ∠1 and ∠7   c) ∠1 and ∠5   d) ∠1 and ∠3
      ∠4 and ∠6   ∠2 and ∠8   ∠2 and ∠6   ∠2 and ∠4
                                       ∠3 and ∠7   ∠5 and ∠7
                                       ∠4 and ∠8   ∠6 and ∠8

3. ∠1 = ∠3 = ∠5 = ∠7   5. ∠1 is suppl. to ∠2, ∠4, ∠6, and ∠8
   ∠2 = ∠4 = ∠6 = ∠8   ∠3 is suppl. to ∠2, ∠4, ∠6, and ∠8
                            ∠5 is suppl. to ∠2, ∠4, ∠6, and ∠8
                            ∠7 is suppl. to ∠2, ∠4, ∠6, and ∠8

7. 67°   9. ∠a = ∠c = ∠e = ∠g = 67°54'   11. ∠c = 53°,
            ∠b = ∠d = ∠f = 112°6'   ∠f = 127°

13. ∠a = 63°   15. ∠ACB = 56°   ∠ABC = 81°   17. ∠1 = 92° ∠5 = 49° ∠8 = 131°
    ∠f = 117°   ∠ADE = 137° ∠DEC = 124°   ∠3 = 92° ∠6 = 43° ∠10 = 88°
                   ∠EDB = 43°   ∠4 = 88° ∠7 = 49°

19. yes, ASA   21. yes, ASA   23.

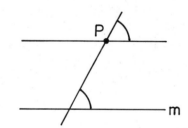

Exercise 3.2 (p. 153)

		Statement	Reason			Statement	Reason
	1.	$\overline{AD}$ ∥ $\overline{BC}$	Given		1.	$\overline{AC}$ & $\overline{DE}$ intersection at B	Given
	2.	$\overline{AB}$ ∥ $\overline{DC}$	Given				
A	3.	∠1 ≅ ∠3	1 & Alt.int. ∠s	S	2.	$\overline{AB}$ ≅ $\overline{CB}$	Given
					3.	$\overline{AD}$ ∥ $\overline{CE}$	Given
A	4.	∠2 ≅ ∠4	2 & Alt.int. ∠s	A	4.	∠ABD ≅ ∠EBC	1 & Vertical ∠s
S	5.	$\overline{AC}$ ≅ $\overline{AC}$	Common side	A	5.	∠DAC ≅ ∠ACE	3 & Alt.int. ∠s
	6.	△ADC ≅ △ABC	3,4,5 & ASA		6.	△ABD ≅ △BCE	2,4,5 & ASA
	7.	$\overline{AB}$ ≅ $\overline{CD}$	6 & CPCTC		7.	$\overline{AD}$ ≅ $\overline{CE}$	6 & CPCTC

	5.	Statement	Reason	7.		Statement	Reason
	1.	$\overline{DC}, \overline{AB}, \overline{EF}$ intersect at R	Given		1.	Segment $\overline{AC}$	Given
					2.	$\overline{AB} \parallel \overline{CD}$	Given
S	2.	$\overline{DR} \cong \overline{CR}$	Given	S	3.	$\overline{AB} \cong \overline{CD}$	Given
S	3.	$\overline{ER} \cong \overline{FR}$	Given		4.	$\overline{AF} \cong \overline{EC}$	Given
A	4.	$\angle DRE \cong \angle CRF$	1 & Vert $\angle$s		5.	$\overline{EF} \cong \overline{EF}$	Same segment
	5.	$\triangle DRE \cong \triangle CRF$	2,3,4 & SAS		6.	CF + FE = CE	Whole = sum
	6.	$\angle D \cong \angle C$	5 & CPCTC			FE + EA = FA	of parts
	7.	$\overline{DA} \parallel \overline{CB}$	6 & Alt.int $\angle$s are $\cong$		7.	CF + FE = FE + EA	4,6 & Trans.
				S	8.	$\overline{CF} \cong \overline{EA}$	7 & Subt. of =s
				A	9.	$\angle BAE \cong \angle FCD$	2 & Alt.int. $\angle$s
					10.	$\triangle BAE \cong \triangle FCD$	3,8,9 & SAS
					11.	$\overline{BE} \cong \overline{FD}$	10 & CPCTC

	9.		Statement	Reason
		1.	0 midpoint of $\overline{AB}$	Given
		2.	0 midpoint of $\overline{CD}$	Given
S		3.	$\overline{BO} \cong \overline{OA}$	1 & Def of midpoint
S		4.	$\overline{CO} \cong \overline{OD}$	2 & Def of midpoint
A		5.	$\angle COB \cong \angle AOD$	Vertical $\angle$s
		6.	$\triangle COB \cong \triangle AOD$	3,4,5 & SAS
		7.	$\angle BCD \cong \angle CDA$	6 & CPCTC
		8.	$\ell_1 \parallel \ell_2$	7 & Alt.int. $\angle$s $\cong$

11.

Given: $\angle 1 \cong \angle 2$ by construction

Prove: $\ell \parallel m$

	Statement	Reason
1.	$\angle 1 \cong \angle 2$	Given
2.	$\ell \parallel m$	1 & corresp. $\angle$s $\cong$

Exercise 3.3 (p. 166)

1. neither; angle not included between $\cong$ sides
3. neither; sides not proportional
5. congruent; SAS
7. Similar; corresponding sides proportional

9. $\angle A = \angle B$ (both are right $\angle$s)
$\angle AED = \angle BEC$ (vertical $\angle$s)
$\angle D = \angle C$ (If 2 pairs of $\angle$s in 2 $\triangle$s are $\cong$, then the 3rd pair is $\cong$.)
$\triangle ADE \sim \triangle CBE$ (AAA)

Corresponding sides: $\overline{AD}$ and $\overline{BC}$
$\overline{AE}$ and $\overline{BE}$
$\overline{DE}$ and $\overline{EC}$

11. $\angle B \cong \angle B$ (same $\angle$)
$\angle BDE \cong \angle BAC$ (corresp. $\angle$s)
$\angle BED \cong \angle ECA$ (corresp. $\angle$s)
$\triangle ABC \sim \triangle DBE$ (AAA)

Corresponding sides: $\overline{BE}$ and $\overline{BC}$
$\overline{BD}$ and $\overline{BA}$
$\overline{DE}$ and $\overline{AC}$

13. 12 ft   15. 3 m   17. 16 ft

19. $6\frac{2}{3}$ ft   21. RS = 6 in., PR = 18 in.   23. BD = 4 cm; AD = 2.4 cm

25. $22\frac{1}{2}$ ft   27. $s = \frac{7t}{2}$ ft

Exercise 3.4 (p. 173)

1. $5\frac{5}{7}$ cm   3. $3\frac{3}{5}$ cm   5. CE = 12 ft   AD = 10 ft   7. $8\frac{8}{9}$ in.
EB = 18 ft   DB = 15 ft

9. 9 mm
11. AC = 4 in.   BD = 6 in.
     CE = 2 in.   DF = 3 in.
13. $S = \dfrac{3D}{5}$ ft

15.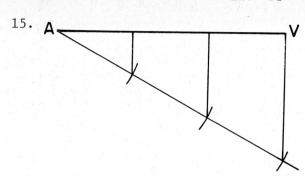

17.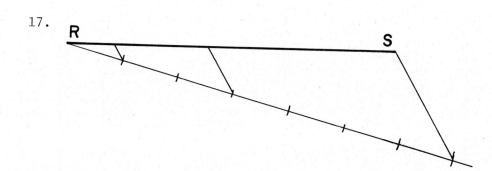

19.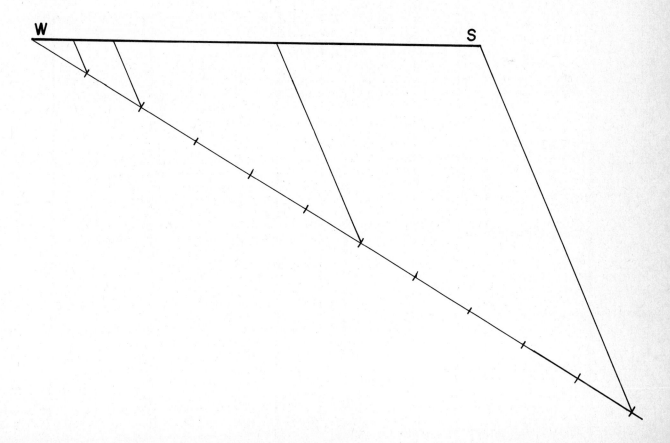

Exercise 3.5 (p. 176)

1.
		Statement	Reason
	1.	$\overline{CE}$ and $\overline{AB}$ intersect at D	Given
	2.	$\overline{EA} \perp \overline{AB}$	Given
	3.	$\overline{DC} \perp \overline{BC}$	Given
	4.	$\angle EAB = 90°$	2 & Def of $\perp$
	5.	$\angle DCB = 90°$	3 & Def of $\perp$
A	6.	$\angle EAB \cong \angle DCB$	4,5 & Trans.
A	7.	$\angle EDA \cong \angle BDC$	Vertical $\angle$s
A	8.	$\angle DEA \cong \angle DBC$	If 2 pair of $\angle$s in 2 $\triangle$s are $\cong$, then 3rd pair is $\cong$.
	9.	$\triangle ADE \sim \triangle CDB$	6,7,8 & AAA

3.
		Statement	Reason
	1.	$\overline{AC} \parallel \overline{DE}$	Given
A	2.	$\angle B \cong \angle B$	Common $\angle$
A	3.	$\angle BDE \cong \angle BAC$	1 & Corresp. $\angle$s
A	4.	$\angle BED \cong \angle BCA$	1 & Corresp. $\angle$s
	5.	$\triangle ABC \sim \triangle DBE$	2,3,4 & AAA

5.
		Statement	Reason
A	1.	$\angle BDC \cong \angle ACB$	Given
A	2.	$\angle B \cong \angle B$	Common $\angle$
A	3.	$\angle BAC \cong \angle BCD$	If 2 pair of $\angle$s in 2 $\triangle$s are $\cong$, then 3rd pair is $\cong$.
	4.	$\triangle ABC \sim \triangle BCD$	1,2,3 & AAA

Chapter 3 Review (p. 178)

1. a) $\angle C$ and $\angle E$ c) $\angle A$ and $\angle E$ $\angle C$ and $\angle G$ 2. $\angle C, \angle E, \angle G$ 3. $\angle B, \angle D, \angle H$
   $\angle D$ and $\angle F$ $\angle B$ and $\angle F$ $\angle D$ and $\angle H$
   b) $\angle A$ and $\angle G$ d) $\angle A$ and $\angle C$ $\angle E$ and $\angle G$
   $\angle B$ and $\angle H$ $\angle B$ and $\angle D$ $\angle F$ and $\angle H$

4. $57.2°$ 5. $\frac{5\pi}{9}$ rad   6. $\angle F = 100°$ 7. 1.71 rad

8. a) $\angle PQC = \angle ABC$ b) $\angle CAB = \angle QRB$ c) $\angle SPQ = \angle PQC$
   corr. angles   corr. angles   alt. int. angles

9. $55°$ 10. $\angle 2 = \angle 3 = 62°$ 11. $104°$ 12. a) $m_1 \parallel m_2$
               $\angle 4 = \angle 5 = 118°$ b) $\ell_1 \parallel \ell_2$
13. $\angle 9 = 60°$ c) $\ell_1 \parallel \ell_2$ and $m_1 \parallel m_2$
    $\angle 10 = 70°$

14. (figure showing two parallel lines cut by a transversal through Q with marked angles)

15.
		Statement	Reason
	1.	$\overline{AB} \parallel \overline{CD}$	Given
	2.	$\overline{AD} \parallel \overline{BC}$	Given
A	3.	$\angle BAC \cong \angle ACD$	1 & Alt.int. $\angle$s
A	4.	$\angle DAC \cong \angle ACB$	2 & Alt.int. $\angle$s
S	5.	$\overline{AC} \cong \overline{AC}$	Common side
	6.	$\triangle ABC \cong \triangle CDA$	3,4,5 & ASA

16.
		Statement	Reason
	1.	$\overline{BC} \parallel \overline{AD}$	Given
	2.	$\overline{AB} \parallel \overline{DC}$	Given
	3.	$\overline{BD}$ and $\overline{AC}$ intersecting at 0	Given
A	4.	$\angle BAC \cong \angle ACD$	2 & Alt.int. $\angle$s
A	5.	$\angle CAD \cong \angle BCA$	1 & Alt.int. $\angle$s
S	6.	$\overline{AC} \cong \overline{AC}$	Common side
	7.	$\triangle ABC \cong \triangle ADC$	4,5,6 & ASA
	8.	$\overline{AD} \cong \overline{BC}$	7 & CPCTC

17.
		Statement	Reason
	1.	Segment $\overline{AD}$	Given
	2.	$\overline{GC} \parallel \overline{ED}$	Given
	3.	$\overline{AB} \cong \overline{CD}$	Given
S	4.	$\overline{GC} \cong \overline{ED}$	Given
A	5.	$\angle GCA \cong \angle EDB$	2 & Corresp. $\angle$s
	6.	$\overline{BC} \cong \overline{BC}$	Same segment
	7.	AB + BC = AC BC + CD = BD	Whole = sum of parts
	8.	AB + BC = BC + CD	3,6 & Add.of =s
S	9.	$\overline{AC} \cong \overline{BD}$	7,8 & transitive
	10.	$\triangle ACG \cong \triangle BDE$	4,5,9 & SAS
	11.	$\angle G \cong \angle E$	10 & CPCTC
	12.	$\overline{AG} \parallel \overline{BE}$	11 & Corresp. $\angle$s are $\cong$

18. x = 2 cm, y = 54 cm

19. y = 6"   x = 4"

20. a) 20 mm   b) 28 mm

21. 27.5 ft   22. 4.2 m   23. DC = 8.75 m AB = 28.8 m   24. $6\frac{2}{3}$ ft

25. $5\frac{1}{3}$ ft   26. $\frac{20-4t}{3}$ ft

29.
		Statement	Reason
	1.	$\overline{AE}$ intersects $\overline{BD}$ at C	Given
	2.	$\overline{AB} \parallel \overline{DE}$	Given
A	3.	$\angle ABC \cong \angle CDE$	2 & Alt.int. $\angle$s
A	4.	$\angle ACB \cong \angle DCE$	Vertical $\angle$s
A	5.	$\angle BAC \cong \angle CED$	2 & Alt.int. $\angle$s
	6.	$\triangle ABC \sim \triangle EDC$	3,4,5 & AAA

27.

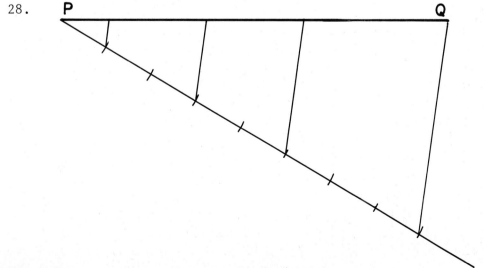

28.

30. | | Statement | Reason | 31. | | Statement | Reason |

	1.	$\overline{ST} \parallel \overline{PR}$	Given
A	2.	$\angle PRQ \cong \angle STQ$	1 & corresp. $\angle$s
A	3.	$\angle RPQ \cong \angle TSQ$	1 & corresp. $\angle$s
A	4.	$\angle Q \cong \angle Q$	Common $\angle$
	5.	$\triangle PQR \sim \triangle SQT$	2,3,4 & AAA

	1.	$\angle ABC \cong \angle EDG$	Given
A	2.	$\angle C \cong \angle C$	Common $\angle$
A	3.	$\angle BAC \cong \angle DEC$	If 2 pairs of $\angle$s in 2 $\triangle$s are $\cong$, then 3rd pair is $\cong$
	4.	$\triangle ABC \sim \triangle EDC$	1,2,3 & AAA

CHAPTER 4

Exercise 4.1 (p. 188)

1. (a) none
   (b) T, P, R
   (c) T

3. (a) opp. sides: $\overline{AB}$ & $\overline{DC}$
       $\overline{AD}$ & $\overline{BC}$
       opp. vert: A & C
                  B & D

   (b) opp. sides: $\overline{FI}$ & $\overline{GH}$
       $\overline{FG}$ & $\overline{IH}$
       opp. vert: F & H
                  I & G

   (c) opp. sides: $\overline{LM}$ & $\overline{ON}$
       $\overline{LO}$ & $\overline{MN}$
       opp. vert: L & N
                  M & O

5. (a) 160°
   (b) $\frac{7\pi}{30}$ rad

7. (a) 132.71°
   (b) 53°41'

9. (a) $\frac{3\pi}{7}$ rad
   (b) 1.75 rad

11. 69°, 91°
    107°, 93°

13. 57°, 125°
    89°, 89°

15. 108°, 72°
    144°, 36°

Exercise 4.2 (p. 197)

1. 6.61 cm
3. 20 mm
5. $10 + 12\sqrt{3}$ ft
7. $\angle A = \angle C = 107°$
   $\angle B = \angle D = 73°$

9. $\angle A = \angle C = 111°$
   $\angle B = \angle D = 69°$
11. $3\sqrt{3}$ cm

Exercise 4.3 (p. 207)

1. (a) i) trapezoid, parallelogram, rhombus
       ii) $\overline{AD} \cong \overline{DC} \cong \overline{BC} \cong \overline{AB}$
       iii) $\angle A \cong \angle C$; $\angle B \cong \angle D$
       iv) not congruent

   (b) i) trapezoid
       ii) none
       iii) none
       iv) not congruent

   (c) i) trapezoid, parallelogram, rectangle
       ii) $\overline{KL} \cong \overline{JM}$; $\overline{KJ} \cong \overline{LM}$
       iii) $\angle K \cong \angle L \cong \angle M \cong \angle J$
       iv) congruent

   (d) i) trapezoid
       ii) none
       iii) none
       iv) not congruent

   (e) i) trapezoid; parallelogram
       ii) $\overline{ST} \cong \overline{RU}$; $\overline{SR} \cong \overline{TU}$
       iii) $\angle S \cong \angle U$; $\angle T \cong \angle R$
       iv) not congruent

   (f) i) none
       ii) none
       iii) none
       iv) not congruent

g) i)   isosceles trapezoid   h) i)   trapezoid, parallelogram, rectangle
    ii) $\overline{AB} \cong \overline{CD}$ square
    iii) $\angle B \cong \angle C; \angle A \cong \angle D$ ii) $\overline{EF} \cong \overline{FG} \cong \overline{GH} \cong \overline{EH}$
    v)   congruent   iii) $\angle E \cong \angle F \cong \angle G \cong \angle H$
                                       v)   congruent

3. $5\sqrt{13}$ in.   5. 2.20 m   7. $x \neq 0$; $x = 8$ ft   9. 29 in.

11. 13 in.   13. 9.42 cm   15. 1 cm   17. 12 cm & $12\sqrt{3}$ cm   19. 76 in.

21. $18\sqrt{2}$ cm   23. 91.36 in.   25. 15 in. by 8 in.

Exercise 4.4 (p. 213)

1. | | Statement | Reason |
    |---|---|---|
    | | 1. ABCD is an isosceles trap. | Given |
    | | 2. $\overline{AB} \parallel \overline{DC}$ | 1 & Def of trapezoid |
    | A | 3. $\angle AEB \cong \angle DEC$ | Vertical $\angle$s |
    | A | 4. $\angle ABC \cong \angle BDE$ | 2 & Alt. int. $\angle$s |
    | A | 5. $\angle BAC \cong \angle ACD$ | 2 & Alt. int. $\angle$s |
    | | 6. $\triangle AEB \sim \triangle CED$ | 3,4,5 & AAA |

3. | | Statement | Reason |
    |---|---|---|
    | | 1. Parallelogram PQRS | Given |
    | S | 2. $\overline{SP} \cong \overline{RQ}$ | 1 & Opp. sides of parallelogram are $\cong$ |
    | | 3. $\overline{SP} \parallel \overline{RQ}$ | 1 & Def of parallelogram |
    | A | 4. $\angle TSP \cong \angle TQR$ | 3 & Alt. int. $\angle$s |
    | A | 5. $\angle TPS \cong \angle TRQ$ | 3 & Alt. int. $\angle$s |
    | | 6. $\triangle TPS \cong \triangle TRQ$ | 2,4,5 & ASA |
    | | 7. $\overline{ST} \cong \overline{TQ}$ | 6 & CPCTC |

5. | | Statement | Reason |
    |---|---|---|
    | | 1. Rectangle JKLM | Given |
    | | 2. N midpoint of $\overline{JK}$ | Given |
    | S | 3. $\overline{JN} \cong \overline{NK}$ | 2 & Def of midpoint |
    | S | 4. $\overline{JM} \cong \overline{KL}$ | 1 & opp. sides of a parallelogram are $\cong$ |
    | | 5. $\angle MJK = 90°$ $\angle LKJ = 90°$ | 1 & Def of rectangle |
    | A | 6. $\angle MJK \cong \angle LKJ$ | 5 & Def of $\cong$ |
    | | 7. $\triangle JNM \cong \triangle KNL$ | 3,4,6 & SAS |
    | | 8. $\angle JMN \cong \angle KLN$ | 7 & CPCTC |

7. | | Statement | Reason |
    |---|---|---|
    | | 1. Rectangle ABCD | Given |
    | | 2. $\overline{EC}$ & $\overline{DF}$ intersect at G | Given |
    | S | 3. $\overline{ED} \cong \overline{FC}$ | Given |
    | | 4. $\angle EDC = 90°$ $\angle FCD = 90°$ | 1 & Def of rectangle |
    | A | 5. $\angle EDC \cong \angle FCD$ | 4 & Def of $\cong$ |
    | S | 6. $\overline{DC} \cong \overline{DC}$ | Common side |
    | | 7. $\triangle EDC \cong \triangle FCD$ | 3,5,6 & SAS |
    | | 8. $\angle GDC \cong \angle GCD$ | 7 & CPCTC |
    | | 9. $\overline{DC} \cong \overline{GC}$ | 8 & if two $\angle$s are $\cong$ in a $\triangle$, then opp. sides are $\cong$. |
    | | 10. $\triangle DGC$ is isosceles | 9 & Def of isosceles |

9. | | Statement | Reason |
    |---|---|---|
    | | 1. $\triangle ABC$ | Given |
    | | 2. $\overline{AD} \parallel \overline{BC}$, $\overline{DC} \parallel \overline{AB}$ | Given |
    | | 3. ABCD is a parallelogram | 2 & Def of parallelogram |
    | A | 4. $\angle CAD \cong \angle BCA$ | 2 & Alt. int. $\angle$s are $\cong$ |
    | A | 5. $\angle BAC \cong \angle ACD$ | 2 & Alt. int. $\angle$s are $\cong$ |
    | S | 6. $\overline{AC} \cong \overline{AC}$ | Common side |
    | | 7. $\triangle ABC \cong \triangle CDA$ | 4,5,6 & ASA |

523

11. 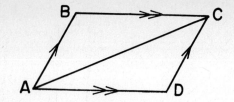   13.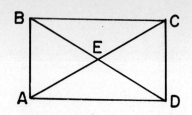

	Statement	Reason
1.	Parallelogram ABCD with diagonal AC	Given
2.	$\overline{BC} \parallel \overline{AD}$	1 & Def of parallelogram
3.	$\overline{AB} \parallel \overline{DC}$	1 & Def of parallelogram
A 4.	$\angle BCA \cong \angle CAD$	2 & Alt.int $\angle$s are $\cong$
A 5.	$\angle BAC \cong \angle ACD$	3 & Alt.int $\angle$s are $\cong$
S 6.	$\overline{AC} \cong \overline{AC}$	Common side
7.	$\triangle ABC \cong \triangle ACD$	4,5,6 & ASA
8.	$\angle B \cong \angle D$	7 & CPCTC

A similar proof using diagonal $\overline{BD}$ can be used to show that $\angle BAD \cong \angle BCD$.

	Statement	Reason
1.	Rect. ABCD with diag. AC & BD intersecting at E	Given
2.	$\overline{BA} \parallel \overline{CD}$	1 & Def of rect.
3.	$\overline{BC} \parallel \overline{AD}$	1 & Def of rect.
A 4.	$\angle BAC \cong \angle ACD$	2 & Alt.int. $\angle$s are $\cong$
A 5.	$\angle CAD \cong \angle BCA$	3 & Alt.int. $\angle$s are $\cong$
S 6.	$\overline{AC} \cong \overline{AC}$	Common side
7.	$\triangle ABC \cong \triangle ADC$	3,5,6 & ASA
8.	$\overline{BA} \cong \overline{CD}$	7 & CPCTC
9.	$\angle BAD = 90°$ $\angle ADC = 90°$	1 & Def of rectangle
10.	$\angle BAD \cong \angle ADC$	9 & Def of $\cong$
11.	$\overline{AD} \cong \overline{AD}$	Common side
12.	$\triangle BAD \cong \triangle ADC$	8,10,11 & SAS
13.	$\overline{BD} \cong \overline{AC}$	12 & CPCTC

15.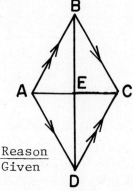

	Statement	Reason
1.	Rhombus ABCD with diag. AC & BD intersecting at E	Given
S 2.	$\overline{AB} \cong \overline{CD}$	1 & Def of rhombus
3.	$\overline{AB} \parallel \overline{DC}$	1 & Def of rhombus
A 4.	ABE $\cong$ EDC	3 & Alt.int. $\angle$s are $\cong$
A 5.	$\angle BAE \cong \angle ECD$	3 & Alt.int. $\angle$s are $\cong$
6.	$\triangle ABE \cong \triangle EDC$	2,4,5 & ASA
7.	$\overline{AE} \cong \overline{EC}$ $\overline{BE} \cong \overline{ED}$	6 & CPCTC
8.	$\overline{AC}$ and $\overline{BD}$ bisect each other	7 & Def of bisect

17.

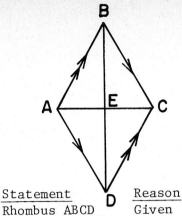

	Statement	Reason
1.	Rhombus ABCD with diag. AC and BD intersecting at E	Given
S,S 2.	AB ≅ BC ≅ AD ≅ DC	1 & Def of rhombus
S 3.	AC ≅ AC	Common side
4.	△ABC ≅ △ADC	2,3 & SSS
5.	Diagonal AC divides the rhombus into two ≅ △s	Restatement of 4
A 6.	∠EAD ≅ ∠EAB	4 & CPCTC
S 7.	AD ≅ AB	1 & Def of rhombus
S 8.	AE ≅ AE	Common side
9.	△EAD ≅ △EAB	6,7,8 & SAS
10.	BE ≅ DE	9 & CPCTC
11.	∠AED ≅ ∠AEB	9 & CPCTC
12.	∠AED + ∠AEB = 180°	BD a strt ∠
13.	∠AED = ∠AEB = 90°	11,12 & two ≅ suppl. ∠s are rt. ∠s
14.	ED ⊥ AC, EB ⊥ AC	13 & Def of ⊥
15.	ED altitude of △ADC, EB altitude of △ABC	14 & Def of altitude
16.	E is midpoint of BD	10 & Def of midpoint
17.	The altitudes of △ADC & △ABC are 1/2 of diagonal BD	15 & 16 & Def of midpoint

19.

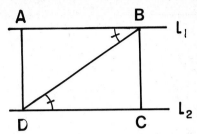

	Statement	Reason
1.	$\overleftrightarrow{AB} \parallel \overleftrightarrow{DC}$	Given
2.	AD ⊥ AB, BC ⊥ AB	Construction
3.	∠DAB = 90° ∠ABC = 90°	2 & Def of ⊥
4.	∠DCB + ∠CBA = 180°	1 & int. ∠s on same side of transv. are suppl.
5.	∠DCB = 90°	3,4 & Subt. of =s
A 6.	∠ABC ≅ ∠DCB	3,5 & Def of ≅
S 7.	DB ≅ DB	Common side
A 8.	∠ABD ≅ ∠BDC	1 & Alt.int. ∠s are ≅
9.	△ABD ≅ △BDC	6,7,8 & ASA
10.	AD ≅ BC	9 & CPCTC
11.	The ⊥ distance between $\overleftrightarrow{AB}$ & $\overleftrightarrow{DC}$ is the same	3,10

21.

	Statement	Reason
1.	Isosceles trap. ABCD with diag. AC & BD intersecting at E	Given
S 2.	AB ≅ CD	1 & Def of isosceles trap.
A 3.	∠BAD ≅ ∠CDA	1 & Base ∠s of isosceles trap. are ≅
S 4.	AD ≅ AD	Common side
5.	△BAD ≅ △CDA	2,3,4 & SAS
6.	AC ≅ BD	5 & CPCTC

Exercise 4.5 (p. 219)

1.

No. of sides	Sum of int. ∠s	Each int ∠
3	180°	60°
5	540°	108°
8	1080°	135°
9	1260°	140°
12	1800°	150°

3. 1980°   5. 18π rad   7. 152.31°

9. $\frac{9\pi}{10}$ or 2.83 rad   11. 7   13. 9   15. 10

17. 9   19. 72°   21. $\frac{\pi}{3}$ rad   23. 8

25. 5   27. 136°, 90°, 102°, 140°, 72°

Chapter 4 Review (p. 222)

1. (a) A, C, E (b) D, E (c) D, E (d) B-E (e) B-E (f) C, E (g) A-F

2. 127°52'   3. 2.06 rad   4. $\frac{\pi}{3}, \frac{\pi}{3}, \frac{\pi}{2}, \frac{5\pi}{6}$ rad   5. 73°, 98°, 63°, 126°

6. 5.2 m   7. $\sqrt{15}$ in.   8. ∠A = ∠C = 71°, ∠B = ∠D = 109°   9. $12\sqrt{2}$ cm

10. 21.4 cm   11. 42 ft by 56 ft; diag. = 70 ft   12. 7 cm   13. 53 in.

14. 15 in. & $15\sqrt{3}$ in.   15. $7\sqrt{2}$ in.   16. 6.08 mm   17. 6.36 ft by 10.18 ft

18.

	Statement	Reason
	1. △ABC	Given
	2. $\overline{DE} \parallel \overline{AC}$	Given
	3. ∠DAF ≅ ∠EFC	Given
	4. $\overline{DA} \parallel \overline{EF}$	3 & corr. ∠s are ≅
	5. Construct $\overline{DF}$	Construction
S	6. $\overline{DF} \cong \overline{DF}$	Common side
A	7. ∠EDF ≅ ∠DFA	2 & Alt. int. ∠s are ≅
A	8. ∠ADF ≅ ∠EFD	4 & alt. int. ∠s are ≅
	9. △ADF ≅ △EFD	6, 7, 8 & ASA
	10. $\overline{DA} \cong \overline{EF}$	9 & CPCTC

19.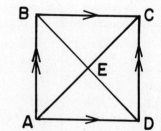

	Statement	Reason
	1. Parallelogram ABCD with diag. AC & BD intersecting at E	Given
	2. $\overline{BD} \perp \overline{AC}$	Given
	3. $\overline{AD} \parallel \overline{BC}$, $\overline{AB} \parallel \overline{DC}$	1 & Def of parallelogram
A	4. ∠BAC ≅ ∠DCA	3 & Alt. int. ∠s are ≅
A	5. ∠ACB ≅ ∠CAD	3 & Alt. int. ∠s are ≅
S	6. $\overline{AC} \cong \overline{AC}$	Common side
	7. △ADC ≅ △ABC	4, 5, 6 & ASA
S,S	8. $\overline{BC} \cong \overline{AD}$, $\overline{AB} \cong \overline{DC}$	7 & CPCTC
A	9. ∠ABD ≅ ∠BDC	3 & Alt. int. ∠s are ≅
	10. △ABE ≅ △CED	4, 8, 9 & ASA
S	11. $\overline{AE} \cong \overline{EC}$	10 & CPCTC
	12. ∠AED = ∠DEC = 90°	2 & Def of ⊥
A	13. ∠AED ≅ ∠DEC	12 & Def of ≅

20. 140°   21. 12   22. 0.90 rad

23. 16   24. 101°, 89°, 123°, 112°, 115°

	Statement	Reason
S 14.	$\overline{ED} \cong \overline{ED}$	Common side
15.	$\triangle AED = \triangle DEC$	11,13,14 & SAS
16.	$\overline{AD} \cong \overline{DE}$	15 & CPCTC
17.	$\overline{BC} \cong \overline{DC} \cong \overline{AD} \cong \overline{AB}$	8, 16 & Trans.
18.	Figure ABCD is a rhombus	1,17 & Def. of rhombus

CHAPTER 5

Exercise 5.1 (p. 232)

1. (a) 23 cm   (b) 24 cm
3. $A = 225 \text{ ft}^2$, $P = 60 \text{ ft}$
5. $A = 4.18 \text{ m}^2$, $P = 8.2 \text{ m}$
7. 144 in^2
9. 59.2 cm
11. 1072.3 cm^2
13. 15 in. by 7 in.
15. 779 ft^2
17. 7 ft by 24 ft
19. 15 ft by 35 ft
21. (a) 76 cm^2 (b) 120 ft^2
23. $A = x(160-x)$
25. $A = \Delta x(3x^2 + 2)$
27. $A = 44 + 2x + \dfrac{132}{2x-3}$

Exercise 5.2 (p. 241)

1. 80 ft^2; 46 ft
3. 144 in^2; 56 in.
5. $45\sqrt{2} \text{ in}^2$; 42 in.
7. 18 in^2; 22.5 in.
9. 30 cm^2; 30 cm
11. 16.25 cm^2; 27 cm
13. 29.70 in^2; 27.61 in
15. 12.9 mm^2; 23.9 mm
17. $49\sqrt{3} \text{ cm}^2$
19. 16 ft
21. 90 m^2
23. 220 in^2
25. 19 cm
27. 36 in^2
29. $49\sqrt{3} \text{ in}^2$
31. 189 in^2
33. 63 m^2
35. $\left(\dfrac{9 + 4\sqrt{3}}{9}\right)x^2 - \left(\dfrac{64\sqrt{3}}{9}\right)x + \dfrac{256\sqrt{3}}{9} \text{ cm}^2$

Exercise 5.3 (p. 252)

1. 117 cm^2
3. 44.41 cm
5. 336 in^2
7. 168.61 cm^2
9. 14 ft
11. 40 in.
13. 13 mm & 16 mm
15. 41.23 mm
17. $98\sqrt{3} \text{ cm}^2$
19. 30 in.
21. 50 in^2
23. $34\frac{5}{16} \text{ in}^2$
25. 6 ft
27. 16 ft
29. 17.1 yd
31. 72 cm^2
33. 420 in^2
35. $50 + 40\sqrt{2} \text{ cm}^2$
37. 9 in. & 23 in.
39. longer base = 19 in., shorter base = 14 in., alt. = 4 in.
41. $\dfrac{1}{2} \Delta x(y_i + y_{i+1})$

Exercise 5.4 (p. 258)

1. 480 in^2 3. 6 ft   5. 236.6 cm^2 7. 11.01 cm   9. $24\sqrt{3}$ in^2

11. $200\sqrt{3}$ m^2 13. 96 cm^2 15. 80 mm^2 17. 69 cm^2

Chapter 5 Review (p. 261)

1. 138 cm^2, 66 cm   2. 28 ft   3. 81 ft^2 4. 22.96 cm   5. 13 in. by 20 in.

6. 5 ft by 18 ft.   7. A = x(300-2x) ft^2 8. $A = 450 + 8x + \frac{2700}{x-6}$ cm^2

9. (a) 264 mm^2, 74 mm   10. (a) 28.28 in^2, 27.2 in.   11. $36\sqrt{3}$ mm^2
   (b) 630 in^2, 126 in.   (b) 12 cm^2, 27.5 cm

12. 144 cm^2 13. b = 18 in.   14. 28 ft^2 15. $x^2 + \frac{\sqrt{3}}{2}\left(\frac{30-2x}{3}\right)^2$
                h = 10 in.
                A = 180 in^2

16. 147 cm^2 17. 50.48 cm   18. 16 in.   19. 18 cm   20. 20 m & 23 m

21. 60.96 m   22. $\frac{81\sqrt{3}}{2}$ in^2 23. 108 in^2 24. 48 in., 57 in^2

25. 7 mm   26. 31 ft   27. 36.6 mm^2 28. 102 in^2; $34 + 12\sqrt{2}$ in.

29. a = 4 yd; b$_1$ = 13 yd, b$_2$ = 9 yd   30. $A = \frac{1}{2}\Delta x[g(x_i) + g(x_{i+1})]$

31. 58.59 cm^2 32. 84.35 in^2 33. $24\sqrt{3}$ cm^2 34. $98\sqrt{3}$ in^2 35. 10 in.

36. $\frac{9\sqrt{3}}{2}$ in.   37. 208 cm^2 38. $24\sqrt{3} - 16$ cm^2 = 25.57 cm^2 39. 264 in^2

40. $121 - 8\sqrt{3}$ in^2 = 107.14 in^2

CHAPTER 6

Exercise 6.1 (p. 276)

1. (a) $\overline{OE}, \overline{OB}, \overline{OA}$, (b) $\overline{BE}$ 3. $\angle B - \stackrel{\frown}{CDA}$; $\angle D - \stackrel{\frown}{CBA}$
   (c) $\stackrel{\frown}{AB}, \stackrel{\frown}{AE}$, (d) $\stackrel{\frown}{ABE}, \stackrel{\frown}{CBA}$ $\angle A - \stackrel{\frown}{BCD}$; $\angle C - \stackrel{\frown}{BAD}$
   (e) $\overline{CD}$, (f) $\angle AOB$ (g) $\stackrel{\frown}{EB}$

5. 102°   7. $\frac{3\pi}{5}$ rad   9. 53°   11. 30°   13. $\frac{5\pi}{12}$ rad   15. 51°

17. 216°   19. 58°   21. 50°   23. 93°   25. 74°   27. 1.20 rad

29. 85°   31. 13 cm   33. 14 in.   35. 5 in.   37. 7 cm   39. 14π in.

41. 11 mm   43. $\frac{12\pi}{5}$ = 7.54 mm   45. π cm   47. $\sqrt{170}\,\pi$ cm   49. $\frac{14\pi}{3}$ cm

51. $\frac{84\pi}{45} = 5.86$ ft   53. 3 in.   55. $\frac{13\pi}{44} = 0.93$ mi/hr   57. $\frac{2640}{\pi} = 840.34$ rev

Exercise 6.2 (p. 295)

1. 6.89 in.   3. 48 in^2 5. 84°   7. $7\sqrt{3}$ in.   9. $\frac{14\pi}{3}$ in.

11. $11\sqrt{2}$ cm   13. $12\sqrt{2}$ in.   15. $\frac{\pi}{4}$ rad   17. $54\sqrt{3}$ in^2 19. 12.96 cm

21. 8 in.   23. 72 in.   25. $8\frac{1}{2}$ m   27. DC = 6 in., AE = 11 in., BF = 13 in.

29. $242\sqrt{3}$ in^2 31. $6\sqrt{3} - 6$ cm   33. $14 - 7\sqrt{2}$ cm

Exercise 6.3 (p. 304)

1. 49π cm^2 3. 3.66 m^2 5. 49π in^2 7. 16.73 mm^2 9. $\frac{49\pi}{20}$ cm^2

11. 9.67 in.   13. $\frac{\pi}{12}$ rad   15. $150\pi - 225\sqrt{3}$ cm^2 17. 1451.25 in^2

19. 40π cm^2 21. $\frac{63\pi}{10}$ in^2 23. $9\pi + 78$ cm^2 25. 64 mm^2

27. $225 - 56.25\pi$ in^2 29. $100\pi - 150\sqrt{3}$ in^2 31. $24\pi + 16\sqrt{3}$ mm^2

33. $50\sqrt{3} - 4\pi$ mm^2 35. $288 - \frac{27\pi}{2}$ in^2 37. $256 - 64\pi$ mm^2 39. 36π in^2

41. 9π cm^2 43. $L = \frac{2P - \pi w}{4}$

Exercise 6.4 (p. 326)

1.

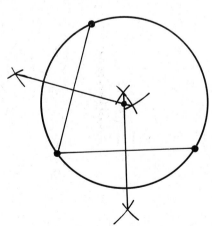

3.

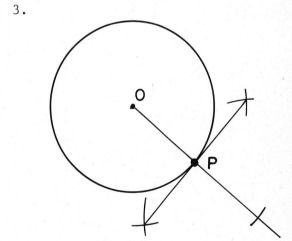

5.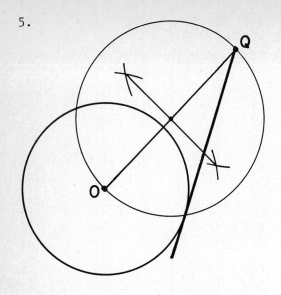

7.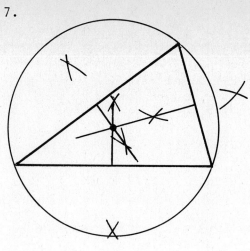

9.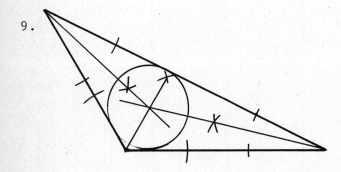

11.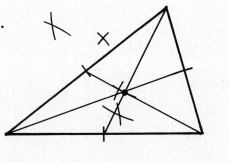

13.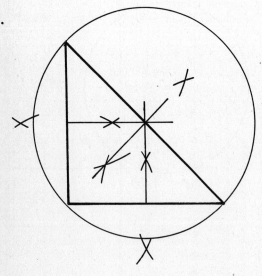

15.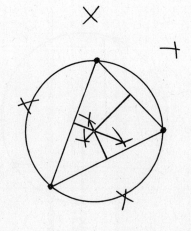

17.
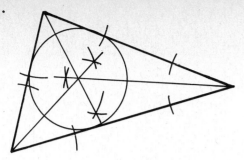

19.
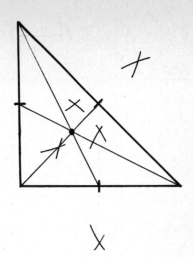

21.
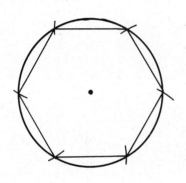

23.
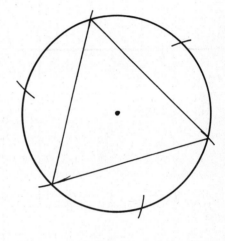

25.
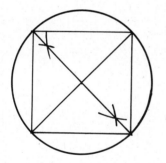

27.
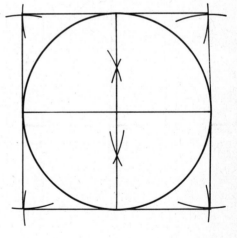

29.
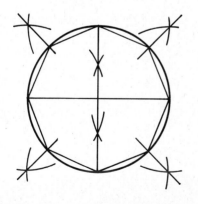

31.

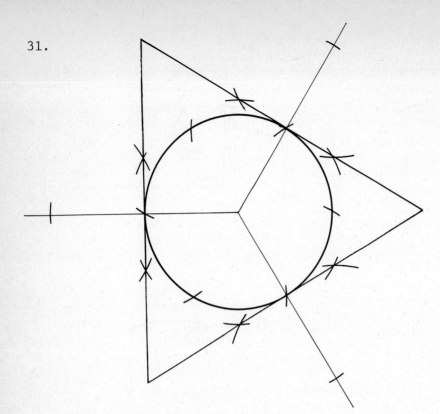

33.

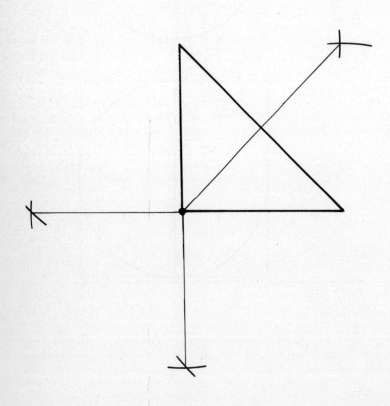

532

Exercise 6.5 (p. 330)

1.

	Statement	Reason
1.	$\triangle AOB$ with $\angle AOB$ a central $\angle$	Given
2.	$\angle AOB = \pi/3$	Given
3.	$\overline{AO}$ & $\overline{BO}$ are radii	Def of radii
4.	$\overline{AO} \cong \overline{BO}$	3 & Def of radii
5.	$\triangle AOB$ is isosceles	4 & Def of isosceles
6.	$\angle OAB \cong \angle OBA$	5 & Base $\angle$s of isosceles $\triangle$s are $\cong$
7.	$\angle OAB + \angle OBA + \angle AOB = \pi$ rad	1 & Sum of $\angle$s in $\triangle = \pi$ rad
8.	$\angle OAB + \angle OBA = 2\pi/3$ rad	2,7 & Subts. of =s
9.	$\angle OAB + \angle OAB = 2\pi/3$ rad	1 & Def of $\cong$
10.	$\angle OAB = \frac{\pi}{3}$ rad = $\angle OBA$	6,9 & Div. of =s
11.	$\triangle AOB$ is equilateral	2,10 & Def of equilateral

3.

		Statement	Reason
A,A	1.	$\widehat{AB} = \widehat{DC}$	Given
	2.	$\angle CDB \cong \angle CAB$ $\angle DCA \cong \angle DBA$	Inscribed $\angle$s of same arc are $\cong$
S	3.	$\overline{DC} \cong \overline{AB}$	1 & Equal chords have = arcs.
	4.	$\triangle DEC \cong \triangle BEA$	2,3 & ASA
	5.	$\overline{DE} \cong \overline{AE}$	4 & CPCTC
	6.	$\triangle AED$ is isosceles	5 & Def of isosceles

5.

	Statement	Reason
1.	$\widehat{RST} = 180°$	Given
2.	$\overline{QR} \parallel \overline{ST}$	Given
3.	Const. $\overline{TR}$	Construction
4.	$\widehat{RST} + \widehat{TQR} = 360°$	Arcs of a circle = 360°
5.	$\widehat{TQR} = 180°$	1,4 & Subt. of =s
6.	$\angle TQR = \angle TSR = 90°$	Inscribed $\angle = \frac{1}{2}$ intercepted arc
7.	$\angle QRT \cong \angle RTS$	2 & Alt. int. $\angle$s are $\cong$
8.	$\angle QTR \cong \angle TRS$	6,7 & if 2 pairs of $\angle$s in $\triangle$ are $\cong$, then 3rd pair are $\cong$
9.	$\overline{QT} \parallel \overline{RS}$	8 & Alt. int. $\angle$s are $\cong$
10.	QRST is a rectangle	2,6,9 & Def of rectangle

7.

		Statement	Reason
	1.	$\triangle ABC$	Given
	2.	$\overline{GD} \perp$ bisector of $\overline{AB}$, $\overline{GE} \perp$ bisector of $\overline{BC}$, $\overline{GF} \perp$ bisector of $\overline{AC}$	Given
S	3.	$\overline{AD} \cong \overline{DB}$, $\overline{AF} \cong \overline{FC}$, $\overline{BE} \cong \overline{EC}$	2 & Def of bisect
A	4.	$\angle ADG = \angle BDG = 90°$ $\angle BEG = \angle CEG = 90°$ $\angle AFG = \angle CFG = 90°$	2 & Def of $\perp$
S	5.	$\overline{GF} \cong \overline{GF}$, $\overline{CE} \cong \overline{GE}$, $\overline{GD} \cong \overline{GD}$	Common side

533

6. △AGF ≅ △CGF   3,4,5 & SAS
   △CGE ≅ △BGE
   △BGD ≅ △AGD
7. $\overline{AG} \cong \overline{BG}$, $\overline{BG} \cong \overline{CG}$,   6 & CPCTC
   $\overline{CG} \cong \overline{AG}$
8. $\overline{AG} \cong \overline{BG} \cong \overline{CG}$ 7 & Transitive

Chapter 6 Review (p. 332)

1. (a) F (b) ∠FAD (c) $\overarc{FD}$, $\overarc{FC}$ (d) $\overline{FA}$ (e) $\overline{AD}$, $\overline{FA}$ (f) $\overline{FD}$ (g) $\overarc{FAC}$
   (h) ∠CPF (i) $\overline{PC}$ (j) $\overline{FE}$, (k) $\overline{AB}$, $\overline{DA}$, $\overline{PC}$, $\overarc{FA}$, $\overarc{FD}$, ∠FAD

2. 2.61 rad   3. 66°   4. 10π in.   5. 22.14 cm   6. $\frac{\pi}{5}$ rad

7. 61.26 cm   8. 0.44 rad   9. 52°   10. 1.92 rad   11. 94.5°

12. 10π cm   13. 7.78 cm   14. 100π cm²   15. 18π in.   16. $\frac{7\pi}{3}$ in.

17. 100π − 25√3 = 270.86 in²   18. $\frac{12\pi}{5}$ cm   19. 20 cm   20. 5√3 mm

21. 12π cm²   22. 400π − 800 = 456.64 in²   23. 4.90 mm   24. 46.6 in.

25. 12 ft   26. AB = 15.5 mm; BC = 8.5 mm; DE = 2.5 mm   27. 600√3 in²

28. 49π cm²   29. 32 − 16√2 = 9.37 in.   30. 57.02 cm²   31. 17.28 cm

32. 144°   33. 297.26 mm²   34. $\frac{25\pi}{2}$ − 13.44 = 25.81 in²   35. 56π in²

36. $\frac{216\pi}{10}$ in²   37. 144 − 36π in²   38. 240 in²   39. 180 − 8π cm²

40. 900 − 225π mm²   41. 30π − 36√3 cm²   42. 294√3 − 49π cm²

43.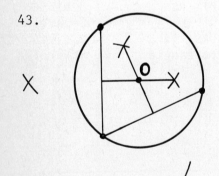

44.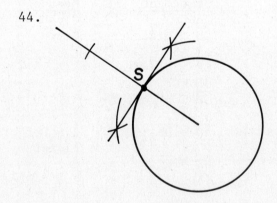

45.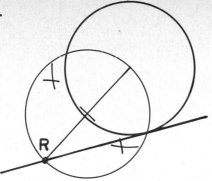

46.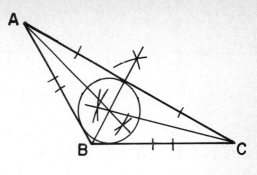

47.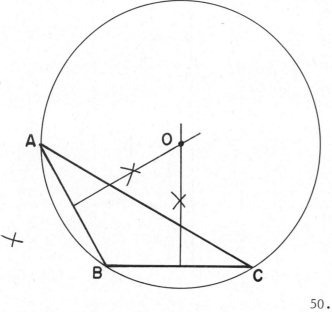

48.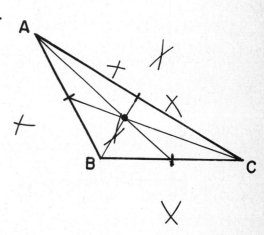

49.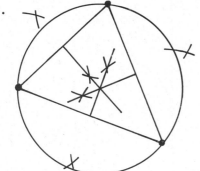

50.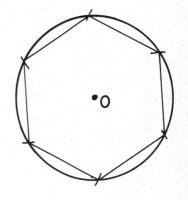

51.

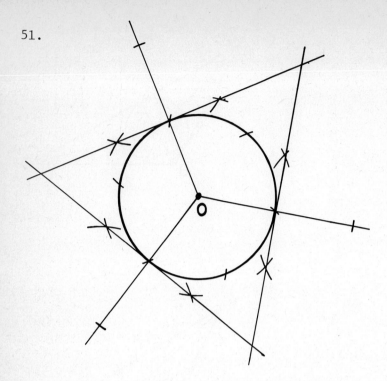

52.

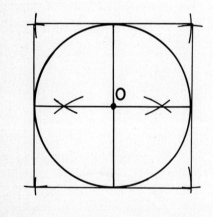

53.

54.

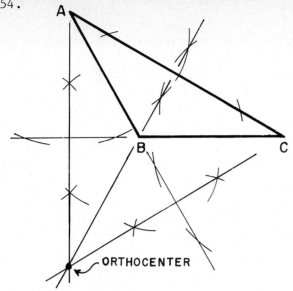
ORTHOCENTER

55.

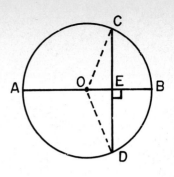

	Statement	Reason
1.	Circle O with diameter $\overline{AB}$	Given
2.	$\overline{DC} \perp \overline{AB}$	Given
3.	$\angle OED = \angle OEC = 90°$	2 & Def of $\perp$
S 4.	$\overline{OD} \cong \overline{OC}$	Def of radii
5.	$\triangle ODC$ is isosceles	Def of isosceles
A 6.	$\angle OCE \cong \angle ODE$	5 & Base $\angle$s of isosceles $\triangle$ are $\cong$
A 7.	$\angle DOE \cong \angle COE$	3,6 & if 2 pairs of $\angle$s in 2 $\triangle$s are $\cong$, then 3rd pair is $\cong$
8.	$\triangle DOE \cong \triangle COE$	4,6,7 & ASA
9.	$\overline{DE} \cong \overline{EC}$	8 & CPCTC
10.	$\overline{AB}$ bisects $\overline{DC}$	9 & Def of bisect

56.

	Statement	Reason
1.	Circle O with chords $\overline{AB}$ and $\overline{CD}$	Given
2.	$\overline{OE} \perp \overline{AB}$ $\overline{OF} \perp \overline{CD}$	Given
S 3.	$\overline{OE} \cong \overline{OF}$	Given
S 4.	$\overline{OA} \cong \overline{OC} \cong \overline{OB} \cong \overline{OD}$	Def of radii
5.	$\triangle AEO, \triangle BEO, \triangle COF, \triangle DOF$ are right $\triangle$s	2 & Def of right $\triangle$s
6.	$\triangle AEO \cong \triangle BEO \cong \triangle COF \cong \triangle DOF$	3,4,5 & if hypotenuse & a leg of 2 right $\triangle$s are $\cong$, then the $\triangle$s are $\cong$
7.	$\overline{AE} \cong \overline{EB} \cong \overline{CF} \cong \overline{FD}$	6 & CPCTC
8.	AE + EB = AB CF + FD = CD	Whole = sum of parts
9.	$\overline{AB} \cong \overline{CD}$	7,8 & Add of =s

537

CHAPTER 7

Exercise 7.1 (p. 350)

1. (a) 0.86 (b) 0.61 (c) 1.46
3. (a) 0.92 (b) 2.94 (c) 1.78
5. (a) 0.35 (b) 1.36 (c) 1.38
7. (a) 49.00° (b) 11.01° (c) 5.00°

9. 22.62°, 67.38°   11. 36.99 in.   13. 54.32 cm

15. 56.68°   17. 0.46   19. 64.79 mm   21. 8.04 in.   23. 7.90 cm

25. 23.42 cm   27. 9.88 in.   29. $\sin \theta = \frac{3}{5}$, $\tan \theta = \frac{3}{4}$ 31. $\cos \beta = \frac{7}{25}$, $\tan \beta = \frac{24}{7}$

Exercise 7.2 (p. 358)

1. 256.51 cm^2 3. 64.20 in^2 5. 167.80 cm^2 7. 2.98 mm & 3.74 mm

9. 139.36 in^2 11. 85.17 mm^2 13. 44.80 ft   15. 32.04 ft

17. 58.06 ft   19. 109.91 ft   21. 2.45 mi/hr   23. 341.48 ft

25. 36.87°   27. $\theta = \arcsin \frac{1+t}{5} = \sin^{-1}(\frac{1+t}{5})$ 29. 20.56°

31. $\arctan \frac{t}{40} = \tan^{-1}(\frac{t}{40})$

Exercise 7.3 (p. 362)

1. csc 60° = 1.15; sec 60° = 2.00; cot 60° = 0.58   3. 0.65   5. 1.00

7. 1.24   9. 0.34   11. 1.28   13. 1.28   15. 19.98   17. 8.21

19. 1.12

Exercise 7.4 (p. 369)

in degrees	0°	30°	45°	60°	90°	120°	135°	150°	180°
in radians	0°	$\frac{\pi}{6}$	$\frac{\pi}{4}$	$\frac{\pi}{3}$	$\frac{\pi}{2}$	$\frac{2\pi}{3}$	$\frac{3\pi}{4}$	$\frac{5\pi}{6}$	π
1. $\sin \theta$	0	$\frac{1}{2}$	$\frac{\sqrt{2}}{2} = 0.71$	$\frac{\sqrt{3}}{2} = 0.87$	1	$\frac{\sqrt{3}}{2} = 0.87$	$\frac{\sqrt{2}}{2} = 0.71$	$\frac{1}{2}$	0
3. $\tan \theta$	0	$\frac{\sqrt{3}}{3} = 0.58$	1	$\sqrt{3} = 1.73$	undef.	$-\sqrt{3} = -1.73$	-1	$\frac{-\sqrt{3}}{3} = -.58$	0
5. $\sec \theta$	1	$\frac{2\sqrt{3}}{3} = 1.15$	$\sqrt{2} = 1.41$	2	undef.	-2	$-\sqrt{2} = -1.41$	$\frac{-2\sqrt{3}}{3} = -1.15$	-1

in degrees	180°	210°	225°	240°	270°	300°	315°	330°	360°
in radians	π	$\frac{7\pi}{6}$	$\frac{5\pi}{3}$	$\frac{4\pi}{3}$	$\frac{3\pi}{2}$	$\frac{5\pi}{3}$	$\frac{7\pi}{4}$	$\frac{11\pi}{6}$	2π
7. $\sin\theta$	0	$-\frac{1}{2}$	$\frac{-\sqrt{2}}{2}$ -0.71	$\frac{-\sqrt{3}}{2}$ -0.87	-1	$\frac{-\sqrt{3}}{2}$ -0.87	$\frac{-\sqrt{2}}{2}$ -0.71	$-\frac{1}{2}$	0
9. $\tan\theta$	0	$\frac{\sqrt{3}}{3}$ 0.58	1	$\sqrt{3}$ 1.73	undef.	$-\sqrt{3}$ -1.73	-1	$\frac{-\sqrt{3}}{3}$ -0.58	0
11. $\sec\theta$	-1	$\frac{-2\sqrt{3}}{3}$ -1.15	$-\sqrt{2}$ -1.41	-2	undef.	2	$\sqrt{2}$ 1.41	$\frac{2\sqrt{3}}{3}$ 1.15	1

13.

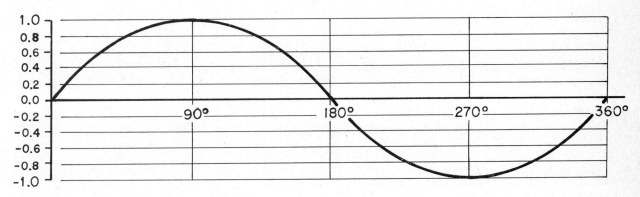

15.

17.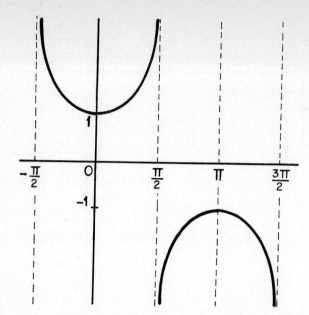

19. Values of θ greater than 180° but less than 360°

21. Values of θ greater than 90° but less than 180°, and values of θ greater than 270° but less than 360°

23. Values of θ greater than 90° but less than 270°

25. 90°, 270°   27. 90°, 180°

Chapter 7 Review (p. 371)

1. (a) E   (b) I   (c) M
    (d) F   (e) C   (f) L

2. 18.16 cm   3. 12.44 in.   4. 50.10°

5. 6.84 in.   6. 98.14 in^2 7. sec θ = $\frac{41}{9}$; csc θ = $\frac{41}{40}$; cot θ = $\frac{9}{40}$

8. 120.71 mm^2 9. 407.97 in^2 10. 8.45 ft & 18.13 ft   11. 64.20 ft

12. 37.68 ft   13. 228.5 ft   14. 23°   15. 18 mm by 13 mm

16. (a) 98 ft   17. (a) 56 ft   18. (a) 145 ft
    (b) 222 ft   (b) 100 ft   (b) 26 ft

CHAPTER 8

Exercise 8.1 (p. 381)

1. A, D, E, G, I, K, L

3.

Figure	No. of faces (or surfaces)	No. of bases	No. of lateral faces (or lateral surfaces)	No. of edges	No. of Vertices
A	7	1	6	12	7
B	2	1	1	1	1
C	3	2	1	2	NA
D	4	1	3	6	4
E	6	2	4	12	8
F	3	2	1	2	NA
G	7	2	5	15	10
H	1	0	1	NA	NA
I	9	2	7	18	11
J	2	1	1	1	1
K	6	2	4	12	8
L	5	1	4	8	5

5. (a) $\overline{DG}$ (b) $\overline{DF}$ (c) $\overline{AD}, \overline{BD}, \overline{CD}$ (d) $\overline{BE}$ (e) $\overline{DG}$

7. F   9. T   11. F   13. F   15.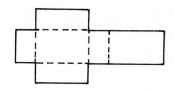

17. 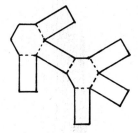   19.   21.

23.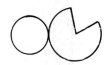

541

Exercise 8.2 (p. 396)

1. 214 cm^2 3. 501.5 in^2 5. 864 cm^2 7. $1\frac{1}{2} \text{ yd}^2$ 9. 21 yd^2

11. $43\frac{3}{4} \text{ in}^2$ 13. 792 in^2 15. 96 in^2 17. $10{,}512 \text{ ft}^2$

19. LSA = 24 ft^2 or 3456 in^2
    TSA = 24.57 ft^2 or 3538.08 in^2

Exercise 8.3 (p. 405)

1. 66 m^2 3. 27.2 cm^2 5. $64\pi \text{ ft}^2$ 7. 15π or 47.12 in^2

9. 3.125π or 9.82 in^2 aluminum 11. 64π or 201.06 in^2 13. $\frac{2(2r+h)}{r+h}$
   11.25π or 35.34 in^2 cardboard

15. LSA = $36\pi \text{ in}^2$ 17. LSA = $156\pi \text{ cm}^2$ 19. $33\frac{1}{3}\pi$ or 104.72 yd^2
    TSA = $54\pi \text{ in}^2$ TSA = $300\pi \text{ cm}^2$

21. 54.25π or 170.43 in^2 23. 4 cm 25. $s = 40 \text{ cm}$, $r = 10 \text{ cm}$ 27. 324 ft^2

29. $2600 + 1200\sqrt{2}$ or 4297.06 ft^2 31. $4.72°$ 33. $d = 12 \text{ mm}$, $h = 36 \text{ mm}$ 35. $216 - 4x^2 \text{ in}^2$

37. $2\pi\sqrt{x} \cdot \Delta x$ 39. $\left(\frac{18-x}{4}\right)^2 + \frac{x^2\sqrt{3}}{36} \text{ cm}^2$ or $\frac{(9+4\sqrt{3})x^2}{144} - \frac{9x}{4} + \frac{81}{4} \text{ cm}^2$

Exercise 8.4 (p. 421)

1. 286 mm^2 3. 693 in^3 5. 163.28 in^3 7. $360\pi \text{ mm}^3$ 9. 250 in^3

11. 300 mm^3 13. 16.39 cm^3 15. 16.39 cm^2 17. Container with greater diameter holds $60\pi \text{ in}^3$ more.

19. 124 cm^3 21. 4791 mm^3 23. 226 ft^3 or 8 yd^3 25. 220 in^3

27. 206.46 cm^3 29. $V_{prism} = 3 \cdot V_{pyramid}$ 31. $\frac{125\pi}{6} \text{ ft}^3$, $25\pi \text{ ft}^2$

33. 524 in^3 35. 2828 in^3 37. 8 to 1 39. 160.39 in^2, 116.91 in^3

41. $630\pi \text{ in}^3$ or   43. $\frac{64000\pi}{3} \text{ cm}^3$ 45. $r = 3 \text{ ft}$, $h = 12 \text{ ft}$ 47. $\pi x \cdot \Delta x$
    1979.2 in^3

49. $SA = 2\pi r^2 + \frac{2V}{r}$ 51. $V = 4x^3 - 46x^2 + 120x \text{ in}^3$ 53. $V = \frac{4\pi x^3}{75}$

Chapter 8 Review (p. 425)

1. (a) $\overline{AD}$, $\overline{AE}$, (b) $\overline{AB}$,   2. 628 cm^2 3. $\frac{3}{8} \text{ in}^3$
   (c) $\overline{CB}$, (d) $\overline{AC}$, (e) $\overline{AB}$ 1040 cm^3

4. $(4 + 2\sqrt{29})\pi \text{ in}^2$ 5. $16\pi \text{ in}^2$ 6. $22\frac{3}{4} \text{ in}^3$ 7. 63.125 m^3
   $\frac{20\pi}{3} \text{ in}^3$ $\frac{32\pi}{3} \text{ m}^3$

8. $13\frac{1}{3}$ in.   9. $88\pi \text{ mm}^2$, 112 mm^3 10. 23365.6 mm^2 11. Must buy 5 qts.

12. $18\pi \text{ in}^3$ 13. 20.1 in^2 14. $w = 4 \text{ m}$, $\ell = 8 \text{ m}$, $h = 6 \text{ m}$ 15. 363.62 in^3

16. 469.95 ft^2 17. 288 in^3 18. 384 in^3 19. $72\pi\sqrt{3} \text{ in}^3$

20. $60\pi \text{ in}^2$ 21. 1157 balls   22. 251.12 in^2 23. $\frac{115\pi}{18} \text{ ft}^3$

24. 959.7 mm^2 25. $36\pi\sqrt[3]{9} \text{ in}^2$ 26. $V = 2w^2x - 6wx^2 + 4x^3$

27. $V = \pi(x^4 + 3x^2 - 4)^2 \cdot \Delta x$ 28. $V = \frac{\pi x^3}{3}$ 29. 7516.8 in^3

30. $72\left[\frac{32\pi \cos^{-1}(\frac{16-x}{16})}{45} - (16 - x)\sqrt{x^2 + 32x}\right]$ 31. $\frac{(L-x)^2}{4\pi} + \frac{L^2\sqrt{3}}{36} \text{ cm}^2$

32. $\frac{100\sqrt{119}}{3}$ or 363.62 mm^3 33. 1305π or 409.98 mm^2 34. 296 in^2
    405π or 1272.3 mm^3 208 in^3

35. 93π or 292.17 in^2
    120π or 376.99 in^3

CHAPTER 9

Exercise 9.1 (p. 434)

1. S   3. NS   5. NS   7. S   9. S

11. (a) Three is an even number or twenty-five is a perfect square.
    (b) Three is an even number and twenty-five is a perfect square.
    (c) T   (d) F

13. (a) $x^2 - 3x - 4 = (x-4)(x+1)$ or $(15)^2 + (20)^2 = (25)^2$
    (b) $x^2 - 3x - 4 = (x-4)(x+1)$ and $(15)^2 + (20)^2 = (25)^2$
    (c) T   (d) T

15. p: He is studying English. $p \wedge q$
    q: He is studying geometry

17. p: He may go with me this summer. $p \vee q$
    q: He may get a job (this summer).

19. p: Quadrilaterals are two-dimensional figures. $p \wedge q \wedge r$
    q: Triangles are two-dimensional figures.
    r: Circles are two-dimensional figures.

21. Susan will be president, and Bobby or Joe will be secretary.

23. This book belongs to either Mary, Bill, or John.

25. false   27. false   29. cannot be determined

Exercise 9.2 (p. 443)

1. This book is not heavy.   3. This pen writes.   5. Some vertical angles are not equal.

7. No number is divisible by 4.
   (All numbers are not divisible by 4.)

9. Some circle is round.   11. All rectangles are congruent.

13. $\sim p \vee r$ 15. $p \vee t$

17. Some squares are not 2-dimensional figures or some circles are not 2-dimensional figures.

19. The weight of Bill cannot be measured in feet and the weight of Bill cannot be measured in pounds.

21. Some squares are not round and some circles do have a center.

23. True   25. False   27. $\sim p \wedge q$ 29. $(\sim p) \vee (\sim q)$

Note: Answers to 31 & 33 will vary depending on the class. These answers assume more than 10 students not meeting on Tuesday.

31. F   33. T   35. T   37. Cannot be determined

Exercise 9.3 (p. 450)

1. p: The number is even.
   q: It is divisible by two.

   $p \rightarrow q$

3. r: You will pass.
   s: You take the final examination.

   $r \rightarrow s$

5. r: $x^2 + 4 \neq (x + 2)^2$
   s: $(x + 2)^2 = x^2 + 4x + 4$

   $s \rightarrow r$

7. p: Two lines form a right angle.
   q: They are perpendicular.

   $p \rightarrow q$

9. s: It is a square.
   t: It has four equal sides.

   $s \rightarrow t$

11. $p \rightarrow q$ 13. $\sim(p \rightarrow \sim r)$

15. $(q \vee \sim r) \rightarrow p$

17. a) Charlie is not a basketball player, and he is not tall.
    b) If Charlie does not like to read, then he is a basketball player.
    c) If Charlie is tall and agile, then he is a basketball player.

19. a) I like math and some homework is difficult.
    b) It is false that all polygons are beautiful and some homework is difficult. OR Some polygons are not beautiful or all homework is easy (not difficult).
    c) If all homework is easy, then I like math.

21. F   23. T   25. T   27. T   29. T   31. T   33. F

35. T   37. T   39. F   41. T   43. T   45. F   47. F

49. T   51. F   53. T   55. F, F   57. T

59. $\angle D \cong \angle E$: true; $\angle E \cong \angle F$: cannot be determined.

Exercise 9.4 (p. 456)

1. p: You are 10 years old.
   q: You are very wise.

   $p \rightarrow \sim q$

   Contrapositive: $q \rightarrow \sim p$
   If you are very wise, you are not 10 years old.
   Converse: $\sim q \rightarrow p$
   If you are not very wise, you are 10 years old.
   Inverse: $\sim p \rightarrow q$
   If you are not 10 years old, you are very wise.

3. p: People drink beer.
   q: People are thirsty.

   $p \rightarrow q$

   Contrapositive: $\sim q \rightarrow \sim p$
   If some people are not thirsty, they will not drink beer.
   Converse: $q \rightarrow p$
   If people are thirsty, they drink beer.
   Inverse: $\sim p \rightarrow \sim q$
   If some people do not drink beer, they are not thirsty.

5. p: We will go home for the holidays.   7. p: I go home.
    q: It snows.   q: I cook.

    $p \rightarrow \sim q$ $p \rightarrow \sim q$

    Contrapositive: $q \rightarrow \sim p$ Contrapositive: $q \rightarrow \sim p$
        If it snows, we will not go   If I cook, I do not go home.
        home for the holidays.   Converse: $\sim q \rightarrow p$
    Converse: $\sim q \rightarrow p$ If I do not cook, I go home.
        If it does not snow, we will   Inverse: $\sim p \rightarrow q$
        go home for the holidays.   If I do not go home, I cook.
    Inverse: $\sim p \rightarrow q$
        If we do not go home for the
        holidays, it will snow.

9. p: You think these are hard.   11. $q \rightarrow p$
    q: You should see the ones I   Conv: $p \rightarrow q$
       omitted.   Inv: $\sim q \rightarrow \sim p$
                                               Contrap: $\sim p \rightarrow \sim q$
    $p \rightarrow q$

    Contrapositive: $\sim q \rightarrow \sim p$ 13. $p \rightarrow \sim q$
        If you do not see the ones   Conv: $\sim q \rightarrow p$
        I omitted, you do not think   Inv: $\sim p \rightarrow q$
        these are hard.   Contrap: $q \rightarrow \sim p$
    Converse: $q \rightarrow p$
        If you see the ones I omitted,   15. $p \rightarrow q$: Inv.
        you think these are hard. $\sim p \rightarrow \sim q$: Stmt.
    Inverse: $\sim p \rightarrow \sim q$ $\sim q \rightarrow \sim p$: Conv.
        If you do not think these $q \rightarrow p$: Contrap.
        are hard, you should not see
        the ones I omitted.

17. $q \rightarrow \sim r$: Contrap.   19. Equivalent.   21. Not equivalent.
    $r \rightarrow \sim q$: Stmt.   Second is   Second is converse of
    $\sim q \rightarrow r$: Conv.   contrapositive   first.
    $\sim r \rightarrow q$: Inv.   of first.

23. Not equivalent.   25. Not equivalent.   27. (a) Conv.-not equiv.
    Second is inverse   Second is inverse   (b) Inv.-not equiv.
    of first.   of first.   (c) Same-equiv.
                                                                      (d) Contrap.-equiv.

29. (a) Same - equiv.
    (b) Inv. - not equiv.
    (c) Contrap. - equiv.
    (d) No relation - not equiv.

Exercise 9.5 (p. 461)

1. q   3. t   5. $\sim s \rightarrow q$ 7. I will learn about logic.

9. If the figures are congruent   11. The altitude does not pass through
    then $\overline{AB} \cong \overline{GH}$.   the center of the base.

13. $\angle ABC \cong \angle BCD$. The lines are   15. No conclusion   17. No conclusion
    parallel.

Exercise 9.6 (p. 466)

1. No concl.   3. t   5. No concl.   7. No concl.   9. Not a logical concl.

11. Not a logical concl.   13. Logical concl.   15. It does not rain.

17. N is divisble by 2.   19. No concl.   21. No concl.

23. Squares do not have 5 sides.   25. (a) No (b) No (c) Yes (d) No

Exercise 9.7 (p. 472)

1. $j \to t$ 3. $r \to \sim q$ 5. $\sim q \to \sim r$ OR $r \to q$ 7. $\sim q \to \sim r$ OR $r \to q$

9. No concl.   11. No concl.   13. All squares are quadrilaterals.

15. No concl.

17. Dr. Gauss is not an infant.

19. If Molly does her homework, she will be appointed to an important position.

21. People who have a driver's license have to carry out their garbage.

23. If the fruit is green, it is not an apple.

Chapter 9 Review (p. 473)

1. (a) NS   (b) S   (c) S   (d) NS   (e) S

2. (a) $x^2 + 5x + 4 = (x + 1)(x + 4)$ and 81 is a prime number.
   (b) $x^2 + 5x + 4 = (x + 1)(x + 4)$ or 81 is a prime number.
   (c) F   (d) T

3. (a) G: Geometry is a hard course.
       A: Algebra is a hard course.
       $G \wedge A$.
   (b) A: Most Freshmen take English 10 their second semester.
       B: Most Freshmen take English 20 their second semester.
       C: Most Freshmen take Math 4 their second semester.
       D: Most Freshmen take Math 5 their second semester.
       $A \vee B \vee C \vee D$

4. (a) Logic is fun, and some math is not exciting.
   (b) Logic is fun or some math is not exciting.

5. (a) The geometry book is not green.
   (b) The door is closed.
   (c) Jack will not sing and I will play the piano.
   (d) Some mathematics courses are easy.
   (e) Every day the sun shines. (The sun shines every day).
   (f) Some doors or some windows should not be closed.

6. a) p: We go tomorrow. b) r: The streets are slick.
      q: We will call you.   s: It rains.

      p → q   s → r

   c) t: A rectangle is a square.
      u: All sides are the same length.

      t → u

7. Original: (a) (~r) → t   (b) Two triangles are congruent if
                                        corresponding sides are the same.

   Inverse: r → (~t)   Two triangles are not congruent if
                                       some corresponding sides are not the
                                       same length.

   Converse: t → (~r)   Corresponding sides are the same
                                       length if two triangles are congruent.

   Contrapositive: ~t → r   Some corresponding sides are not the
                                       same length if two triangles are not
                                       congruent.

8. (a) F   (b) T   (c) Cannot be   9. a) F   b) T   c) F
                       determined

10. (a) 1. Same 2. Inv. 3. Conv. 4. None 5. Contrap. (b) 1 & 5

11. (a) p   (b) no concl. (c) no concl. (d) t   (e) ~s → ~v   (v → s)

12. a) You understand logic. b) No concl. c) You study.
    d) No concl. e) If you have a good weekend, you talked to Mrs. Moneybags.

13. People who like rain are not students at State University.
    (Students at State University do not like rain.)

Cumulative Review (p. 476)

1. (a) Yes (b) Yes (c) No   2. $\dfrac{3\pi}{10}$

3.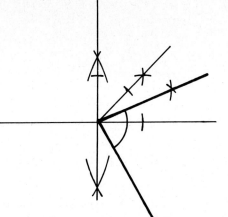

4. 73°27'   5. 45°, 60°, 75°

6. 75 n mi   7. 25t n mi

8.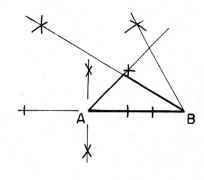

9. (a) $x = y = 120°$, $z = 60°$
   (b) $6x - 3° = 117°$, $3x + 3° = 63°$
   (c) $4x + 30° = 70°$, $x^2 - 2x + 30° = 110°$

10. $w = y = 109°$; $x = 23°$; $z = 21°$

11. (a) $6\sqrt{2}$ cm   (b) 12 cm   (c) $6\sqrt{3}$ cm   (d) $6 + 6\sqrt{3}$ cm

12. (a) $y = z = 2$ cm   (b) $x = 4\sqrt{2}$ ft, $y = 4$ ft
    (c) $y = z = \dfrac{3\sqrt{2}}{2}$ in.   (d) $x = 4$ m, $y = 2\sqrt{2}$ m

13. (a) 15"   (b) 16'   (c) 8'   14. 156°   15. 10'
    (d) $2\sqrt{6}$ in.   (e) 4'   16. 25 ft   17. $\dfrac{25t}{3}$ ft

18.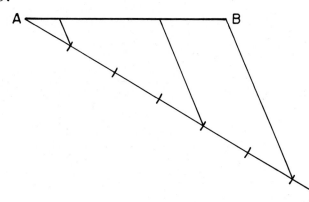

19. | Statement | Reason |
    |---|---|
    | 1. $\overline{CG} \parallel \overline{DE}$ | Given |
    | A 2. $\angle CFG \cong \angle DFE$ | Vertical $\angle$s |
    | A 3. $\angle C \cong \angle E$ | 1 & Alt. int. $\angle$s |
    | A 4. $\angle G \cong \angle D$ | 1 & Alt. int. $\angle$s |
    | 5. $\triangle CFG \sim \triangle DFE$ | 2,3,4 & AAA |

20.

	Statement	Reason
S 1.	$\overline{AS} \cong \overline{BR}$	Given
2.	$\overline{AS} \perp \overline{BC}$ $\overline{BR} \perp \overline{AC}$	Given
3.	$\angle ARB = 90° = \angle BSA$	2 & Def of $\perp$
A 4.	$\angle ARB \cong \angle BSA$	3 & Def of $\cong$
S 5.	$\overline{AB} \cong \overline{AB}$	Common side
6.	$\triangle ARB \cong \triangle BSA$	1,4,5 & If hypotenuse & one leg of rt. $\triangle$s are $\cong$, then $\triangle$s are $\cong$
7.	$\angle RAB \cong \angle SBA$	6 & CPCTC
8.	$\overline{AC} \cong \overline{CB}$	7 & Side opp. $\cong \angle$s are $\cong$.
9.	$\triangle ABC$ is isosceles	8 & Def of isosceles

21.

	Statement	Reason
1.	W is the midpoint of $\overline{AT}$ & $\overline{SR}$	Given
S 2.	$\overline{AW} \cong \overline{WT}$	1 & Def of midpoint
S 3.	$\overline{SW} \cong \overline{RW}$	1 & Def of midpoint
A 4.	$\angle SWT \cong \angle RWA$	Vertical $\angle$s
5.	$\triangle SWT \cong \triangle RWA$	2,3,4 & SAS
6.	$\angle WTS \cong \angle RAW$	5 & CPCTC
7.	$\overline{AR} \parallel \overline{ST}$	6 & Alt. int. $\angle$s are $\cong$

22.

	Statement	Reason
S 1.	$\overline{CE} \cong \overline{EB}$	Given
S 2.	$\overline{DE} \cong \overline{EF}$	Given
A 3.	$\angle DEC \cong \angle BEF$	Vertical $\angle$s
4.	$\triangle DEC \cong \triangle BEF$	1,2,3 & SAS
5.	$\overline{DC} \cong \overline{BF}$	4 & CPCTC

23. $x - 5 = 95°$ $x + 8 = 108°$ 24. (a) 40 ft   25. $100\sqrt{5}$ ft^2
    $x + 30 = 130°$ $x - 3 = 97°$ (b) 168°
                 $x + 10 = 110°$

26. 24 in., 26 in^2 27. (a) 5 in.   (b) 60 in^2 (c) 13 in.   28. 16 cm

29. (a) $6\sqrt{2}$ in.   (b) $8\sqrt{2}$ cm   (c) $\sqrt{6}$ m   30. (a) $\sqrt{73}$ m   (b) $\sqrt{85}$ cm   (c) $\sqrt{41}$ in.

31. (a) $\frac{49\pi}{12}$ ft^2, $\frac{7\pi}{6}$ ft   (d) 29.4π m^2 4.2π m   32. (a) 60°
                                                                        (b) 120°
     (b) 2π in^2, π in.   (e) 12.7π ft^2 $4.2\overline{3}\pi$ ft   (c) 14'
     (c) 11.2π cm^2, $1.8\overline{3}\pi$ cm   (f) $41.\overline{7}\pi$ yd^2 $10.\overline{4}\pi$ yd

33.

	Statement	Reason
S 1.	$\overline{AD} \cong \overline{DC}$	Given
2.	$\angle ADE \cong \angle CDE$	Given
3.	$\angle ADE + \angle ADB = 180°$ $\angle CDE + \angle CDB = 180°$	Def of strt. $\angle$
4.	$\angle ADE + \angle ADB = \angle CDE + \angle CDB$	3 & Trans.
A 5.	$\angle ADB \cong \angle CDB$	2,3 & Subt. of =s
S 6.	$\overline{DB} \cong \overline{DB}$	Common side
7.	$\triangle ADB \cong \triangle CDB$	1,5,6 & SAS
8.	$\overline{AB} \cong \overline{BC}$	7 & CPCTC

34.

	Statement	Reason
1.	KOFA a rectangle	Given
2.	OUFA a parallelogram	Given
3.	$\overline{KF} \cong \overline{OA}$	1 & Diagonals of rectangle are $\cong$
4.	$\overline{UF} \cong \overline{OA}$	2 & Opp. sides of parallelogram are $\cong$
5.	$\overline{KF} \cong \overline{UF}$	3,4 & Trans.
6.	$\triangle KUF$ is isosceles	5 & Def of isosceles

35. 0.8π rad   36. 35°, 145°   37. $8 + 8\sqrt{3}$ cm^2, $12 + 4\sqrt{2} + 4\sqrt{3}$ cm   38. $420.75

39.

	Statement	Reason
S 1.	$\overline{AC} \cong \overline{CE}$	Given
S 2.	$\overline{CD} \cong \overline{BC}$	Given
A 3.	$\angle C \cong \angle C$	Common side
4.	$\triangle ACD \cong \triangle ECB$	1,2,3 & SAS
5.	$\angle CDA \cong \angle CBE$	4 & CPCTC
6.	$\angle ABE + \angle CBE = 180°$ $\angle EDA + \angle ADC = 180°$	Def of supplementary $\angle$s
7.	$\angle ABE + \angle CBE = \angle EDA + \angle ADC$	6 & Trans.
8.	$\angle ABE \cong \angle EDA$	5,7 & Subt. of =s

41.

	Statement	Reason
1.	$\angle 2 + \angle 3 = 180°$ $\angle 3 + \angle 5 = 180°$	Def of supplementary $\angle$s
2.	$\angle 2 + \angle 3 = \angle 3 + \angle 5$	1,2 & Trans.
3.	$\angle 2 \cong \angle 5$	3 & Subt. of =s
4.	$\ell_1 \| \ell_2$	4 & Corresp. $\angle$s are $\cong$

43.(a)

	Statement	Reason
1.	ABCD is a parallelogram	Given
A 2.	$\angle CPB \cong \angle BQA = 90°$	Given
A 3.	$\angle A \cong \angle C$	1 & Opp. $\angle$s in a parallelogram are $\cong$
A 4.	$\angle PBC \cong \angle ABQ$	2,3 & If 2 pairs of $\angle$s in 2 $\triangle$s are $\cong$, then 3rd pair are $\cong$
5.	$\triangle PBC \sim \triangle BQA$	2,3,4 & AAA

(b) $8\frac{1}{3}$ in.

40.

	Statement	Reason
1.	$\triangle ABC$ is isosceles	Given
2.	$\angle A \cong \angle CDE$	Given
3.	$\angle A \cong \angle B$	1 & Def of isosceles
4.	$\angle B \cong \angle CDE$	2,3 & Trans.
5.	$\overline{AB} \| \overline{ED}$	4 & Corresp. $\angle$s are $\cong$

42.

	Statement	Reason
1.	ABCD is a parallelogram	Given
S 2.	$\overline{AM} \cong \overline{NC}$	Given
3.	$\overline{DC} \| \overline{AB}$	1 & Def of parallelogram
A 4.	$\angle MAB \cong \angle NCD$	3 & Alt. int. $\angle$s are $\cong$
S 5.	$\overline{DC} \cong \overline{AB}$	1 & Opp. sides of parallelogram are $\cong$
6.	$\triangle MAB \cong \triangle NCD$	2,4,5 & SAS
7.	$\angle AMB \cong \angle DNC$	6 & CPCTC
8.	$\angle AMB + \angle BMN = \angle DNC + \angle DNB = 180°$	Def of supplementary $\angle$s
9.	$\angle BMN \cong \angle DNM$	7,8 & Subt. of =s
10.	$\overline{BM} \| \overline{DN}$	9 & Alt. int. $\angle$s are $\cong$
11.	$\overline{AD} \| \overline{BC}$	1 & Opp. sides of parallelogram are $\cong$
A 12.	$\angle NCB \cong \angle MAD$	11 & Alt. int. $\angle$s are $\cong$
S 13.	$\overline{AD} \cong \overline{BC}$	1 & Opp. sides of parallelogram are $\cong$
14.	$\triangle BCN \cong \triangle DAM$	2,12,13 & SAS
15.	$\angle CNB \cong \angle AMD$	14 & CPCTC
16.	$\angle CNB + \angle BNM = \angle AMD + \angle DMN = 180°$	Def of supplementary $\angle$s
17.	$\angle BNM \cong \angle DMN$	15,16 & Subt. of =s
18.	$\overline{DM} \| \overline{BN}$	17 & Alt. int. $\angle$s are $\cong$
19.	MBND is a parallelogram	10,18 & Def of parallelogram

44. 4.2 ft   45. y = 5 mm, x = 10 mm

46. 39"   47. $\sqrt{4930}$ ft

48. 5 in. & $5\sqrt{3}$ in.

49. x = 5 cm, y = 2.5 cm, A = $2r^2$

50. x = 10 ft, y = 10 ft

51. $20\frac{2}{7}$ in., $41\frac{4}{7}$ in., $80\frac{1}{7}$ in.   52. $13\frac{1}{3}$ m   53. x = 11.75 cm, y = 14 cm

54. $\frac{8\sqrt{3}}{3}$ cm   55. 5 cm   56. 4 ft   57. 52 m   58. $6\frac{6}{19}$ in.

59. 9 ft²   60. x = 14 cm, y = 4 cm   61. 144 ft   62. 18 ft by 24 ft

63. (a) 20', (b) 60°   64. 55°   65. (a) 16'   (b) $4\sqrt{5}$ in.

66. $2\sqrt{3}$ cm

                                                      (c) $20\sqrt{3}$ in. (d) $\frac{11\sqrt{3}}{3}$ ft

67. (a) 10'   68. (a) 143°   69. (a) & (b) 60°
    (b) 14'   (b) 42°   (c) 90°
    (c) 26'   (c) 71.5°   (d) 30°
    (d) 46°   (d) 87.5°   (e) 120°
    (e) $3\sqrt{7}$'   (e) 73°   (f) 30°

70. AF = 8 m, CE = 36 m, BD = 12 m   71. Two angles = 80°

72. $405 + 124.5\sqrt{2} - 6\sqrt{3}$ in²   73. 16 in²

74.
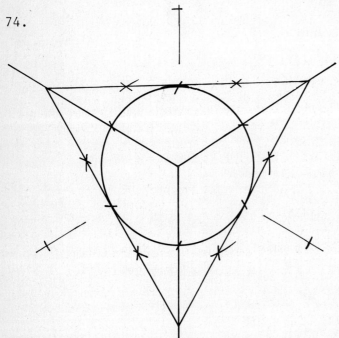

75.
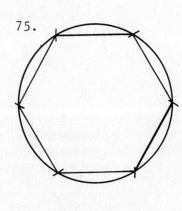

76.
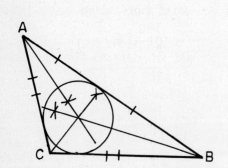

77.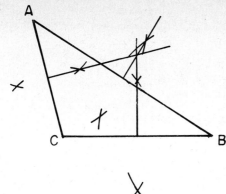

78. 61°   79. 257.07 ft

80. 1764.23 in^2 81. 32.87 ft

82. 162°   83. 70 cm

84. (a) $16\pi - 24\sqrt{3}$ in^2

(b) $25\pi - \dfrac{75\sqrt{3}}{2}$ in^2

85. $\dfrac{950\pi}{3}$ cm^3 86. $54\sqrt{3}$ ft^3

160π cm^2 $108 + 54\sqrt{3}$ ft^2

87. $114\sqrt{3}$ in^3 88. (a) NS
(b) S
$108\sqrt{3}$ in^2 (c) NS
(d) S
(e) S
(f) NS
(g) S

89. (a) Conv: If you go to jail,
then you break the law.
Inv: If you don't break the law,
then you don't go to jail.
Contrap: If you don't go to jail, then
you don't break the law.

(b) Conv: If you are happy, then the sun shines.
Inv: If the sun does not shine, then you are not happy.
Contrap: If you are not happy, then the sun does not shine.

(c) Conv: If I go on Saturday, then I got paid.
Inv: If I do not get paid, then I will not go on Saturday.
Contrap: If I do not go on Saturday, then I did not get paid.

90. (a) Greedy creatures can fly.
(b) Some antelopes do not delight the eye.
(c) Some black rabbits are not young.
(d) Some Englishmen are Frenchmen.
(e) All oysters are amusing.
(f) Some line is not a set of points.
(g) John will not buy a calculator, and Bob will not buy a calculator.
(h) John or Bob will buy a calculator.
(i) Some day is not Christmas, or cats are not fickle.
(j) Some nice guys do not finish last, and all students will pass this course.

91. (a) T (b) T (c) T

92. (a) If you like mathematics, all students like this text.
(b) You do not like mathematics or all students like this text.
(c) If some students do not like this text then you do not like mathematics.
(d) If you do not like mathematics then all students like this text.
(e) If you like mathematics and some students do not like this text then all students like this text.

93. (a) You don't brush your teeth with Brylcreem.
    (b) No conclusion.
    (c) If you have time to eat breakfast, fleas are on your bed.
    (d) If you smoke pot, you will have to move to Bellefonte.
    (e) If Harry throws a party, he will have a miserable time.
    (f) If John goes to the store then Sam doesn't come with Jane.

INDEX

A

AAA similarity, 155
Alternate interior angles, 139
Angle(s)
    acute, 30
    alternate exterior, 139
    alternate interior, 139
    bisecting, 41
    complementary, 35
    constructing, 41
    corresponding, 103, 139
    depression, of, 354
    elevation, of, 354
    inscribed, 269
    obtuse, 30
    reflex, 30
    right, 30
    straight, 30
    supplementary, 34
    vertex, 27
    vertical, 35
Annulus, 302
Apothem, 255
Arc, of a circle, 267
Arc Cosine, 346
Arc Sine, 346
Arc Tangent, 346
Area
    annulus, 302
    circle, 298
    equilateral triangle, 240
    parallelogram, 235
    polygon, 255
    rectangle, 226
    rhombus, 245
    square, 226
    trapezoid, 248
    triangle, 238
Arguments
    invalid, 462
    valid, 458
ASA congruence, 106

B

Bisect, 15
    angle, 47
    line segment, 20

C

Centroid, 317
Chord, 269
Circle(s), 2, 265
    arc, 267
    area, 298
    central angle, 266
    chord, 269
    circumference, 272
    concentric, 265
    constructions, 308
    diameter, 266
    inscribed angle, 269
    proofs, 329
    radius, 265
    sector, 299
    segment, 299
    tangent, 268
Circular cylinder, 378
    surface area, 399
    volume, 409
Circumcenter, 313
Circumference, 272
Circumscribed polygon, 290
Coincident lines, 5
Coincident planes, 7
Coincident points, 1
Collinear points, 3
Compass, 17
Complementary angles, 35
Conclusion, 444
Conditional statements, 444
Cone, 379
    surface area, 400
    volume, 414
Congruent
    angles, 34
    line segments, 15
    triangles, 102
Constructions
    angles, 41-53
    circles, 308-326
    lines, 17-22
    parallel, 145
    perpendicular, 44-47
    proportional parts
        of line segments, 171-173
    triangles, 91-98
Conjunction, 429
Contrapositive, 454

Converse, 454
Coplanar, 7
Corresponding angles, 103, 139
Cosecant, 361
Cosine, 341
Cos^{-1}, 346
Cotangent, 361
Cube, 377
    surface area, 388
    volume, 409
Curve, 1
    simple closed, 1, 61
Cylinder, 378
    surface area, 399
    volume, 409

D

Degree, 30
Degree mode, 343
Depression, angle of, 354
Diameter, 266
Disjunction, 430

E

Edge of a polyhedron, 375
Elevation, angle of, 354
Equilateral triangle
    area, 240

F

Face of a polyhedron, 375

H

Hexagon, 215

I

Incenter, 316
Implications, 444
Inscribed, 269, 284
Intersecting lines, 6
Invalid arguments, 462
Inverse of trigonometric
    functions, 346
Isosceles triangle, 63

L

Lateral faces, 376
Lateral surface area, 385

Line, 3
    coincident, 5
    constructions, 17
    intersecting, 5
    parallel, 6
    segment, 4
    skew, 6
Logic, 428

M

Median, 317
Midpoint, 15
Minute, 32
Mode
    radian, 343
    degree, 343

N

Negations, 435
    conjunction, of a, 439
    disjunction, of a, 440

O

Orthocenter, 317

P

Parallel lines
    constructions, 44-47
    proofs, 151-153
Parallelepiped, 377
Parallelogram, 183
    properties of, 192-197
    area and perimeter, 235-237
Perpendicular
    construction, 44-47
Plane(s), 5
    coincident, 7
    intersecting, 7
    parallel, 7
Point(s), 1
    coincident, 1
    collinear, 3
Polygon(s), 61, 214-218
    area and perimeter, 255-258
Polyhedron, 374
    edge, 375
    face, 375
    vertex, 375
Premise, 444
Prisms, 376
    surface area, 386
    volume, 409

Proofs
    circles, 329
    congruent triangles, 115-122
    logic, 468
    parallel lines, 151-153
    similar triangles, 175-176
    quadrilaterals, 211-212
Proportional segments, 171-173
Protractor, 30
Pyramid, 377
    surface area, 391
    volume, 414
Pythagorean theorem, 70
    triples, 84

Q

Quadrilaterals, 61, 183-212
    proofs, 211-212

R

Radian, 274
Radian mode, 341
Radius
    circle, of, 265
    sphere, of, 374
Ray, 4
    opposite, 5
Rectangle, 183, 199-202
    area and perimeter, 226-232
Rectangular prism, 376
Regular prism, 376
Regular pyramid, 377
Rhombus, 183, 202-205
    area and perimeter, 245-248
Right angle, 30

S

SAS congruence, 109
Secant, 361
Segment, length of, 15
Similar triangles, 155-166
Sin^{-1}, 346
Sine, 341
Sphere, 374
    surface area, 403
    volume, 418
Square, 183, 206-207
    area and perimeter, 226-232
SSS congruence, 104
Statement, 428
Straightedge, 17
Supplementary angles, 34

Surface area, 384
    cones, 400
    cylinders, 399
    prisms, 386
    pyramids, 391
    sphere, 403

T

Tan^{-1}, 346
Tangent, 268, 391
Tetrahedron, 376
Transversal, 138
Trapezoid, 183, 190-192
    area and perimeter, 248-252
Triangle(s), 61
    acute, 64
    area and perimeter, 238-241
    centroid, 317
    circumcenter, 313
    congruence, 102-110
    equilateral, 63
    incenter, 316
    isosceles, 63
    median, 317
    obtuse, 64
    orthocenter, 317
    proofs, 115-122
    right, 64
    scalene, 63
    similar, 155-166
    proofs, 175-176
Trigonometry, 340
    applications, 352
Total surface area (TSA), 385

V

Valid arguments, 458
Vertex of a polyhedron, 375
Vertical angles, 35
Volume, 409
    cone, 414
    cylinder, 409
    prism, 409
    pyramid, 414
    sphere, 418